Mark S. Wilson 7.50

op

Forest and Shade Tree Entomology

Forest and

ROGER F. ANDERSON
School of Forestry
Duke University

Shade Tree Entomology

John Wiley & Sons, Inc., New York · London · Sydney

10

Library of Congress Catalog Card Number: 60-11714

Printed in the United States of America

ISBN 0 471 02739 1

Preface

This book is both a textbook and a manual of forest, shade tree, and wood products entomology that is designed to help forestry and entomolgy students obtain an understanding of forest and tree insect problems and the methods used to prevent damage in forests, trees, and wood products.

In the first section of the book I have dealt with basic facts about insects and insect life. These general aspects include insect structure, physiology, development, classification, and ecology. After this introduction I have discussed the more specialized subjects of forest insect surveys and control methods.

The second and larger section of the book treats individually the more important forest, shade tree, and wood products insect pest species. Identification aids consist of descriptions, figures, and tables that stress characteristics observable in the field. These include types of tree injury together with the characteristics of the tree-infesting stages. Other biological information presented for each species includes hosts, distribution, insect appearance, life cycle, and applied control methods. I have always emphasized silvicultural control methods, whenever this approach has appeared feasible.

The field of forest entomology has not been studied intensively; therefore, much of what is known consists of observational data, which are often scanty. Much additional information is sorely needed, especially experimentally obtained data. Consequently, forest entomology is a virgin field with respect to research opportunities. This means that much of the information presented here may be amplified and modified by subsequent research.

I am very grateful to my predecessor, Dr. J. A. Beal, who accumulated much of the fine collection of insects and insect damage that I used for study and for photographic models. Although the majority of the photographic work in this book is my own, many of the photographs were taken by Dr. Beal and his students, and the sources of the remaining ones are indicated in the legends.

Durham, North Carolina
July, 1960

ROGER F. ANDERSON

Contents

SECTION I

SECTION II

SECTION *I*

General Principles of Tree, Forest, and Wood Product Entomology

SPECIAL INSTRUCTIONS:

COMPLIMENTS OF WILEY 052

IF SHIPMENT ARRIVES SHORT OR DAMAGED, ADVISE WITHIN 30 DAYS.

JOHN WILEY & SONS, INC.
1530 So. Redwood Road - Salt Lake City, Utah 84104

REASON *CODE	READY DATE	QUANTITY	AUTHOR AND SHORT TITLE	CODE	C	PRICE
		1	ANDERSON ENTOMOLOGY	W02739		

PACKING LIST

J87572	1
	TOTAL QUANTITY

* REASON NOT SHIPPED

1 IN PREPARATION. READY ABOUT - BACK ORDERED.
2 OUT OF STOCK. READY ABOUT - BACK ORDERED.
3 TEMPORARILY OUT OF PRINT. NEW EDITION PENDING. READY ABOUT - BACK ORDERED.
4 IN PREPARATION. PUBLICATION INDEFINITE. CANCELLED.
5 OUT OF PRINT. CANCELLED.
6 NOT OUR PUBLICATION. CANCELLED.
7 IMPORT. CANNOT SUPPLY. OUR MARKET RESTRICTED.
8 WE WILL ADVISE IN A FEW DAYS.

CHAPTER 1

Introduction

THE FIELD OF FOREST, SHADE TREE, AND WOOD PRODUCT ENTOMOLOGY

THIS BRANCH OF ENTOMOLOGY DEALS WITH THE PROTECTION OF TREES, forests, and forest products from damage caused by insects. Foresters, aboriculturists, and others who constantly need information on this subject frequently are handicapped by lack of sufficient training and by a lack of readily available information. These deficiencies are especially evident to those who become involved with tree and wood protection problems created by insects.

The primary reason for studying forest, shade tree, and wood products entomology is to obtain information necessary to carry out the known methods for preventing insect-caused damage. In order to apply these corrective measures intelligently, certain basic information must be known about insects and insect life. When a problem involving tree injury or wood destruction occurs, the first thing that must be done is to ascertain its cause. This decision requires the ability to identify the various types of insect-caused damage and to differentiate these damages from injury caused by other agencies such as weather, fungi, birds, or mammals. In making a diagnosis it is often necessary to identify the insects; therefore, some knowledge of insect structure is needed to make the insect descriptions comprehensible.

Another aspect of forest entomology is the evaluation of insect outbreaks. This evaluation includes determining the amount of damage already caused and estimating possible future damage if the epidemics continue. For this work there are special survey techniques that should be familiar to the field man. Some general knowledge of insect ecology is also desirable if one is to understand why great fluctuations of insect populations occur so frequently. Finally, students should become familiar with the more important forest pest species that illustrate the way in which seasonal development, general biology, host, and distribution influence the application of control methods.

DAMAGE CAUSED BY FOREST INSECTS

Insects have always played an important role in the life of forests. Even before the time that wood was of great value to man, insect activity was a prominent factor influencing the development and destruction of forests. Insects often have been and still are the principle harvesters of mature trees. Sometimes this harvesting is a very gradual and continuous process, whereas at other times timber over large areas is killed rapidly. Such forest destruction simply removes the old and makes room for the young. When insects are the killing agent and the new generation consists of tree species different from the old, the ecological succession of the stand is usually advanced. In the more stable climax types, however, the tree species may remain the same, but the age of the stand is reduced. The amount of this reduction is dependent on the rate at which the mature stand is destroyed—the more rapid the destruction the greater the reduction in stand age.

Mature trees usually are of greater economic value to man because harvesting and product manufacture can be accomplished more efficiently, and more high quality products are obtainable from these larger trees. This mature timber can be stored most economically by being left standing. In this case however, it is exposed to attack and vulnerable to destruction by insects as well as other injurious agencies. Of course, other stages of forest growth are also susceptible to insect attack, and the resulting destruction of part or all of the trees influences the ecological succession, rate of stand growth, tree form, and species composition of the stand. In these ways insects compete with man for the use of forest trees, with the result that man frequently has to fight the insects in order to prevent loss.

Most tree injury caused by insects occurs when they use various parts of the trees for food. Many insects live more or less exposed on the outside of the plants, where they feed on the leaves or other succulent parts. The most troublesome of these have mouthparts which are adapted for chewing so that the solid leaf parts can be ingested. Others have elongated slender tube-like mouthparts which enable them to feed by piercing and sucking the liquid sap from the plant tissues. Some soil-inhabiting species also live external to the host plant, but of course they are concealed by the soil. These root-infesting insects feed in either of the previously mentioned ways, depending on the species. Other insects always feed inside the trees by boring their way through the plant tissues.

All parts of trees and all stages of tree growth are subject to insect attack. Different insect species infest different parts, but all stages of growth are subject to all types of insect attacks. Some types of injury, however, are more common or more serious in certain age groups. For example, insects that bore mostly within the inner bark usually cause most trouble to mature trees, whereas the root-feeding insects are most injurious to young trees. Defoliators, considered as a group, probably cause most damage to pole-size and older trees.

The losses insects cause to forests are either directly the result of primary attack or only indirectly an effect of it. The direct losses include killing the tree or parts thereof, reduction of woody growth, wood destruction, and seed destruction. The chief indirect effects are increased fire hazards, changes in stand composition, reduced recreational values, and the providing of entrance courts for or by the actual transmission and inoculation of disease producing organisms.

Killing of trees or parts thereof is usually the result of heavy, sustained feeding on the leaves, twigs, or roots by concentrated boring in the inner bark cambium-phloem region or by boring in the growing tips. When the aerial terminals of young trees are killed, the lateral branches that develop into the new leaders often do not completely straighten; therefore, such trees develop crooked boles. Whenever two or more of these laterals on the same tree become leaders and grow at about the same rate, multiboled or excessively branched trees result. Both of these characteristicis cause the woody growth to be produced in an undesirable form and thereby reduce the value of affected trees. When leaf destruction is heavy, but not sufficiently so to kill the tree, wood production is reduced. Direct damage also occurs when insects eat and/or bore through the seeds or the wood. Wood products containing insect holes, borings, or stains are less useful, hence less valuable.

The susceptibility of an area to fire may be greatly increased by the presence of large amounts of dead timber; therefore, unsalvaged insect-killed timber may contribute greatly to the fire hazard of an area. As soon as much of this dead timber falls, the tangle of fallen trees makes fire fighting almost impossible.

When insects attack only some of the tree species in mixed stands, these hosts may be reduced in number or eliminated from the stands. This selective type of attack may change the stand composition either for better or worse, depending on the value of the trees eliminated and the value of those left. If the trees destroyed are inferior weed or cull trees, the stand is benefited, whereas the opposite is true if the more valuable trees are killed.

The recreational value of forests can also be influenced by insects. Dead timber is unsightly and depressing. Likewise, the mere presence of insects is objectionable to many people even though the insects do not attack them directly. Consequently, when large numbers of certain forest pests are present many people prefer to vacation elsewhere. Insects and related animals which attack man, such as mosquitoes, flies, and ticks, are of interest to foresters because they prevent people from enjoying forested areas and from working comfortably and safely in them. Therefore, foresters should know how to protect themselves from the attacks of these pests.

A fourth type of indirect loss due to insects is their involvement in transmitting, inoculating, or providing a means of entrance for organisms that cause tree diseases. These diseases may kill the trees, destroy the wood, or only discolor it. Insect transmission of tree-killing pathogens has been of more importance in shade trees; nevertheless, there is a close relationship between boring insects and the spread of tree-killing, wood-staining, and wood-rotting fungi.

BENEFICIAL FOREST INSECTS

Not all insect attacks on trees and wood result in economic damage. Sometimes tree killing by insects may benefit man. It has already been stated that the composition of the stand may be improved when undesirable or less desirable species are killed. Insects may also speed the rate of pruning and increase the rate of slash decomposition. The destruction of defective cull trees is also beneficial, especially when such trees are in inaccessible stands or in forests not intensively managed. Other insects are of value in forests because they attack and destroy destructive insect species.

LOSSES CAUSED BY FOREST INSECTS

The best estimates available of forest insect-caused losses for the United States are presented in the latest U. S. Forest Service *Forest Resource Report* 14. Growth impact loss (tree mortality plus the reduction of growth) resulting from insect attack are estimated for an average year to be about 3 billion cubic feet of timber. This amount is over 14 per cent of the net annual growth. The total annual tree mortality caused by insects was estimated to be over 1¾ billion cubic feet, or more than 1½ times that caused by diseases and over 4 times that caused by fire. Total insect-caused losses (growth impact), however, are not proportionally as great; they are only slightly more (11 per cent) than those caused by fire and only 4/10 as much as those caused by diseases. Much (about 70 per cent) of the disease losses are due to heart rots. One fact regarding the growth impact loss caused by insects, which was not mentioned in the *Forest Resource Report* 14, is that there were almost no data available concerning the losses caused by borers in eastern hardwoods. Recently, however, several studies have been made which show that borer damage to hardwoods is extremely great (Hay and Wootten, 1955; Bryan, 1958).[1] Thus, insect-caused losses may rival those caused by wood rots for the eastern United States. The relative importance of the various destructive agencies is shown in Table 1.1.

TABLE 1.1 RELATIVE IMPORTANCE OF THE VARIOUS DESTRUCTIVE AGENCIES AFFECTING FORESTS IN THE UNITED STATES AND COASTAL ALASKA*

Cause	Growth Impact Loss,† per cent	Tree Mortality Loss, per cent
Disease	45	26
Insects	17	41
Fire	16	10
Weather	9	4
Mammals	8	2
Miscellaneous	5	17

* Calculated from *U.S.D.A. Forest Resource Report* 14. 1958.
† Tree-mortality loss plus growth loss.

In Canada the annual timber depletion caused by insects and diseases combined is about two times greater than that caused by fire. This drain by both insects and diseases has varied from 15 to 33 per cent of the amount of timber utilized by man during the past 30 years.

Most of the insect-caused sawtimber mortality (90 per cent) in the United States has been caused by bark beetles which bore in the inner bark. Only about 1 per cent of the tree mortality has been caused by leaf feeders. Bark beetles also have been responsible for most of the growth impact loss (over 60 per cent), with the leaf feeders causing only about 15 per cent. Inasmuch as such a large proportion of the insect-caused loss is due to bark beetles, and most bark beetles injury occurs in the West, it is not surprising that about 60 per cent of the growth-impact loss also has occurred there. In Canada, on the other hand, the insect-caused loss pattern is different. Here most of the damage has

[1] Parenthetical names and dates refer to bibliographical entries at the end of each chapter.

occurred in the East and has been caused by a defoliator, the spruce budworm.

Fire is extremely dangerous to forests. Through the years increasing amounts of money have been spent for forest fire control so that at present over 63 million dollars are spent annually in the United States by both public and private agencies for this purpose. In contrast to this amount, only about 5 or 6 million dollars are spent annually for the control of forest insects. Even though insects have caused about the same amount of injury to our forests as fire has, the total amount spent is only about one tenth as much for insect control. This statement does not imply that forest insect control needs funds as extensive as the amount spent for fire control; nevertheless, it does suggest that in years past too little attention has been directed toward protecting forests from the ravages of forest insects. Currently, however, more attention is being given by everyone concerned to the study and control of both forest insects and diseases. This attention is evidenced by the fact that the United States government is now spending about half as much for each of these destructive biotic agents as for forest fire control.

SELECTED REFERENCES

Annual reports on forest insect conditions in various regions of the U. S. *U.S.D.A. Forest Serv. Exp. Sta. Papers.*

Annual reports of the forest insect and disease survey. *Canada Dept. Agri. Forest Biol. Div.*

Anonymous. 1957. Canada's forests 1946–1950. *Can. Dept. Northern Affairs and Natl. Resources Bull.* 106.

Balch, R. E. 1942. On the estimation of forest insect damage, with particular reference to *Dendroctonus piceaperda* Hopk. *J. Forestry* 40:621–629.

Brown, A. W. A. 1940. A note on the gross estimate of forest damage in Canada. *Forestry Chronicle* 16:249–254.

Bryan, W. C. 1958. Defect in Piedmont hardwoods. *U.S.D.A. Forest Serv. S. E. Forest Exp. Sta. Research Note* 115.

Craighead, F. C. 1942. The influence of insects on the development of forest protection and forest management. *Smithsonian Inst. Publ. Rept. No.* 3665:367–392. 1941.

Gobeil, A. R. 1941. *Dendroctonus piceaperda Hopk.:* A detrimental or beneficial insect. *J. Forestry* 39:632–640.

Hay, C. J., and J. F. Wootten. 1955. Insect damage in hardwood sawlogs. *U.S.D.A. Forestry Serv. Tech. Paper* 148.

Hepting, G. H., and G. M. Jemison. 1958. *U. S. Forest Serv. Resources Rept.* 14:185–220.

Orr, L. W., and R. J. Kowal. 1956. Progress in forest entomology in the South. *J. Forestry* 54:653–656.

Prebble, M. L. 1954. Review of forest entomology, 1948–1953 (Canada). *Rept. 6th Commonwealth Entomological Conference* 206–224.

CHAPTER 2

Structure, Physiology, and Development of Insects

EXTERNAL STRUCTURE

It is necessary to identify correctly the insect involved, if we are to be sure of the cause of the injury observed. Often this identification can be accomplished most easily if we can examine not only specimens of the insects responsible but also samples of the damage. When collecting these samples, the field man must be sure that the insects taken are the cause of the injury, for frequently many insects found at or near the site of the damage are only secondary invaders, parasites, or predators. After we have determined that the injury has been caused by a certain insect, the insect must be identified. To accomplish this, we must be familiar with certain structural characteristics; therefore, elementary insect morphology will be considered first.

THE BODY WALL

The external covering of insects, as well as of other arthropods, forms the main skeletal structure. It consists mostly of a thin microscopic single layer of cells, the *epidermis* (epi = over, dermis = skin), and a thick outer visible noncellular layer, the *cuticle,* which is secreted by the epidermis. This body wall covers the insect so completely that even the eyes and the body hairs are thinly covered. It also lines the fore and hind intestines as well as the breathing tubes (*tracheae*). The cuticle is composed of many substances but the most characteristic constituent is a polysaccharide, *chitin.* Chitin is present in the exoskeleton of both soft, fleshy insects and those which are hard and horny. The hard characteristic, however, is due to a protein named *sclerotin.* Any part of the body may have slender to stout, or sharp to rounded projections which are named variously according to location and form. These include spines, horns, spurs, claws, tubercles, and warts. Body hairs are named *setae* and depressed circular or oval places in the cuticula are called *pores.* The body wall serves as the main skeletal structure to which most of the muscles of locomotion are attached. In addition, it performs the usual functions of protection and secretion and contains numerous sensory receptor nerve cells.

Most adult insects have three main body divisions: the head, thorax, and abdomen. These are composed of a number of ring-like parts called *segments* which, in turn, are made up of plate-like areas named *sclerites.* These sclerites are commonly delimited by impressed lines (*sutures*) and/or by membranous areas. The abdominal segments usually show segmentation clearly, whereas the head segments have coalesced so that they can not be distinguished. The thoracic segmentation is somewhat confused being less distinct than that of the abdomen.

THE HEAD

The insect head usually is a capsule-like globular or flattened structure. On adults and some immature insects there is a pair of large, *compound eyes.* These are so named because they are composed of a great many similar tube-like units (*ommatidia*). The ommatidia extend from the surface of the eye inward, so that only the numerous minute outer hexagonal facets can be seen in the cornea. On some insects small simple eyes (*ocelli*) are present on the side, top, or front of the head. On immature insects, which lack compound eyes, there may be several ocelli-like eyes. These eyes are actually ommatidia.

A pair of large segmented feelers (*antennae*) are attached to the head, usually near the compound eyes. The shapes of the antennae vary and are therefore used to help describe certain insects. These shapes are as follows: thread-like (*filiform*), bristle-like (*setaceous*), like a string of beads (*moniliform*), segments shaped like saw teeth (*serrate*), terminal segments gradually or abruptly enlarged (*clubbed*), clubbed antennae with flattened plate-like terminal segments that are enlarged and extend out from only one side of the antennal axis (*lamellate*), one part of each antenna attached to a second part by a sharp angle bend (*elbowed*), antennae that somewhat resemble the teeth of a comb (*pectinate*), hairy or feathery (*plumose*), and a small stout globular antenna with a bristle attached to the side (*aristate*). The function of antennae is chiefly sensory, for numerous olfactory and tactile receptor nerve cells are concentrated in these structures. Setae are usually associated with

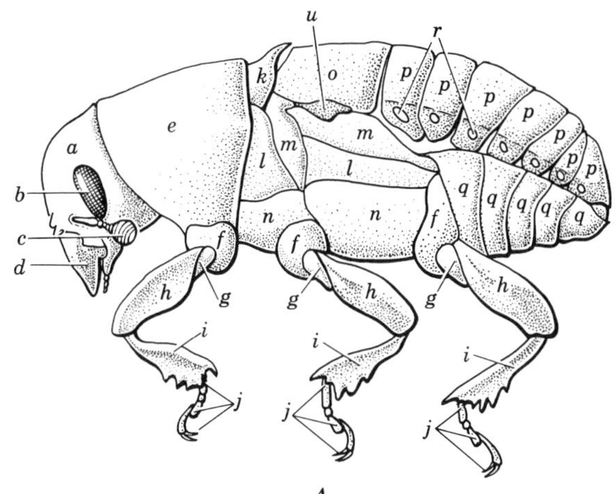

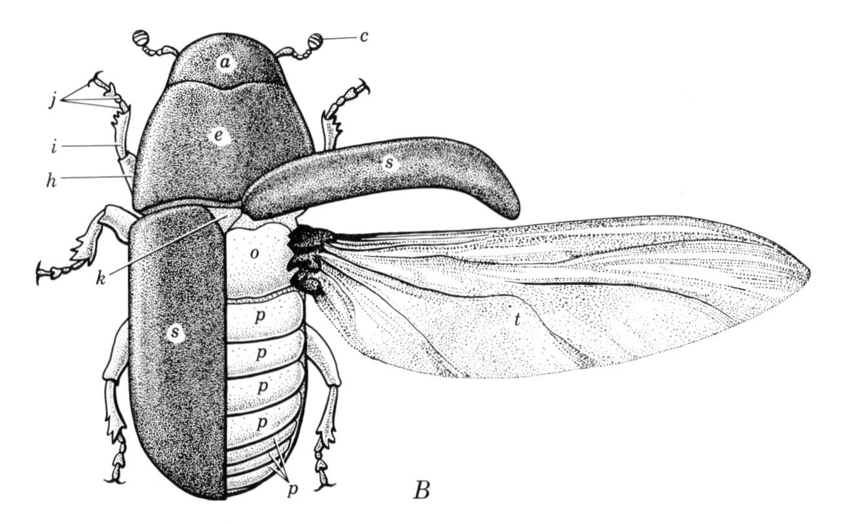

Fig. 2.1. External insect morphology. *A,* side view of a bark beetle with elytra (wing covers) and membranous wings removed; *B,* dorsal view (Adapted from Hopkins, A.D. 1909. *U.S.D.A. Entomol. Bull.* 83, Part 1.) LEGEND: *a,* head, *b,* compound eye, *c,* antenna, *d,* mouthparts, *e,* pronotum, *f,* coxa, *g,* trochanter, *h,* femur, *i,* tibia, *j,* tarsus, *k,* mesonotum, *l,* episternum (pleuron), *m,* epimeron (pleuron), *n,* sternum, *o,* metanotum, *p,* terga of abdomen, *q,* sterna of abdomen, *r,* spiracles, *s,* elytra (wing covers), *t,* membranous wings, *u,* area where membranous wing is attached.

tactile receptors, whereas those that are stimulated by odors are located beneath pores.

Mouthparts of insects are modified in various ways. The most important type found in forest pests is that which is adapted for chewing or masticating solid plant parts. Other insects can feed only on plant sap because they have piercing and sucking mouthparts. Nectar feeding species also have tubular sucking type mouthparts since they feed by lapping, sponging, and/or siphoning the liquid food. Some adult insects have only rudimentary mouth structures, so that it is impossible for them to feed; consequently, these live as adults for only the short period of a day or two.

Insects with chewing mouthparts may have an upper lip (*labrum*) which extends over a pair of main jaws (*mandibles*) but sometimes the labrum may be small or absent. Each mandible consists of one segment which usually is a stout, toothed structure. They operate or move sideways rather than up and down. Beneath the pair of mandibles is a second pair of lateral-moving accessory jaws (*maxillae*). Each of these consists of many segments often having three free parts on each. One of these parts is a short segmented feeler (*maxillary palpus*). The other two may be flattened, spoon-like, pointed, or toothed. The mandibles are the main chewing structures, whereas the maxillae function for holding, manipulating, and tasting the food being eaten. The underside of the mouth is closed by the lower lip (*labium*). The labium also consists of several segments including a pair of segmented feelers (*labial palpi*).

Insect mouthparts which are adapted for piercing plant tissue and sucking the sap usually appear as a rod-like beak. This visible portion of the beak consists mostly of an enlarged labium which has become elongate and tube shaped. Inside the labium are housed the slender, delicate needle-like stylets. The inner pair, the maxillae, are united laterally so as to form a canal through which the sap is ingested. The second or outer pair are the mandibles. During feeding, the mandibular stylets cut a minute opening in the plant tissue into which the maxillary stylets are inserted immediately. This insertion continues until suitable sap-containing tissues are reached. Insects with lapping, sponging, and/or siphoning types of mouthparts are modified in various other ways to form beak-like structures. These may be coiled, folded, or project out from the underside of the head.

The head capsule also contains a number of sclerites. The more important of these are the *front,* the *clypeus* (located between the front and the labrum), the *genae* (cheeks), the *occiput* (back of head), and sometimes a *gula* (throat).

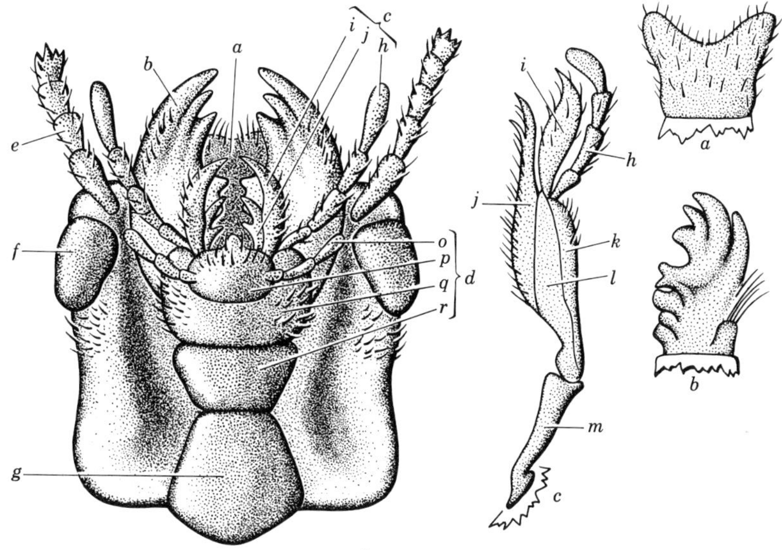

A

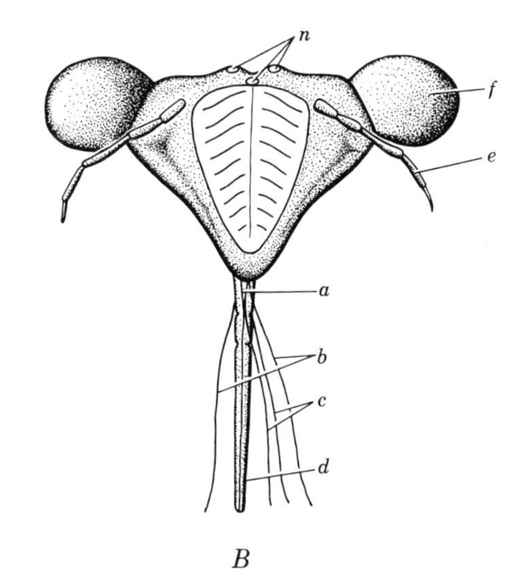

B

Fig. 2.2. Mouthparts. *A*, ventral view of chewing-type mouthparts on a *Passalus cornutus* Fab. beetle; *B*, ventral view of a cicada head showing piercing and sucking-type mouthparts. The bristle-like mandibular and maxillary stylets are shown removed from the labium where they are normally retained. LEGEND: *a*, labrum (upper lip); *b*, mandible (main jaw); *c*, maxilla (accessory jaw); *d*, labium (lower lip); *e*, antenna; *f*, compound eye; *g*, gular (neck) sclerite of head; *h*, maxillary palpus; *i*, galea; *j*, lacinia; *k*, palpifer; *l*, stipes; *m*, cardo; *n*, ocelli; *o*, labial palpus; *p*, ligula; *q*, mentum; *r*, submentum.

It is obvious that the function of the head is to ingest food and serve as a center for several important sensory organs.

THE THORAX

The middle body region to which the jointed legs are attached is called the thorax. It usually consists of three segments which are the anterior *prothorax* (pro = first), the middle *mesothorax* (meso = middle), and the posterior *metathorax* (meta = after). These segments can be most easily recognized by observing where the legs are attached, for they usually join the body close to the posterior region of each segment. Each of these segments consists of several sclerites. The top of each thoracic segment is the *notum* and the bottom the *sternum.* Sometimes, however, these regions may contain several sclerites. There may be one, two, or no recognizable sclerites on the sides. When present, the one which is closer to the sternum is named *episternum* (epi = upon), whereas the sclerite located closer to the leg attachment is the *epimeron* (mer = leg). The prefixes *pro-*, *meso-*, and *meta-* are frequently added to the names of all these thoracic sclerites to indicate specific parts. For example, *pronotum* designates the notal or top sclerite of the prothorax. In beetles and bugs part of the *mesonotum* is usually a small, triangular sclerite (*scutellum*) between the bases of the fore wings. In beetles the *metanotum* is a soft sclerite completely covered by the wing covers.

One pair of legs is attached to the posterior area of each sternum, and on many adult insects, one pair of wings commonly articulates with each of the mesonotum and metanotum. The forewings of the beetles and the earwigs are hardened to form a pair of wing covers (*elytra* = cover). These elytra protect the membranous hindwings and part or all of the mesonotum, metanotum, and dorsum of the abdomen. Elytra lack veins, and when they are in place, they cover the back so that the inner margins meet in a straight line along the middle. The basal two thirds or three fourths of the forewings of some bugs are also thickened, but the terminal part is membranous. These are named *hemelytra* (hem = half). Only the membranous wings are used for flight. They usually have branching veins which form the strengthening framework for the thin, fragile, membranous portions. These veins have names, and their relative positions are used to identify some insects.

Each leg consists of several parts. The segment attached to the sternum is the *coxa.* Sometimes it is large and appears to be more closely associated with the sternum than with the rest of the leg. Close observation, however, will show that it is movable, whereas the rest of the sternum is rigid. A small, often triangular segment, the *trochanter,* is attached to the coxa, but in a few insects this structure consists of two small segments. Sometimes the trochanter can be seen from only one side because it is attached to the side of the following sclerite, the *femur*. The femur is usually the largest segment of the leg and is followed by a more slender elongate *tibia*. Frequently the tibia is armed with spines, especially at the distal end. On adult insects each foot or *tarsus* commonly consists of three to five segments, but sometimes only one or two are present, with the terminal segment bearing sharp claws. Adhesives secreted by the tarsal pads together with the claws make it possible for many insects to walk or cling to smooth or rough surfaces while in any position. Some immature insects have legs as described above, some are legless, and others have both jointed thoracic legs and fleshy, unjointed abdominal legs (*prolegs*). Sometimes the ends of the prolegs have rows or circles of hook-like small spines (*crochets*), which function for holding onto the silk webbing these insects spin. When walking or running the fore- and hindlegs on one side of the thorax and the middle leg on the opposite side are either in motion or at rest simultaneously; consequently, the insects are always supported on a tripod consisting of 3 of the 6 legs.

There may be one or two pairs of small circular, oblong, or crescent-shaped structures containing slit-like openings located in the membranous side areas of one or more of the thoracic segments. These are the external openings to the respiratory system and are named *spiracles*.

From the above description of the various structures present on the thorax it is evident that the function of locomotion is centered there.

THE ABDOMEN

The abdomen is the posterior body region and consists of a number of rather similar segments. Wings often cover the dorsum of many adult insects; therefore, abdominal segmentation usually can be observed more readily on the underside. Each of the segments always consists of a back sclerite (*tergum*) and a "belly" sclerite (*sternum*), but sometimes side sclerites (*pleura*) are also present. On each side of most abdominal segments there is a breathing pore or spiracle.

Certain insects have a pair of structures which may be of various shapes and are attached to the posterior of the abdomen. These are *cerci*. Another terminal abdominal structure, the *ovipositor* (ovi = egg, posit = placer) may also be present on females. It functions for the placement or deposition of the eggs. This ovipositor may be either stout or slender, short or long but always consists of several closely united parts which arise from the posterior abdominal sterna. In wasps and bees it is frequently modified to form a sting. Structures somewhat similar to ovipositors sometimes occur on males, but then they are called male genetalia. Abdominal legs (prolegs), which are present on some immature insects, have been described previously.

INTERNAL STRUCTURE AND PHYSIOLOGY

Nearly all insect-caused damage results from their feeding activities; consequently, the elementary aspects of food utilization by insects should be understood. In addition, other physiological processes of insects will be presented briefly in order to complete the picture of how insects function.

ALIMENTARY SYSTEM

A tube-like structure, often dilated in various places and consisting of several thin layers of muscles, extends from the head to the posterior end of the abdomen. The food is collected and prepared for ingestion by the mouthparts. It then passes through the adjacent mouth opening into the alimentary canal. From here the food is conveyed backward by peristalsis into the *fore–gut,* which consists of a slender tube, the *esophagus,* and often an enlarged storage portion which is called the *crop.* Immediately behind the crop is another enlarged part, the *mid–gut,* which secretes many of the digestive enzymes. The posterior part of this stomach-like organ opens into the smaller but commonly longer *hind–gut* which consists of both large and small intestines. Most of the digestion and absorption of food takes place in the hind-gut. Finally the short somewhat enlarged *rectum* stores the undigested food residue for a short time as it accumulates before being excreted through the anus.

Digestion usually starts in the mouth where the saliva enters and mixes with the food. Other digestive enzymes are secreted by the mid-gut so that digestion continues as the food passes through most of the intestinal tract. Those parts of the food that are liquefied by digestion are absorbed, whereas the undigested residue continues on and accumulates in the rectum until finally excreted through the anus as feces.

CIRCULATORY SYSTEM

As soon as the absorbed nutrients pass through the wall of the digestive tract they become incorporated into the blood. The blood bathes all the tissues; consequently, the needed food materials can be absorbed readily. The *heart* is a slender, muscular tube which extends along the mid-dorsal region of the abdomen. Blood enters the heart through numerous slit-like valves (*ostia*) and is then pumped forward through an arterial prolongation to the head where it is discharged over the brain. Through the remainder of the circulation cycle the blood is not enclosed

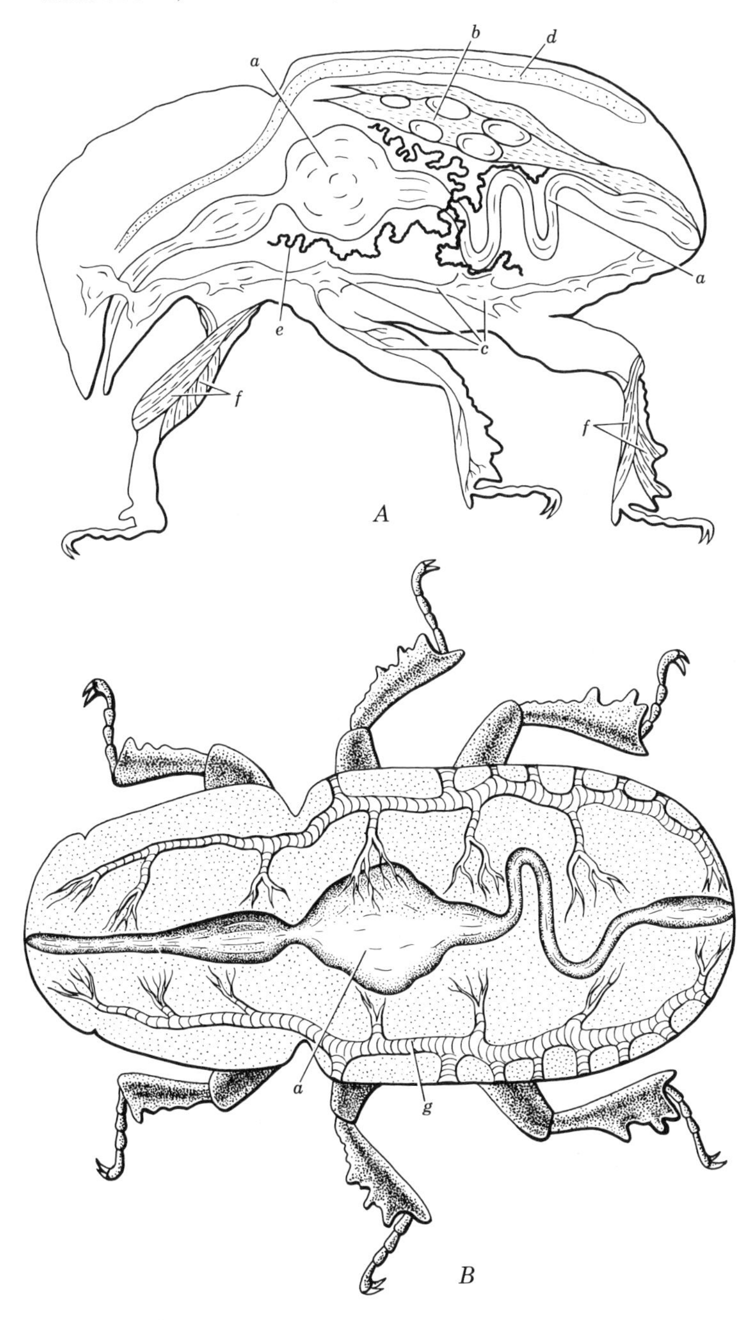

Fig. 2.3. Internal insect morphology. *A,* side view; *B,* dorsal view. LEGEND: *a,* alimentary canal; *b,* reproductive system; *c,* nerve cord, nerve, and ganglia; *d,* heart: *e,* malpighean tubule; *f,* pairs of opposing muscles; *g,* trachea of respiratory system.

in vessels but simply circulates through the small spaces of the head, thorax, legs, and abdomen, thereby bathing and nourishing all the internal body tissues. The cycle is completed when the blood again re-enters the heart.

RESPIRATORY SYSTEM

The oxygen needed for the respiration is usually obtained from the air which enters the respiratory system through external openings known as *spiracles.* The number of spiracles varies greatly, but commonly there are many pairs located on the sides of the abdomen and two pairs on the sides of the thorax. These openings are connected to common air tubes (*tracheae*) which extend along both sides of the insect just inside the body wall. In some insects the lateral tracheae are dilated in places so as to form large air sacs. From these lateral tubes and/or sacs arise many secondary tracheal branches. These, in turn, branch repeatedly and thereby become progressively smaller and increasingly numerous until all tissues of the body are supplied by the ultimate minute tips. Most of these tracheae are strengthened and held open by cuticular spiral thickenings. The air passes through the tracheae directly to the cells where oxygen is needed; thus, the blood is not an intermediate carrier of oxygen. In some aquatic forms, however, parts of the body wall are thin, evaginated, and richly supplied with tracheae. Oxygen diffuses from the water in which the insects live through the cuticle and into the trachea. These are called tracheal gills.

EXCRETORY SYSTEM

The utilization of food materials by the tissues results in the production of liquid metabolic waste products which consist chiefly of sodium and potassium salts of uric acid. These degradation products are first released in the blood and from there absorbed by the *Malpighian* tubules. This organ consists of numerous minute slender tubes which extend throughout the various spaces in the insect body. One end of each tube extends into the body cavity, whereas the other end is attached to and opens into the digestive tract at the junction of the mid-gut and the intestine. The metabolic wastes are absorbed and transported through the Malpighian tubules and then finally discharged into the intestine. From this point the wastes are carried along and discharged through the anus together with the undigested food residue.

MUSCULATORY SYSTEM

Insects have a great number of muscles, with the head and thorax containing most. Movement of the appendages is accomplished by muscular contraction in conjunction with appropriate fulcrum points at each articulation. Flight, on the other hand, involves movement of the thoracic nota to which the wings are attached. Contraction of strong dorso-ventro muscles depress the notal sclerites and causes the wings to go up. This occurs because the meso- and metanota of winged insects are movable, whereas the sterna and pleura are fused together to form a rigid unit. The upper parts of the pleura form the fulcrum points over which the wing bases swing. When the longitudinal muscles contract as the vertical muscles relax the nota are arched up again forcing the wings to beat down. This occurs because the wing bases are attached to and swing over the pleural fulcrums. Many additional muscles also are involved when the wings and legs are tilted or twisted in various ways. This provides all degrees and variations of the simple, straight movements mentioned. The muscular system of insects is very complex having many more muscles than are found in man.

ORGANS OF SECRETION

Many of the secretory organs are located in the epidermis of the body wall. In certain insects, wax, odoriferous compounds, or spittle are secreted from special glandular epidermal body cells, whereas poison secretion is commonly associated with setae. The digestive enzymes are, of course, secreted by the various parts of the digestive tract.

REPRODUCTIVE SYSTEM

Bisexuality is the usual condition for insects, so that both males and females are needed to produce offspring. There are many exceptions, however, in which females reproduce without mating. This type of reproduction is called *parthenogenesis* (parthen = maiden, genesis = born). Most insects deposit eggs which later hatch into active young (*oviparous* reproduction), but in some species the eggs are withheld in the genital tract until they have hatched so that active young are born (*viviparous* reproduction).

The reproductive organs are located in the abdomen. Here the eggs develop in the two ovaries. After the eggs mature they pass through the genital ducts to the outside. As each egg approaches the genital opening it is fertilized by a sperm that is released from a storage sac (*spermatheca*) opening into the vagina. The sperm is stored in this sac from the time of copulation until the time it is needed to fertilize the passing eggs. Certain glands may also have openings into the genital tract through which secreted substances such as egg adhesives or egg covering materials are discharged. The ovipositor aids in the placement of the eggs.

Each egg consists of a shell (*chorion*), yolk, and a nucleus. There are one or more minute openings (*micropyles*) in the chorion through which the sperm enters. The color, shape, size, and markings of insect eggs varies greatly with species so that frequently it is possible to identify many insects by these characteristics.

Sperm is produced in the paired testis of males and conducted to the outside through genital ducts. The copulating organs, or genitalia, which may be present, are homologous with the ovipositors of females.

NERVOUS SYSTEM

The *main brain* is located in the head near the eyes and the antennae. It is connected to the *subbrain* and the main *nerve cord,* which extends along the lower median part of the body, by a pair of nerves which pass one on either side of the digestive tract. The ventral nerve trunk consists of a series of ganglia connected by a double nerve cord. One of these ganglionic masses is located in the head, close to the mouthparts, and is sometimes referred to as the subbrain as indicated above. The other ganglia are usually arranged segmentally, but in some insects the number of ganglia in the thorax and abdomen has been reduced as a result of several adjacent ganglia coalescing to form common masses. From each ganglion thread-like nerves radiate to all parts of the particular segment or segments enervated. These nerves consist of many axons of both sensory and effector nerve cells. A third type of nerve cell is the interconnecting nerve. These cells are located in the ganglia and the central nerve cord.

INSECT DEVELOPMENT

When a sperm enters the egg via a micropyle it immediately unites with the nucleus to form a zygote. Fertilization has then been accomplished. Growth, divisions and redivisions of the zygote, soon forms an embryo which is nourished by the food in the yolk. After the embryo has developed sufficiently, the insect escapes from the egg shell in various ways which depend on the species. These include chewing, rasping with special spines, or exerting fluid body pressure against the chorion shell. Following hatching the insect enters a stage in which feeding and growth are the primary functions. Often the fully grown insect may be several thousand times larger than it was when first hatched. Such tremendous growth, together with the fact that the body wall is relatively inelastic, requires some mechanism to adjust for this apparent incompatability. This adjustment is accomplished by period-

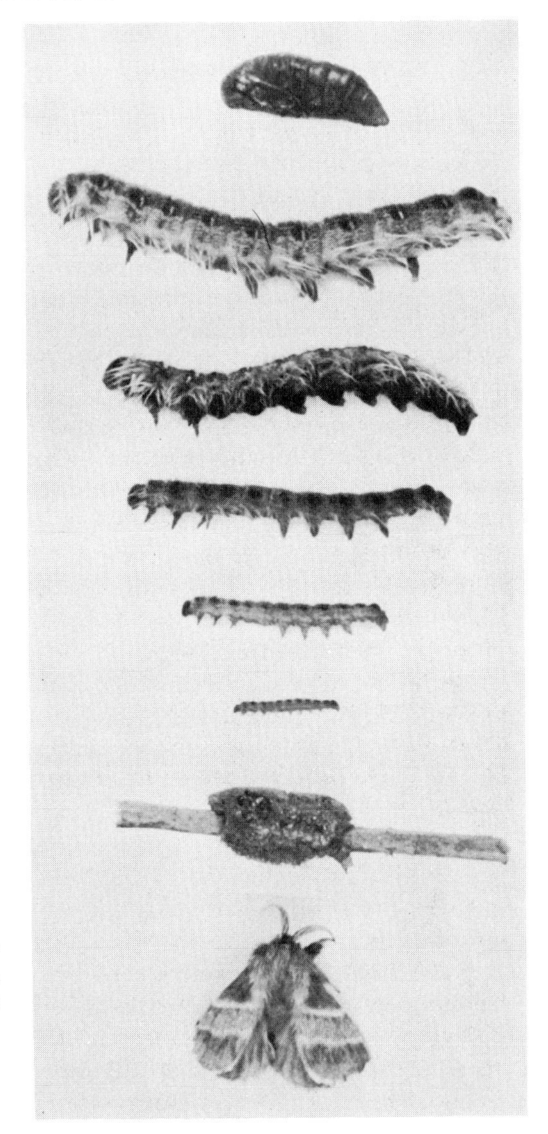

Fig. 2.4. Complete metamorphosis showing adult, egg mass, five larval instars, and a pupa (X1).

ically reconstructing the body wall so as to provide the needed larger size. Part of the old cuticle is digested, and the degradation products absorbed and reused by the epidermis to form the new enlarged cuticle. After the new cuticle has been formed, the remaining old outer layer is shed. This cuticular remnant first splits along the mid-back from the head to the thorax and thereby provides an opening through which the insect can emerge. The number of times insects molt in this way varies from 2 to 20 or more times depending on the species. The usual average is 3 to 6 times. The form of the insect between to successive molts is called an *instar.* For example, if an insect is said to be in the third instar it is understood that the insect has molted twice. Most insects do not molt after they become adults.

In addition to the previously described growth and molting, greater changes often take place when the insects transform into the adult instar. All changes in the form of insects as they develop throughout life are known as *metamorphosis* (meta = change, morph = form). There are four main types of metamorphosis, but the most common with respect to forest and shade tree insect pests are the complete and gradual types. Insects with gradual metamorphosis have three main stages. These are the egg, nymphal, and adult stages. There are a number of nymphal instars but usually only one instar for the adult. Nymphs usually are rather similar to the adults in appearance, except that in winged forms the nymphs lack fully developed wings. Nymphs, however, have compound eyes and visible wing buds if these structures are present in the adults. Bugs, termites, and grasshoppers are examples of insects that have gradual metamorphosis.

Insects with complete metamorphosis have four main stages: the egg, larval, pupal, and adult stages. The larval, or immature feeding stage, is frequently, but not always, worm-like and has neither compound eyes nor external wing buds. Usually there are several larval instars but only one pupal and one adult instar. During the pupal instar the insect is relatively inactive, although great internal changes are occurring because the insect is transforming from the larval to the adult form. It is at this time that many worm-like larvae change into such insects as moths, butterflies, beetles, and wasps. Usually pupae can wiggle their abdomens; otherwise they are immobile. During this pupal stage many of the internal organs and tissues appear to become disorganized and are mostly a viscous liquid. From this liquid, with its disorganized appearance, the adult insects develop. During the pupal stage insects frequently conceal themselves in protected places such as in the ground litter, in wood, in crevices, or in the soil. Many species of Lepidoptera and Hymenoptera construct silk cocoons in which they undergo this transformation.

SELECTED REFERENCES

Comstock, J. H. 1940. *An introduction to entomology.* Comstock Publishing Co., Ithaca, N. Y. 9th edition: 29–205.

Essig, E. O. 1942. *College entomology*. Macmillan, N. Y.: 1–52.
Roeder, K. D. (editor). 1953. *Insect physiology*. John Wiley and Sons, N. Y.
Ross, H. H. 1956. *A textbook of entomology*. John Wiley and Sons, N. Y. 2nd edition : 50–202.
Snodgrass, R. E. 1935. *Principles of insect morphology*. McGraw-Hill, N. Y.
Wigglesworth, V. B. 1947. *The principles of insect physiology*. Methuen and Co., London. 3rd edition.

CHAPTER 3

Insect Classification

Insects belong to the phylum *Arthropoda* (arth = joint, pod = leg). Some of the other major phyla are as follows: Protozoa, the one-celled animals; Porifera, the sponges; Coelenterata, the jellyfish and sea anemones; Platyhelmenthes, the tapeworms and flukes; Nemethelminthes, the unsegmented roundworms; Echinodermata, the starfish, sea urchins, and others; Mollusca, clams, snails, and octopuses; Annelida, the segmented earthworms and leeches; and Chordata, the mammals, birds, reptiles, amphibians, and fish.

Some wood-destroying animals belong to the phyla *Mollusca* and *Chordata,* but most animal-caused tree and wood damage is due to insects. Therefore we are primarily interested in the phylum Arthropoda. Adult members of this phylum are characterized by having a segmented body, jointed legs, and a body wall that forms the chief skeletal structure. The segmented characteristic usually can be seen most easily on the abdominal region, but on some arthropods the external segmentation is not evident. The immature stages of many arthropods also have these characteristics, but in some cases the jointed legs are absent. With a little experience one can learn to recognize these legless stages of insects. Sometimes, however, the external definable characteristics may be similar to those for members of the phylum Annelida; therefore, it might be well to point out that there is a difference in the internal structure of members of the two phyla. Annelids have segmentally arranged transverse septa that divide the inside of each individual, except for the digestive tract, into numerous chambers. Arthropods lack these cross walls. The structural characteristics of the various arthropoda classes are presented in Table 3.1.

TABLE 3.1 CLASSES OF ARTHROPODS COMMONLY FOUND IN FORESTS

	Characteristic	
1.	Jointed legs present	2
1.	Legs absent	(some insect larvae) HEXAPODA
2.	Three pairs of jointed legs	3
2.	More than 3 pairs of jointed legs	4
3.	Head distinct from thorax; antennae present; wings may be present	(insects) HEXAPODA
3.	Head and thorax fused into one structure (cephalothorax); antennae absent; wings never present	(immature ticks and mites) ARACHNIDA
4.	Four pairs of legs	(spiders, ticks, scorpions, and mites) ARACHNIDA
4.	More than 4 pairs of legs	5
5.	Legs of about equal size on each body segment; head distinct from rest of body	6
5.	Legs on posterior segments absent or much smaller than others; head and thorax fused into one structure (cephalothorax)	(pill bugs) CRUSTACEA
6.	One pair of legs attached to each body segment	(centipedes) CHILOPODA
6.	Two pairs of legs appear to be attached to each body segment	(millipedes) DILOPODA

The insect class, *Hexapoda* (hex = six, pod = leg), is comprised of a tremendous number of different kinds or species. Estimates of this number vary from 600,000 to several millions. This is more than all other species of plants and animals combined. Only about 80,000 of these occur in North America north of Mexico, but the number of these which are injurious is only a small part of the total.

Arachnida include the spiders, ticks, mites, and scorpions. Many of these are common in forests, but only certain species of mites attack trees. These phytophagus mites are mostly pests of shade trees, whereas both ticks and some mites (chiggers) are well known pests of man. Chilopoda (centipedes) and Diplopoda (millipedes) are common under dead bark, in ground litter, and in other concealed places in the woods, but none of these are injurious to forests or man. The centipedes are predaceous on other small animals, whereas millipedes are scavengers. Pill bugs (Crustacea) are common in forests where they inhabit decaying wood, the ground litter, and other moist places. They are innocuous scavengers. Certain related marine species (wood lice or gribbles), however, cause considerable damage to wood placed in sea or brackish waters. Other more familiar crustaceans such as shrimp, crabs, and lobsters cause no damage to wood or trees.

The next (lower) taxonomic category to be considered is the *order*. Characteristics of the more important terrestrial insect orders are presented in Tables 3.2 and 3.3. Those which include the more important tree and wood pests are listed in decreasing order of their importance

Fig. 3.1. Common forest arthropods other than insects. Upper left: centipede (X1½); upper right: millipede (X2); center left: engourged tick (X2); lower left: pill bug (X3); lower right: black widow spider (X1½).

TABLE 3.2 COMMON ORDERS OF INSECTS WITH FULLY DEVELOPED WINGS OR WITH WING COVERS

1. Two pairs of wings unlike in structure; hindwings, if present, always membranous; forewings horny, leathery, parchment-like, or somewhat thickened 2
1. All wings membranous, although they may be covered with scales or hairs . . . 5
2. Forewings horny or leathery without veins (elytra); may be ridged but never overlap each other 3
2. Forewings leathery or parchment-like with veins 4
3. Posterior of abdomen terminating in a pair of forcep-like structures (cerci); elytra small and do not cover abdomen . . . . (earwigs) DERMAPTERA
3. Cerci absent; elytra usually, but not always, cover abdomen . . . . (beetles) COLEOPTERA
4. Mouthparts with jaws adapted for chewing . . . . (grasshoppers, walking sticks, and relatives) ORTHOPTERA
4. Mouthparts form a rod-like beak . . . . (bugs) HEMIPTERA
5. Tarsal claws absent; last tarsal segment bladder-like; wings usually very narrow and fringed with hairs; insects less than 1/16 inch long; body slender and elongate . . . . (thrips) THYSANOPTERA
5. Tarsal claws present; last tarsal segment not bladder-like 6
6. Wings partly to completely covered with scales . . . . (moths and butterflies) LEPIDOPTERA
6. Wings completely naked and transparent or thinly clothed with hairs 7
7. One pair of wings 8
7. Two pairs of wings 10
8. Posterior abdominal filament(s) present 9
8. Without posterior abdominal filament(s) . . . . (flies) DIPTERA
9. Small knobbed structures (halteres) present behind wings . . . . (male scales) HEMIPTERA
9. Halteres absent . . . . (Mayflies) EPHEMEROPTERA
10. Mouthparts forming a rod-like beak; tarsi each with 3 or fewer segments . . . . (bugs) HEMIPTERA
10. Mouthparts rudimentary or with jaws adapted for chewing 11
11. Antennae inconspicuous; bristle-like 12
11. Antennae conspicuous, variously shaped but not bristle-like 13
12. Both pairs of wings rather elongate and of nearly same length; tarsi with 3 segments . . . . (dragonflies and damselflies) ODONATA
12. Wings triangular shaped; hindwings smaller than forewings; tarsi with 4 segments . . . . (Mayflies) EPHEMEROPTERA
13. Front of head prolonged into a beak . . . . (scorpionflies) MECOPTERA
13. Front of head not prolonged into a beak 14
14. Forewings distinctly larger than hindwings 15
14. Hindwings almost as large or larger than forewings 16
15. Tarsi usually with 5 segments; head and thorax hard . . . . (ants, bees, wasps, etc.) HYMENOPTERA
15. Tarsi with 2 or 3 segments; small soft bodied insects . . . (booklice) PSOCOPTERA

16. Pair small projections (cerci) on the sides of the posterior abdominal segment; tarsi with 3 to 5 segments 17
16. Cerci absent; tarsi 5 segments; wings usually net veined; rather similar in size, shape and veination . . . . . (nerve-winged flies) NEUROPTERA
17. Four or 5 segmented tarsi 18
17. Three segmented tarsi . . . . . (stone flies) PLECOPTERA
18. Wings clothed with hairs; antennae long and thread-like (filiform) . . . . . (caddis flies) TRICHOPTERA
18. Wings not clothed with hairs; antennal segments rounded and bead-like (moniliform); small blackish ant-like insects; wings shed readily . . . . (termites) ISOPTERA

TABLE 3.3 COMMON ORDERS OF IMMATURE AND WINGLESS ADULT INSECTS*

1. Three pairs of jointed thoracic legs present 2
1. Jointed thoracic legs absent 29
2. Tick-like or lice-like external parasites on larger animals 3
2. Plant feeders, predators or internal parasites on other insects, or scavengers . . 7
3. Body compressed (flattened sideways); legs long and adapted for jumping; rows of short, stout spines often present on various body parts . . (fleas) SIPHONAPTERA
3. Body depressed (flattened dorso-ventrally) 4
4. Each tarsus with 1 large claw 5
4. Each tarsus with 2 claws 6
5. Mouthparts with laterally moving jaws adapted for chewing . . . . . (bird lice) MALLOPHAGA
5. Mouthparts adapted for sucking; without jaws . . . . . (sucking lice) ANOPLURA
6. Abdomen indistinctly segmented . . . . . (sheeptick) DIPTERA
6. Abdomen distinctly segmented . . . . . (bedbugs) HEMIPTERA
7. Appendages encased in membranes so that the joints are indistinct; wing pads, when present, project somewhat toward the underside; quiescent forms often inside cocoons or in cells in wood or soil (pupae) 33
7. Appendages not encased in membranes; joints distinct; wing pads, when present, project straight back 8
8. Compound eyes present 9
8. Compound eyes absent; simple eyes (ocelli) may be present 20
9. Lower lip (labium) jointed and enlarged so that it can be extended greatly; aquatic; some have terminal plate-like or thread-like tracheal gills . . . . . (dragonfly and damselfly naiads) ODONATA
9. Labium characteristic not as in 9 above; aquatic or terrestrial 10
10. Aquatic forms; usually with many pairs of bunched, feathery, plate-like or thread-like tracheal gills on abdomen or thorax 11
10. Terrestrial; tracheal gills absent 12
11. Paired tracheal gills usually present at bases of thoracic legs; sometimes present on head or at posterior end of abdomen . . . . . (stonefly naiads) PLECOPTERA
11. Paired tracheal gills on sides of most of the abdominal segments . . . . . (Mayfly naiads) EPHEMEROPTERA

12. Forcep-like cerci at tip of abdomen . . . . (earwigs) DERMAPTERA
12. Forcep-like cerci never present but cerci of other shapes may be present 13
13. Head prolonged into a beak with small jaws at tip . . . (scorpionflies) MECOPTERA
13. Head not prolonged into a beak 14
14. Body ovate; densely covered with scales and/or hairs; mouthparts rudimentary; if posterior abdominal structures (cerci) present see 21 below . . . . (wingless moths) LEPIDOPTERA
14. Insects without the combination of characteristics in 14 above 15
15. Body strongly constricted between thorax and abdomen . . . . (ants and wingless wasps) HYMENOPTERA
15. Thorax broadly joined to abdomen 16
16. Tarsal claws absent; terminal tarsal segment bladder-like; small elongate insects, less than 1/16 inch long . . . . (thrips) THYSANOPTERA
16. Tarsal claws present 17
17. Mouth a rod-like beak . . . . (bugs) HEMIPTERA
17. Mouthparts are laterally moving jaws 18
18. Pair of short structures on sides of last abdominal segment (cerci) present . . . 19
18. Cerci absent . . . . (booklice) PSOCOPTERA
19. Antennae bead-like (moniliform); whitish to black ant-like insects . . . . (termites) ISOPTERA
19. Antennae thread-like (filiform); body of various forms . . . . (grasshoppers, walking sticks, roaches, and others) ORTHOPTERA
20. Globular, hemispherical, or flattened insects; often covered with woolly, powdery, or scale-like wax; insects stationary on foliage, twigs, or larger parts of trees; mouthparts bristle-like . . . . (adelgids and scales) HEMIPTERA
20. Insects without the combination of characteristics in 20 above 21
21. Mouthparts retracted into head so that only tips show or body clothed with scales; minute insects 22
21. Mouthparts exposed; body not clothed with scales 23
22. Abdomen with more than 6 segments . . . . (bristletails) THYSANURA
22. Abdomen with 6 or fewer segments; usually with a springing structure attached to tip of abdomen . . . . (springtails) COLLEMBOLA
23. Antennae well developed and distinct, with 12 or more bead-like segments; whitish ant-like insects which live in colonies in soil or wood . . . . (termites) ISOPTERA
23. Antennae inconspicuous, consisting of less than 12 segments, often bristle-like; insects usually caterpillar- or worm-like 24
24. Two or more pairs short fleshy unsegmented, abdominal legs (prolegs) and/or paired groups of minute hooks (crochets) on underside of some abdominal segments 25
24. Prolegs and/or crochets absent† 27
25. Crochets present; usually with 2 to 5 pairs prolegs with crochets on ends . . . . (moth and butterfly caterpillars) LEPIDOPTERA
25. Crochets absent 26
26. With only one pair of simple eyes (ocelli) on head; body naked; with 6 to 8 pairs of prolegs . . . . (sawfly caterpillars) HYMENOPTERA
26. With many ocelli on each side of head . . . . (scorpionfly larvae) MECOPTERA

27. Aquatic larvae that live in cases or webs; tracheal gills may be present on both abdomen and thorax; posterior of abdomen usually has hooks . . . . (caddisfly larvae) TRICHOPTERA
27. Insects do not live in cases or webs; either aquatic or terrestrial 28
28. Main jaws (mandibles) and accessory jaws (maxillae) united lengthwise to form a pair of sickle-shaped jaws; 7 pairs abdominal spiracles . . . . (nerve-winged fly larvae) NEUROPTERA
28. Mandibles and maxillae separate; 8 pairs abdominal spiracles . . . . (beetle larvae) COLEOPTERA
29. Head indistinct; mouthparts consist of 1 or 2 parallel dark-colored hook-like structures located at tip or inside pointed end of body; 1 pair breathing pores (spiracles on last abdominal segment large and prominent . . . . (maggots) DIPTERA
29. Head distinct; mouthparts and posterior spiracles not as in 29 above 30
30. Mandibles large and well developed; head harder than body and also darker colored; breathing pores (spiracles) always distinct . . (beetle larvae) COLEOPTERA
30. Mandibles small, often smaller than accessory jaws; mouthparts poorly developed; spiracles distinct or indistinct; head may or may not be colored darker than body 31
31. Several long hairs (setae) present on each body segment; last abdominal segment has a comb-like transverse row of short setae . . . . . (flea larvae) SIPHONAPTERA
31. Long setae absent; comb-like setal structure absent on abdomen 32
32. Larvae live within plant tissue, are parasites in other insects, or live in mud, wax, paper-like cells, or other structures constructed by the adults; body usually saclike, or grub-like . . . . (bee, wasp, and ant larvae) HYMENOPTERA
32. Characteristics not as of Hymenoptera listed in 32 above; many aquatic or semiaquatic species . . . . (fly larvae) DIPTERA
33. With one pair of wing cases; each pupa often enclosed in a brown, segmented, ovoid capsule (puparium) . . . . (fly pupae) DIPTERA
33. With 2 pairs wing cases; insect never enclosed in a puparium but may be inside a cocoon, a cell, or in other protected places 34
34. Appendages so closely appressed to body so that they appear to be lacking; spindle shaped body, anterior blunt, posterior distinctly segmented and pointed . . . . (moth and butterfly pupae) LEPIDOPTERA
34. Appendages are not tightly glued to the body 35
35. Body strongly compressed (flattened sideways); small insects, less than ⅛ inch long . . . . (flea pupae) SIPHONAPTERA
35. Body not strongly compressed 36
36. Prothorax much larger than other thoracic segments; antennae with 11 or fewer segments . . . . (beetle pupae) COLEOPTERA
36. Prothorax usually about same size or smaller than other thoracic segments; antennae with more than 12 segments 37
37. Head and mouthparts elongated (beaked); antennae with 16 or more segments, attached to the head near the eyes . . . . (scorpionfly pupae) MECOPTERA
37. Head and mouthparts not beaked 38
38. Mandibles short, stout, and cylindrical, cross each other; aquatic forms; usually

found in cases; tracheal gills may be present on thorax and abdomen . . . . (caddisfly pupae) TRICHOPTERA

38. Mandibles sometimes large but never overlap each other; aquatic or terrestrial forms 39
39. Antenna with 12 or 13 segments . . . . (bees, wasps, ants, and sawfly pupae) HYMENOPTERA
39. Antenna with more than 13 segments . . . (nerve-winged fly pupae) NEUROPTERA

* Wing pads may be present.

† A few Lepidoptera and Hymenoptera caterpillars lack both prolegs and crochets; therefore, these cannot be identified with this key. See pages 164 and 183.

as follows: Coleoptera (beetles), Lepidoptera (moths and butterflies), Hymenoptera (sawflies, bees, wasps, and ants), Hemiptera (bugs), Isoptera (termites), Orthoptera (grasshoppers and walking sticks), and Diptera (flies). Each of these order names contains the root "ptera" which is derived from a Greek word meaning wing.

The order Coleoptera (col = sheath, ptera = wing) is the largest of the insect orders and also contains the largest number of species that feed on trees or wood. Adults have their forewings modified into horny wing covers (elytra). These wing covers lack veins, and when the insect isn't flying, the two elytra usually cover the back in such a way that the inner margins meet in a straight line along the mid-back region. Membranous hindwings are usually present, which, when not in use, are folded beneath the elytra. Forceps-like structures (cerci) are never present at the posterior end of the abdomen. The mouthparts are well developed and adapted for chewing. Beetle larvae or grubs vary greatly in form. They may either be legless or have jointed thoracic legs, but the head and the chewing mouthparts are always well developed. Frequently they are worm-like with a straight or curved body. Pupae resemble the adults except that they are often light in color and have small wing pads instead of elytra and membranous wings. The prothorax is large and well developed and the appendages are not tightly appressed to the body but usually extend out. Transformation is commonly undergone in cells made in soil or wood.

Among the beetles are leaf, bark, and root eaters, bark and wood borers, and also predaceous forms which feed on other insects. Forest entomologists are chiefly concerned with only about a dozen of the 150 or more families of beetles.

Lepidoptera (lepid = scale, ptera = wing) adults are the commonly observed and well-known moths and butterflies, whereas the larvae, which are always the direct cause of the tree damage, are known as caterpillars. The female moths of a few species are wingless, but the winged members have four membranous wings which are generally more or less covered with minute, overlapping, flattened scales (modified setae). These scales are easily detached so that when moths or butterflies are handled the scales rub off readily and appear as dust. The body, legs, and other appendages are densely clothed with hairs and scales. Mouthparts of the adults sometimes are in the form of an elongated tube so that only liquid food can be digested. When not feeding, this tube is held coiled beneath the head. Many species, however, have nonfunctional mouthparts. They cannot eat and therefore live for only a short time. The caterpillars are usually cylinderical and worm-like. Each has a distinct head, 3 pairs of short, slender, jointed, tapered, peg-like legs on the 3 thoracic segments and usually 2 to 5 pairs of thick, short, fleshy, jointless abdominal legs (prolegs). These prolegs are armed at the extremities with one or more rows or circles of minute hooks (crochets). When 5 pairs of prolegs are present they occur on the third, fourth, fifth, six, and tenth abdominal segments. The hairiness of caterpillars varies according to species. Some appear naked, whereas others resemble small brushes. Lepidoptera pupae are somewhat spindle shaped and usually deeply colored tan to dark brown. The wing pads, legs, mandibles, maxillae, and antennae are tightly appressed to the body surfaces so that the pupae appear legless and resemble mummies. Many caterpillars, when preparing for pupation, spin a cocoon of silk which is secreted from modified salivary glands. The places they select for pupation may be in rolled leaves, in the ground litter, in the soil, or even in wood, for the wood-boring species.

Moths are night fliers. They rest with their wings spread horizontally, or folded roof-like over their abdomen, and have thread-like, comb-like, or feathery antennae. Butterflies, on the other hand, are day fliers that usually hold their wings together vertically over their backs when resting; they have clubbed antennae. The caterpillars are always the tree damaging stage. They are the leaf feeders, the tip borers and the wood borers.

The order Hymenoptera (hymen = membrane, ptera = wing) includes the ants, bees, wasps, sawflies, and horntails. Some adults are always wingless, but those that are winged have four membranous wings with the hindwings smaller than the forewings. A row of hooks on the front margins of the hindwings attach over a fold on the hind margins of the forewings in such a way that the 2 wings on each side are held together when the insects fly. The mouthparts are adapted for either chewing or for both chewing and sucking. Frequently the ovipositor is modified to form a sting, whereas on others this structure has serrate saw-like teeth (*sawflies*). The larvae of the latter group are caterpillars that have chewing mouthparts, 3 pairs of short, tapered, jointed, thoracic legs, and 6 to 10 pairs of abdominal prolegs. These prolegs differ from those on Lepidoptera caterpillars by usually being more numerous and by not having crochets. Most other Hymenoptera larvae are grub-like. They are legless and have poorly developed chewing mouthparts. Just before pupation the larvae commonly spin white to brown, parchment-like, silken cocoons in which they transform to adults. Legs and wing pads of the pupae are not tightly appressed to the body but are somewhat free. They can be distinguished from beetle pupae, which have similar characteristics, by the small size of the hymenopterous prothorax. The habits of these insects vary greatly. Leaf feeders, borers, and many parasitic species belong to the order Hymenoptera.

The order Hemiptera (hem = half, ptera = wing), includes bugs, lacewings, cicadas, tree hoppers, leafhoppers, spittlebugs, aphids, and scale insects. These insects have mouthparts combined into rod-like beaks, which are adapted for piercing plant tissues and sucking sap. The winged forms have 2 pairs of wings with the front pair the larger. Some species have all membranous wings which are held roof-like over the body, whereas others have the basal half or more of the front pair of wings thickened, with only the distal parts being membranous. When resting the wings of the latter species are held approximately horizontal. This is one of the 3 orders important to foresters in which the individuals experience gradual metamorphosis. The immature forms (nymphs) usually appear very much like the adults except that they are smaller and lack fully developed wings. Nearly all Hemiptera are plant sap feeders.

Termites or white ants also have gradual metamorphosis. The order name, Isoptera (iso = equal, ptera = wings) refers to the characteristic that the fore- and hindwings of the winged individuals are very similar with respect to size and appearance. Only at certain times, however, do any of the adults have wings, and these are soon shed after the termites swarm from the parent colonies. The abdomen is broadly joined to the thorax and the mouthparts are adapted for chewing. The somewhat spherical shape of each of the antennal segments make them appear to resemble a string of beads. Short 2 to 6 jointed cerci are present. Colonies of most species contain members which belong to several castes. For the more common species occurring in North America these castes are the primary and secondary reproductives, the workers, and the soldiers. The primary reproductives differ in several respects from the others. They are colored dark brown or black, have compound eyes, and have wings for a short time before they swarm. These insects develop with gradual metamorphosis; therefore, the young resemble the adults. Wing buds develop only on the nymphs which are destined to become primary reproductives. Most of the termites encountered in the United States construct their nests in the soil, under objects lying on the ground or in wood. Commonly they burrow in the soil and then extend their nests into the wood lying nearby on the ground or in adjacent wooden structures. They feed on any type of cellulosic material but are of most interest because of their wood-destroying habits. In the tropics and southernmost portions of the United States, however, there are certain species which do not require contact with moist soil. The more important of these are appropriately named dry-wood termites.

True flies belong to the order Diptera (di = two, ptera = wing). The winged members have one pair of membranous wings located on the mesothorax. A small pair of knobbed structures (*halteres*) are located where the second pair of wings should arise. These are homologous with the second pair of wings present on other insects. On some flies a basal lobe (*squama*) occurs beneath the base of each wing. This lobe is part of the forewing and is not a reduced hindwing. The mouthparts are adapted for sponging, sucking, and sometimes also for piercing. As with the bugs, the labium forms a sheath inside of which the other parts are enclosed. In those which feed by piercing, such as mosquitoes and deer flies, the mandibles and maxillae are stylet-like and function to pierce the tissues fed upon. Flies undergo complete metamorphosis. The larvae (maggots) are legless and in many the head is poorly developed consisting of only a pair of parallel mouth hooks which are invaginated into the pointed anterior end. The pupae may be either naked or enclosed in a brown segmented, cylindrical structure (puparium), which is formed from the last larval skin. Flies are of interest to foresters primarily because many species are pests of man and because some are parasitic on other insects. A few cause minor damage to trees.

Among the remaining groups are the walking sticks, the roaches, and the grasshoppers. These all belong to the order Orthoptera (orth = straight, ptera = wing). The winged members have two pairs of wings. The forewings are narrow and frequently somewhat thickened, whereas the hindwings are broad and membranous. Some species are always wingless. The mouthparts are always of the chewing type. The members of this order also undergo gradual metamorphosis; consequently, they have no pupal stage. The nymphs have compound eyes and wing pads when the latter are present on the adults. Sometimes walking sticks are defoliators of oak forests, whereas grasshoppers may cause injury to seedlings planted in grassy areas.

See the references listed at the end of Chapter 2.

CHAPTER 4

Insect Abundance (Ecology)

It is obvious that large insect populations must be present in order to cause significant tree damage; therefore, it is desirable to understand the principles of insect population development and especially to appreciate the effects various environmental factors may have on insects. If the conditions necessary for excessive insect multiplication are known, valuable predictions and adequate preparations can be made to prevent or reduce the expected economic loss. At the present time, however, the importance and exact role played by many of these factors on even the most destructive forest pests are known only imperfectly. Although much research is being done along these lines at the present time, the great many ways each factor, singly and in concert with others, may influence a population makes the problem very complex. Nevertheless, knowledge of this kind is most important because it may permit development of silvicultural methods which will yield the best practical preventative measures.

The large number of insect species in existence indicates that they have successfully adapted themselves to changing conditions in the past. Many of the resulting pestiferous species also are very successful with respect to producing large insect populations. Nevertheless, none of these injurious forest insects ever develops large destructive populations which continue at a high level until all the host trees are destroyed over continental areas. Even on smaller areas populations of forest insect pests exhibit a cyclic or periodic pattern. Great insect abundance and the resulting damage may continue for a time, but then the insects again become relatively few in number. These large population fluctuations suggest that only small environmental changes may cause the balance to tip in favor of or against the insects. At the peak of these periods of abundance large quantities of timber may be destroyed, whereas during the intervening low periods these same insect species may become so scarce that even trained entomologists have difficulty finding specimens. Some of the most destructive forest insect species exhibit this cyclic characteristic, whereas other tree-infesting species that are constantly present, and always easily found, often are not as troublesome.

The factors that influence insect population development may be either biological or physical. The biological factors include reproductive and survival powers of the animal, the quantity and quality of food, and the various parasites and predators that attack and destroy the host organisms.

BIOTIC FACTORS

INHERENT FACTORS

The size of any animal population is determined partly by various constantly changing factors which are inherent in the organisms and partly by the suitability of external environmental conditions. The inherent qualities any insect species needs in order to be successful

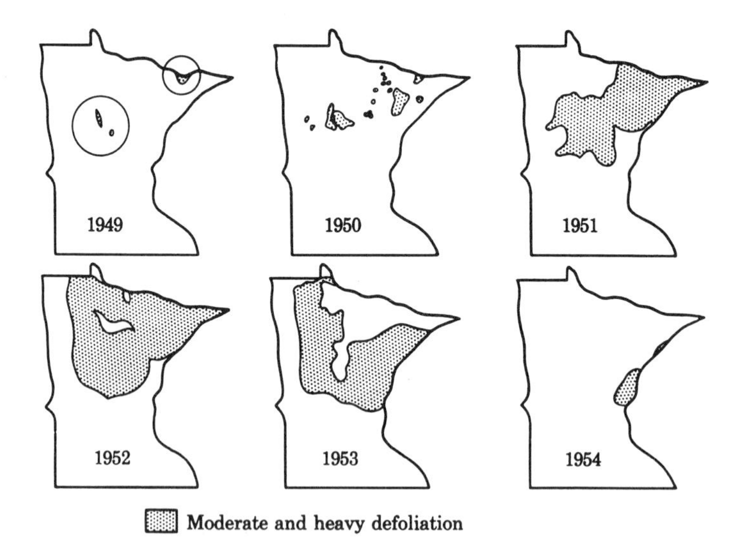

Fig. 4.1. History of an outbreak of the forest tent caterpillar in Minnesota. After J. W. Butcher.

include the following: sufficiently great reproductive powers; a short life cycle; and sufficient adaptability to survive and thrive under a wide range of environmental conditions. Such an inherent property of an organism to reproduce and survive is known as the *biotic potential.* The components of this concept, therefore, are the *reproductive potential* and the *survival potential.* The first of these, the reproductive potential, may be defined as the total population that develops in a one year period for each adult present at the beginning when optimum environmental conditions exist throughout the period. The various factors that comprise the reproductive potential include fecundity, or number of offspring produced per female; length of life cycle; proportion of females in the population; parthenogensis (reproduction without fertilization); and polyembryony (production of 2 or more individuals from 1 egg). Parthenogensis occurs in only a few forest insect pests and polyembryony is characteristic of only a few parasitic wasps.

The fecundity of forest insects is often very high. Frequently the average number of eggs laid per female is between 100 and 300. Sometimes, however, with insects such as termites, ants, and some bees, where the reproductive function is limited to only a few individuals, the number of eggs produced per female is in the thousands. Most forest pest insects complete their life cycle in 2 months to 2 or 3 years. A 1-year cycle is very common, but rarely extremely long cycles of 13 to 20 years may occur. Some wood borers have their life cycles extended greatly when they are living in wood in which the nutrient and/or moisture conditions are unfavorable for the insects. One cicada species normally has a very long life cycle of 13 or 17 years. Males of a few forest insect species are scarce or even absent; consequently, most of these females reproduce parthenogenetically. Usually, however, reproduction is bisexual, and the percentage of females in the population is near 50.

If certain data are available for a species, the reproductive potential can be calculated readily. A few computations made for a species with an assumed fecundity of 100, a sex ratio of one male to each female, and 1, 2, 3, or more generations per year, will indicate the theoretical rapidity with which such populations can increase under most favorable conditions. By continuing these calculations for a few years, the expodential characteristic of the reproductive potential curve can be demonstrated. These theoretical maximal population growth rates never occur, although they may be approached for a time under very favorable conditions. If the theoretical rate of increase continued very long for any insect species the earth would soon be overrun by these animals.

Reproduction data collected on insects reared under carefully controlled optimum conditions are seldom available. In fact, the optimum conditions for most species are not known. Usually the best data available for estimating the reproductive potential of forest insect species is that collected during the time of rapid increase or during the development of an outbreak. Even under these conditions, however, we only have the realized reproductive performance under very favorable conditions. Therefore, inasmuch as the environmental conditions seldom, if ever, are optimum for a species, the full impact of the reproductive potential is seldom, if ever, realized.

The other component of biotic potential is the survival potential. It is the inherent property of a species to survive and thrive. This toughness or vigor of a species is similar to the various other inherent factors in that it is not a constant value for a species but continually changes as the animal population changes. A good example of how this may occur is suggested in a study of Wellington's (1957).

The more important external biotic factors which influence insect population development are food quantity and quality, the population size, and the effectiveness of predators and parasites.

NUTRITIONAL FACTORS

It is obvious that nutritional factors are of great importance for food is a basic need of any type of life. Forest insect pests require certain of the food constituents such as carbohydrates, proteins, fats, minerals, and some of the vitamins. As would be expected, those insects which feed on the more nutritious parts of trees, such as the leaves and inner bark, develop faster and, consequently, have shorter life cycles than those that feed on the woody parts. Nutritional factors that influence forest insect population development are the quantity of suitable hosts trees, the density and species composition of the stands, host age and vigor, and in a few cases, the presence of alternate hosts. Some forest insects are limited to only one host, but usually the most troublesome forest pests feed and develop to maturity on several or many species of trees. Even for these latter insects, however, the hosts are usually limited to groups of related tree species. One reason for this host specificity is the oviposition habits of the adult females. This characteristic of the females to select the same species of trees as that on which they and their ancestors for many generations were reared is known as *Hopkins' Host Selection principle.*[1] The principle applies chiefly under natural conditions, for, even though some species may have several hosts, the

[1] Hopkins was the first chief forest entomologist of the U. S. Department of Agriculture. He did much basic work in forest entomology and elucidated the principle bearing his name.

insects often appear to thrive best on only one of these hosts. In different localities, however, the preferred hosts of the same insect species may be different.

Certain tree species growing in mixed stands are attacked by insects only infrequently; consequently, these trees are thought to be exceptionally resistant. Often, however, when these same trees are grown in pure stands their supposed resistance is not apparent. The reasons for this phenomenon are not completely understood, but the resistance may be partly due to the mechanical barrier of unsuitable trees which separates the preferred trees from one another. It also may be partly due to the larger and more varied parasite and predator population supported by the more diverse endemic insect host population present in mixed stands.

Tree vigor is often of great importance in determining vulnerability of living trees to borer attack. Many of these insects are unsuccessful or develop poorly when they infest either dead trees or trees that are growing vigorously. They live and thrive best only when the hosts are growing poorly or are temporarily in a state of reduced vigor. This relationship between host vigor and insect attack frequently results in selective attacks and tree destruction that can be either detrimental or beneficial to forest stands. If the poorly growing trees thus destroyed are distributed in such a way that the remaining trees are provided with needed growing space, the stands are benefited. On the other hand, should much of the stand be in a state of reduced growth vigor due to temporary unfavorable growing conditions, the resulting tree destruction may reduce the stand below optimum stocking. Economic loss then occurs.

Leaf-feeding insects, on the other hand, are not obviously related to host vigor. Nevertheless, heavy continued attacks by defoliators often cause tree decadence and thereby help produce suitable host conditions for the borers.

These fundamental factors of food quantity and quality are often determined by stand composition. In the past, logging, clearing, burning, and other activities of man have caused great changes in forest stand composition. The resulting more temporary forests, which replaced those destroyed, generally were more insect susceptible. The greater insect resistance of the virgin forests may be one reason why these particular tree communities evolved as climax types. This lower incidence of insect attack in climax types may not be due to any resistance *per se*, but is more likely the result of a complex interrelation of the environmental factors acting to hold the insect populations at low levels so that little damage is caused. Most tree species that make up climax types are not necessarily free of insect attack, but generally they are troubled less than are the early or intermediate successional stages.

Some insects prefer to attack trees of a certain age group (for example, the western pine bark beetle infests mostly overmature trees), with the younger and the older trees usually suffering most. Nevertheless, no age class is immune; consequently, these insect attacks help to reduce the number of trees in stands throughout the life of the forests.

Starvation sometimes becomes an important factor in limiting or causing a decrease in an insect population. If the lack of food is sufficiently prolonged, death results, whereas partial starvation often causes reduced fecundity and reduced vigor. The insects weakened by starvation also are more susceptible to other environmental pressures.

PARASITES AND PREDATORS

The last of the biotic agents to be considered, which often help reduce large insect populations or prevent them from developing, are parasites and predators. *Parasites include those organisms that live in or on a host organism of another species and from which the parasite obtains sustenance.* It is an antagonistic type of symbiosis (sym = together, bios = life) in which one of the partners suffers and the other benefits as a result of the association. The parasite is the attacking organism, whereas the host is the organism attacked. When insect parasites attack insect hosts, death of the latter usually occurs. *Predators are free living organisms which kill the hosts quickly by external attack and always require numerous hosts in order that they may develop and sustain themselves during their lifetime.* Some species, however, can not be classified readily, for they have habits that are intermediate between the two types.

Organisms that are parasitic on insects occur in both the plant and the animal kingdoms. Those that might be considered plants include certain viruses and Thallophytes (bacteria and fungi). Parasitic animal species are found in the phyla Protozoa (one-celled animals), Nemathelminthes (nemas or round worms), and the Arthropoda (mites and insects). The more important predators of forest insect pests are insects, spiders, birds, and mammals. Other animals of minor importance as forest pest predators are fish, amphibians, and reptiles.

Birds are the most abundant and important insectivorous vertebrates present in forests. During the breeding season, when young are being fed, birds consume the greatest quantity of insects. At this time of the year forest pests are frequently the most available and desirable food for the birds. Different bird species naturally have different food habits. It is well known that woodpeckers feed on boring insects; flycatchers catch most of their food when in the air; warblers feed on crawling insects and thrushlike birds eat both crawling animals and vegetable food. Even many adult birds that are primarily seed eaters, such as sparrows and cardinals, feed their young mostly with insects. It is possible, but difficult, to evaluate the importance of predators; consequently, we do not have much precise information on their effectiveness in reducing populations of forest pests. The most intensive work along this line has been with birds feeding on the spruce budworm (Dowden et al., 1953, George and Mitchell, 1948). The effectiveness of woodpeckers in reducing bark beetle populations also has been reported several times.

The most important insect predators of forest insect pests are the larvae and adults of beetles. These predators include checkered beetles (Cleridae) and ostomids (Ostomidae), both of which feed on bark beetles, the aphid-eating lady bird beetles (Coccinellidae), and the ground beetles (Carabidae), which feed on caterpillars. Syrphid fly maggots (Diptera, Syrphidae) and aphid lion larvae (Neuroptera, Chrysopidae) both are voraceous predators of plant lice. Many species of adult ants (Hymenoptera, Formicidae) also commonly prey on insects, but other ants have different habits. These ant species protect and take care of plant lice and insect scales for the purpose of obtaining the excreted honeydew (mutualistic symbioses). Various other predaceous insects including papermaking and digger wasps, dragon and damsel flies (Odonata), certain bugs (Hemiptera), and mantids (Orthoptera) also destroy insects, but these predators never have been effective in reducing large insect host populations.

Many species of ichneumonid wasps (Hymenoptera) are among the most common and important parasitic insects. Of the flies (Diptera) that are parasitic, the tachinid or larvaevorid flies (Tachinidae) are the most important, but sometimes flesh flies (Sarcophagidae) are very effective. Some parasitic insects always attack only one host species, whereas others may attack many. One advantage of those parasites that attack many different species of insects is that they are better able to maintain themselves during periods when the population of the main pest host is low.

Most adult parasitic wasps and flies feed on the nectar of flowers. The habits of some, however, differ. For example, the parasitic flesh flies usually feed on carrion, and some of the parasitic wasps have the habit of feeding on the exudate from body wounds made on the hosts.

The oviposition habits and mode of reproduction vary somewhat for the various species of parasitic insects. Some reproduce parthenogenetically and for a few the eggs always develop by polyembryony. A few parasitic flies are viviparous, but most flies and wasps lay eggs.

Generally the flies glue their eggs to the outside cuticular surface of the hosts, whereas most of the parasitic wasps insert their eggs beneath the insect cuticle. A few simply lay their eggs in places frequented by the host insects so that the newly hatched parasites must either crawl onto or else be ingested by the hosts when they feed on the foliage. This type of oviposition habit is most common for some parasitic wasps that attack wood borers and ocassionally occurs for some flies that parasitize leaf-feeding caterpillars. These female wasps insert their ovipositors into and lay their eggs in galleries inhabited by wood-boring grubs, whereas the flies glue their eggs on foliage. In the latter case, the ingested parasite eggs hatch when stimulated by gastric juices of the hosts, and the young maggots immediately bore through the gut wall. They complete their development in the body cavities of the hosts.

Usually only the larvae are parasitic. They develop either internally or externally, depending on the species. Those living externally attach themselves to the host insects and feed through holes in the body walls. This type of attack occurs most commonly with those wasp parasites that parasitize insects concealed in wood or in soil. Generally, however, the larvae of most parasitic insects live inside the bodies of the hosts where they are immersed in the nutritious blood. Usually the parasitized insects remain alive until the parasite larvae approach maturity. At this time depletion of the host tissues is so great that the parasitized insects die. Some species, however, quickly kill the host insects and then live as saprophytes on the dead tissues. Endoparasitic larvae obtain the oxygen needed for respiration in various ways. The younger ones probably satisfy their small requirements by absorbing the dissolved oxygen from the blood of the hosts; but as the parasites grow, they may rupture the smaller trachea and thereby increase the oxygen supply. Certain parasitic species, on the other hand, establish respiratory connections with the free external air by connecting their spiracles with large tracheal branches or the body walls of the hosts. Most insect parasites leave their hosts before they pupate, but some species pupate inside the dead insects. Those that leave may spin a cocoon or form a puparium close by, or they may drop to the ground and enter the litter or soil to pupate.

There are various relationships between parasites and hosts that are of importance in determining the effectiveness of any parasite. Sometimes two or even more different species parasitize the same individual (multiple parasitism), with the result that one usually destroys the other. Sometimes in this way the effectiveness of an otherwise good parasitic species may be greatly reduced. When one parasite parasitizes another parasite, the phenomenon is called *hyper-parasitism.*

The most effective parasitic control of an insect pest species usually is realized when each of several parasites attack a different stage during host development and thereby keep the host insects under attack throughout much of their life. Thus, each parasite destroys a portion of the pest population and yet does not compete too greatly with the other parasites. Another important requirement for parasites to be effective in reducing pest populations is that they should not be highly specific. Those that are general feeders have the advantage of being able to maintain themselves on other host species at times when the pest population is low.

PHYSICAL FACTORS

CLIMATIC FACTORS

Many physical factors, including temperature, light, precipitation, humidity, and wind, help to regulate animal abundance. Various combinations of intensities of the aforementioned factors are known as *weather* and *climate.* These two words have somewhat similar meanings but differ in that climate defines the average weather conditions—especially average extremes—over a long period of years, whereas weather refers to the atmospheric conditions which prevail at any particular hour, day, week, or month. The climate of an area may be such that individuals of an insect species cannot become sufficiently numerous to cause trouble or perhaps cannot even exist. In such places the greatest deviations in the weather from the expected climatic conditions are never great enough to be favorable for the insect. On the other hand, insects that do become epidemic are generally adapted to the climatic conditions of the areas where they occur. In these locations the weather factors often determine whether the insects thrive.

The atmospheric conditions immediately surrounding individual insects (for example, insects boring inside trees or logs or those living near the ground on young reproduction) may be very different from those occurring in the open atmosphere of the forests (Graham, 1924; Henson, 1952; Wellington, 1950). These localized conditions are called *microclimates.*

The numerous references cited at the end of the chapter indicate that there is much current interest in the physical factors composing weather and climatic as they influence insects and insect population development. Canadian research forest entomologists are especially active in this field.

Insects are characterized as *cold-blooded.* This term signifies that the internal temperatures of animals so constituted are at or very near that of the surrounding atmosphere. *Warm-blooded* animals (mammals and birds) on the other hand, automatically regulate their internal temperature so that it remains rather constant irrespective of the temperature of the environment. Because insects are dependent on the temperature of the surrounding air, they can grow and develop completely from egg to adult only when the temperature is within a certain favorable range for each of the various stages. These suitable developmental temperatures are called *effective temperatures.* As would be expected, the effective temperature zone may vary for different stages of the same insect as well as for the different insect species. Commonly it is between 60° and 100° F. In general, the rate of development throughout much of the effective temperature range is directly proportional to the temperature—the higher the temperature the faster the development. The lower and upper limits of this developmental zone for any stage of any insect species are called the *minimum* and *maximum threshold temperatures.* Somewhere within the effective temperature zone there is a narrower range—the *optimum temperature zone.* Other things being equal, the rate of insect population growth for any species is greatest when the temperatures remain within optimum growth zone. At higher temperatures the rate of development may be greater, but this rate is more than offset by greater mortality; at lower temperatures, survival may be better, but rate of development is slower. Under outdoor conditions, however, narrow temperature ranges seldom occur throughout the developmental period of any stage, but probably the closer the temperatures are to these optimum ranges the faster will be the growth of the insect populations.

Both below and above the effective temperature ranges are *dormancy zones.* Insects exposed to temperatures in these zones become inactive but survive for variable periods of time. This adaptation enables them to live through periods of unfavorable conditions such as commonly occur during the winter. At the upper and lower extremes of the dormancy zones are the *death zones.* In these zones, insect mortality increases as the temperatures approach the extremes, and the exposure time lengthened. At either the high or low temperature extremes two factors are of importance with respect to amounts of mortality. These factors are the intensity of the unfavorable temperature (the actual temperature) and the duration or length of exposure. Commonly these mortality zones for forest insects occur between 100° and 125° F for the upper zone and from freezing to −50° F or below at the lower end. Insect species, stage of development, and physiological condition largely account for the resistance or susceptibility to these extremes.

The exact causes of death due to either high or low temperatures are very complex, and therefore they are only imperfectly understood. Accumulation of metabolic wastes has been suggested as one possible cause for both types of mortality. Some insects die only when the body fluids freeze, whereas those that are very cold hardy can survive after being frozen solid. During the gradual autumnal cooling, those insects that live through the winter commonly become physiologically better adapted to survive the cold. As the amount of water decreases, the quantity of salts increases, and the colloids change so that they incorporate or "bind" part of the water. This bound water cannot be frozen. When dry insects are exposed to freezing temperatures they usually are protected somewhat by the body fluids "undercooling" to a temperature some degrees below the actual freezing point of their body fluids. This phenomenon of undercooling apparently is associated with the absence of suitable nuclei for forming ice crystals. As soon as the first crystals form, however, freezing of the body fluids occurs instanteously. The amount of undercooling varys from a few degrees below 0° C to −20° C or lower. When free water is in contact with the external insect cuticle, the undercooling phenomenon does not occur. The ice crystals form in the external water and in some unknown way initiate ice crystal formation in the body fluids; therefore, exposed wet insects do not undercool. These various mechanisms all help to protect insects from being frozen.

In addition to these protective physiological changes, many insects also avoid unfavorable cold conditions by concealing themselves in protected places such as the ground litter or soil, where the temperature extremes are not as great as they are in the atmosphere.

Light is seldom a limiting factor for insects. Under natural conditions the effects of light on the activities and development of insects are difficult to separate from the effects of temperature. In a few cases, however, it is known that seasonal changes in the amount of daylight induce the production of winged forms and the generation of males for some species of aphids. Insect behavior also commonly is associated with light. For example, most metallic wood borers and some long-horned borers prefer those parts of trees exposed to full sun, because of the higher temperatures and/or the greater amount of light (Graham, 1924). Leaf-eating caterpillars also may move about in the tree crowns for the purpose of finding most suitable light conditions (Wellington, 1948). This positive phototaxic response is very evident with many species that defoliate the tops of the tree crowns first.

The main effects that variations of air pressure have on insects result from the storms associated with or caused by these air-pressure changes. Winds may be either beneficial or detrimental to insects, depending on their direction and velocity. Insect destruction may be great if the animals are blown too violently against solid objects or into areas devoid of food. On the other hand, some forest insect epidemics have been traced to winds that transported the insects from distant breeding place to other stands of suitable host trees.

Moisture in the air may exist as a gas, a liquid, or a solid. In any of these forms it may affect insect populations. The gaseous form is known as humidity and commonly is expressed as relative humidity. This term expresses the percentage of water vapor in the air as compared to the amount that the air can hold at the particular air temperature being observed. Insects have certain humidity requirements. When the atmospheric moisture deviates too greatly from these needs, the insects suffer and often die. Also involved is the rapidity and efficiency with which lost water can be replaced by ingested and/or metabolic water. Much of the water loss that occurs by evaporation escapes through the spiracles because the cuticle itself usually is well waterproofed with a thin outer waxy layer. Insects with thick hard cuticle, such as beetles, are especially resistant to dehydration. The fact that some insects can live in dry environments is not necessarily contingent upon whether they live exposed or concealed, for some borers (powder post beetles) live in rather dry wood.

Sometimes hail, sleet, and rain may be responsible for killing many exposed insects, whereas those better protected (borers, tent makers, and soil insects) may escape. Snow on the other hand, is seldom an important destructive agent, because when snow falls most insects are hibernating. On the other hand, snow may be beneficial for those insects that hibernate in the soil, in litter, or close to the ground. In these locations a flocculent blanket of snow can insulate the insects against the more intense atmospheric cold.

These various climatic and weather factors always are operating and affecting any insect populations present. Sometimes all the physical factors may be favorable for the insect, whereas at other times one or more of the elements may be unfavorable. In either case, the resulting interaction of these various factors frequently presents a rather complex picture. These influences may also act indirectly, especially on those borers that develop more favorably when the radial growth rate of the trees is reduced or even stopped for a period of time (Anderson, 1944). The reduction of tree growth often is due to adverse weather conditions, such as drought and floods.

Generally when physical factors affect an animal population directly, the proportion of the animals destroyed is constant, irrespective of the size or density of this population. Such influences sometimes are referred to as *density independent* factors. Certain biotic factors such as parasites and predators, on the other hand, vary in effectiveness according to the density of the host population; consequently, these are called *density dependent* factors.

SELECTED REFERENCES

Allee, W. C., O. Park, A. E. Emerson, T. Park, and E. P. Schmidt. 1949. *Principles of animal ecology.* W. B. Saunders Co., Philadelphia, Pa.

Anderson, R. F. 1944. Relation between host condition and attacks by the bronzed birch borer. *J. Econ. Entomol.* 37:588–596.

Andrewartha, H. B., and L. C. Birch. 1954. *The distribution and abundance of animals.* University Chicago Press, Chicago, Ill.

Bird, F. T. 1957. A Virus disease and introduced parasites as factors controlling the European spruce sawfly. *Canadian Entomol.* 89:371–378.

Blais, J. R. 1952. The relationship of the spruce budworm to the flowering condition of balsam fir. *Can. J. Zool.* 30:1–29.

———. 1953. Effects of the destruction of the current year's foliage of balsam fir on the fecundity and habits of the spruce budworm. *Can. Entomol.* 85:446–448.

———. 1957. Some relationships of the spruce budworm to black spruce. *Forestry Chron.* 33:364–372.

Buckner, C. H. 1956. Mannalian predators of the larch sawfly. *Proceed. 10th Internal. Congr. of Entomol.* 4:354–362.

Chapman, R. N. 1931. *Animal ecology.* McGraw-Hill, N. Y.

Christensen, C. M., and A. C. Hodson. 1954. Artificially induced senescence of forest trees. *J. Forestry* 52:126–129.

Clausen, C. P. 1940. *Entomophagous insects.* McGrow-Hill, N. Y.

Craighead, F. C. 1921. Hopkins host selection principal as related to certain cerambycid beetles. *J. Agri. Research* 22:189–220.

———. 1925. Bark-beetle epidemics and rainfall deficiency. *J. Forestry* 23:340–354.

Dowden, P. B., H. A. Jaynes, and V. M. Carolin. 1953. The role of birds in a spruce budworm outbreak in Maine. *J. Econ. Entomol.* 46:307–312.

Freeman, J. A. 1945. Distribution of insects by aerial currents. *J. Animal. Ecol.* 14:128–154.

George, J. L., and R. T. Mitchell. 1948. Calculations on the extent of spruce budworm control by insectivorous birds. *J. Forestry* 46:454–455.

Graham, S. A. 1924. Temperature as a limiting factor in the life of subcortical insects. *J. Econ. Entomol.* 17:377–383.

———. 1956. Ecology of forest insects. *Ann. Rev. Entomol.* 1:261–280.

Greenbank, D. O. 1956 and 1957. The role of climate and dispersal in the initiation of outbreaks of the spruce budworm in New Brunswick. *Can. J. Zool.* 34:453–476 and 35:386–403.

Henson, W. R. 1951. Mass flights of the spruce budworm. *Can. Entomol.* 83:240.

——— and R. F. Shepard. 1952. The effects of radiation on the habitat temperatures of the lodgepole needle miner · · ·. *Can. J. Zool.* 30:144–153.

Hodson, A. C. 1941. An ecological study of the forest tent caterpillar. *Univ. Minn. Tech. Bull.* 148.

Hopping, G. R. 1950. Timber types in relation to insect outbreaks in the Canadian Rocky Mountains. *Ann. Rept. Entomol. Soc. Ontario* 81:72–75.

Morris, R. F., C. A. Miller, D. O. Greenbank, and D. G. Mott. 1958. The population dynamics of the spruce budworm in eastern Canada. *Proc. 10th Intern. Congr. Entomol.* (1956) 4:137–149.

Patterson, J. E. 1930. Control of the mountain pine beetle in lodgepole pine by the use of solar heat. *U.S.D.A. Tech. Bull.* 195.

Soloman, M. E. 1957. Dynamics of insect populations. *Ann. Rev. Entomol.* 2:121–142.

Steinhaus, E. A. 1949. Principles of insect pathology. McGraw-Hill, N. Y.

Sullivan, C. R., and W. G. Wellington. 1953. The light reactions of larvae of the tent caterpillars. *Can. Entomol.* 85:297–310.

Wellington, W. G. 1948. The light reactions of the spruce budworm. *Can. Entomol.* 80:56–82.

———. 1949. The effect of temperature and moisture upon the behavior of the spruce budworm. *Sci. Agri.* 29:201–215, 216–229.

———. 1950. Variations in the silk-spinning and locomotor activities of larvae of the spruce budworms . . . at different rates of evaporation. *Trans. Royal Soc. Canada* 44:89–101.

———. 1950. Effects of radiation on the temperatures of insect habitats. *Sci. Agri.* 30:209–234.

———. 1952. Air mass climatology of Ontario . . . before outbreaks of the spruce budworm and the forest tent caterpillar. *Can. J. Zool.* 30:114–127.

———. 1954. Atmospheric circulation processes and insect ecology. *Can. Entomol.* 86:312–333.

———. 1957*a*. Individual differences as a factor in population dynamics: The development of a problem. *Can. J. Zool.* 35:293–323.

———. 1957*b*. The synoptic approach to studies of insects and climate. *Ann. Rev. Entomol.* 2:143–162.

Wellington, W. G., J. J. Fettes, K. B. Turner, and R. M. Belyea. 1950. Physical and biological indicators of the development of outbreaks of the spruce budworm. *Can. J. Research D.* 28:308–331.

Yuill, J. S. 1941. Cold hardiness of two species of bark beetles. *J. Econ. Entomol.* 34:702–709.

CHAPTER 5

Insect Control

"Insect control" refers to any method that can bring about the reduction or prevention of damage caused by insects. This objective may be accomplished naturally without any conscious help from man, or he may apply some direct effort; thus, insect control methods are commonly referred to as either *natural control* or *applied control.*

NATURAL CONTROL

Natural control results when the operation of the naturally occurring environmental factors are such that the pest animals are adversely affected and reduced in numbers so that economic damage does not occur. These various factors include those already discussed such as physical factors, nutrition, and entomophagous organisms. Sometimes a single factor may exert a tremendous pressure on an insect population, whereas at other times a number of factors seem to be equally important. Whether one or many, those factors which are of greatest importance in effectively acting against an insect population are the *limiting factor(s).* Natural control, therefore, may be defined as the equalizing action of naturally occurring factors that balance the population sizes of all plants and animals in a community in a manner favorable to man's economic interests. Frequently, this economic concept of natural control is referred to as the *balance of nature.* Probably a better term would be *balance of nature favorable to man.* Another definition for this idea of the *balance of nature* concept is that it is an assumed tendency of the various environmental factors to act as a series of checks and balances in order to keep the various plant and animal species relatively constant in number. Some people like to think that the most favorable balance of nature usually occurs if man does not interfere with nature. This idea may be true for some communities under some conditions, but frequently complete dependence cannot be placed on this natural control to maintain desirable population levels of organisms in any given community.

Frequently, the population sizes of the various associated species in a community vary greatly and do not remain constant for very long. In times past, the activities of man undoubtedly have accentuated these population fluctuations. In forests, for example, operations such as logging, burning, clearing, and cultivating the land have caused great changes. These operations were conducted on a tremendous scale and resulted in replacement of the more stable climax forest types with the less stable, earlier successional stages of forest growth. The more desirable white and red pine forests of the New England and Lake States regions have been largely replaced with post-fire types consisting of species such as aspen, birch, cherry, and jack pine. In the Southeast Piedmont region the climax oak-hickory forests have been replaced by the more valuable pines; whereas in the West, lodgepole pine and brush are the temporary post-fire types. These sub-climax type forests often are more susceptible to certain types of insect depredations than were the virgin stands.

Overmature stands, even those of the climax type, especially are vulnerable to insect attacks because of conditions associated with the extremely slow radial tree growth. Ordinarily these stands should be replaced by faster growing younger stands, but often this practice is not economically feasible; therefore, man tries to store the timber on the stump until needed at some future time. In these cases, when the timber is destroyed the economic concept of the balance of nature favorable to man (natural control) fails.

When insect damage is great, foresters immediately are concerned with knowing how long the damage will continue if no direct control is applied and whether direct control should be applied. As yet we do not have sufficient knowledge regarding the collection and use of data to compute the exact answers to these questions. Nevertheless, population trends can be obtained, by periodically sampling to determine insect abundance. It is especially important to make some of these observations immediately following any abnormal weather conditions, because, at such times, great reductions in insect numbers frequently occur.

When parasites are involved, their relative importance can be readily determined by periodically collecting insect samples in each stage throughout the developmental period and rearing them in confinement. From these collections host mortality and the identity of the reared parasites can be determined. Sometimes, however, parasitism can be detected on the basis of external appearance of the hosts or by disecting and removing the parasites. The latter is very time consuming and specific identification of the larval parasites is difficult, if not impossible, in most cases.

It is more difficult to evaluate the effectiveness of predators. Reduction in insect numbers not caused by adverse weather conditions or the activity of parasites may be due to predation. Involved time-consuming studies are necessary to determine the efficiency of larger predators such as birds or mammals, but approximate estimations of the effectiveness of the smaller predators can be made more easily. For example, the rate of host consumption by a predator can be determined roughly by feeding the host prey to individually caged predators and then estimating the predator population of the pertinent areas by sampling. By combining these two sets of data, an estimate of the predator effectiveness for an area can be calculated. Of course, caution must be used when evaluating such data, because confined conditions are not exactly comparable to those which occur in forests.

Whenever dealing with mortality figures, one should be careful not to overestimate the effectiveness of high mortalities. As an example, if each insect female produces 100 offspring that have a sex ratio of 1 male to 1 female, the total mortality necessary, from egg to adult, to merely hold the population constant is 98 per cent. Consequently, under these conditions, only when the total mortality is greater than 98 per cent will the next generation start out with fewer numbers than the previous one. Of course, a high mortality during the early developmental stages of the following generation may result in a lower population irrespective of the survival of the parent generation. Similarly, a high mortality of only one stage does not necessarily mean that the population will be reduced the next generation. For example, a species whose biological characteristics are the same as those given previously may have 90 per cent of the pupae destroyed without suffering a reduction in numbers during the following generation. This situation would occur if the total mortality due to all other factors were less than 8 per cent. If the mortality up to the pupal stage was 20 per cent, and then 90 per cent of the pupae were parasitized, the total number of surviving offspring for each pair of parents would be 8. Thus, even with a pupal mortality of 90 per cent the number of offspring would still be four times larger than the parent population.

The total mortality from all causes is difficult to determine, whereas that due to certain factors such as parasites can be estimated more easily. Nevertheless, the sum of all mortalities occurring to each of the stages is needed if accurate predictions are to be made. Students of animal populations always have been interested in total mortalities and total survivals as they occur throughout the life of a species. They are especially interested in knowing when the mortalities are greatest because these most vulnerable times often greatly influence future population trends. Data of this type are now being tabulated in a manner similar to that used by human actuaries, and these tables are also called *life tables* (Morris and Miller, 1954).

Sometimes the natural control factors, which act very rapidly, decimate large insect populations in a seemingly magical manner. More frequently, however, the controlling factors operate inconspicuously by preventing small insect populations from becoming large.

APPLIED CONTROL

Whenever the natural controls of an insect pest become inoperative so that severe tree injury occurs or seems likely to occur, forest managers immediately become concerned and want to know what can be done. Well-informed foresters, therefore, should be familiar with the available control methods so that they can approach the problem at hand more intelligently.

Historically, the almost unlimited forests present in North America made it necessary to use very extensive types of forest management; therefore, applied control methods were used only for emergencies. Recently, however, more intensive management practices are being developed sc that in the future more effort probably will be directed toward applied forest insect control.

In order to be effective, applied control, like natural control, must destroy a sufficiently large proportion of the insect pest population. Therefore, control operations only partially completed may be ineffective unless they are supplemented by a sufficient amount of natural control. Removal of only part of a large population, which has a high reproductive potential, may serve only to temporarily reduce some of the competition for food, and possibly space, so that the remaining insects will thrive better. As a consequence, the population may quickly expand and soon be as large or larger than it was before control was undertaken.

Applied control methods can be classified under five headings:

1. Indirect control by silvicultural and management techniques

2. Direct physical methods
3. Use of entomophagous organisms (biological control)
4. Regulatory control
5. Direct chemical methods (insecticidal control)

SILVICULTURAL CONTROL

Sometimes the most practical type of long-term forest insect control is to develop insect-resistant forests by the use of silvicultural methods. These methods are directed toward preventing the occurrence of insect-caused damage, rather than waiting for trouble to start. One advantage is that the work frequently can be done at the same time other management operations are carried out. This makes it possible to perform the insect control work in a more orderly and less costly manner.

Control of forest insects by silvicultural methods consists of regulating the forest growth so that food conditions are unfavorable for the injurious insects. The quantity and quality of insect food are controlled by regulating the composition, density, vigor, and age of the stands. If certain trees are highly favored by an insect, and these are continually injured, it may be best to reduce the quantity of these hosts (for example, the spruce budworm). Sometimes mixed stands are less susceptible to insect damage than are pure stands. In such cases, mixtures of various tree species should be favored and developed whenever practicable. Sometimes, however, forest managers may consider it desirable to grow insect-susceptible trees under vulnerable conditions; but whenever this is done, the foresters should be fully aware of the dangers involved and be prepared to apply direct control when necessary.

PHYSICAL METHODS OF CONTROL

Mechanical and physical insect control methods include the use of the following methods:

1. Unfavorable temperatures
2. Unfavorable moisture conditions
3. Insect habitat destruction
4. Trapping and destroying the insects
5. Excluding insects by mechanical means
6. Insect destruction by hand or mechanical means

1. ***Unfavorable temperatures.*** The most obvious use of high temperatures for killing insects is to burn them. Often this can be accomplished most easily by burning the insect-infested wood. When large trees are involved, this procedure requires much work, because the trees usually have to be felled and sometimes even moved, to prevent dam-

aging adjacent uninfested trees. Frequently, this use of fire is costly and inconvenient, because the burning must be restricted to certain times of the year when the danger of starting forest fires is low. Nevertheless, much burning has been done in the past, especially for the control of those insects that bore and conceal themselves beneath the bark (bark beetles). Currently, however, this method is being supplanted by insecticide applications.

Destruction of insect-infested wood products by burning is usually easier to accomplish, because the smaller size of the partially manufactured products facilitates handling, and facilities generally are available so that burning can be done safely and cheaply at any time. Destruction by burning, therefore, is an economically feasible method to help control insects such as powder post beetles that infest wood debris around mills and factories. Exposure of lightly infested but still usable materials to high temperatures in kilns also is a practical control method for killing powder post beetles.

If thin-barked logs infested with bark beetles are exposed to full sunlight for one to several days during the summer months, those on the top side are killed. Because only some of the insects are killed during any one period of exposure, this solar treatment necessitates turning the logs so that all surfaces eventually are fully exposed to the hottest rays of the sun. This entails an excessive amount of work that, in combination with the variability of the atmospheric conditions, makes the method costly to use and the results uncertain; therefore, seldom is it practical to use this solar method for controlling inner-bark borers.

2. ***Unfavorable moisture conditions.*** Sometimes the moisture content of wood products can be regulated so as to repell or influence the activity of the infesting insects. Sawing logs soon after they have been cut and rapidly drying the lumber produced, quickly reduces the wood moisture content so that it becomes unfavorable for many boring insects such as ambrosia beetles. Conversely, delayed utilization and slow drying, especially during the warm summer months, are conducive to attacks by these pinhole borers. The principle behind one important method of subterranian termite control is also moisture regulation. When colonies of these termite pests become established in the dry wood of buildings, the insects usually must obtain the needed moisture from the soil; therefore, when a portion of a colony in a building is isolated from that part in the soil, the former eventually dies from dehydration unless other sources of water are available. Moisture regulation in living trees or in other habitats of tree infesting insects generally is not possible. Sometimes, however, it may be feasible to temporarily flood forest lands and thereby destroy the insects present in the soil or litter (for example, larch sawfly prepupae in cocoons).

3. ***Insect habitat destruction.*** The immature stages of most insects that live in the inner-bark region can be destroyed by peeling the bark from infested logs or tree boles. The immature stages of these inner-bark borers are incapable of moving to and reentering a new host; consequently, they quickly die. This type of habitat destruction usually can be most easily accomplished by applying the debarking treatment after the new broods are about half grown, for by this time the boring insects will have loosened the bark considerably. Debarking also results in more rapid drying of the logs so that moisture conditions become unfavorable for those borers located deep in the wood. Unfortunately, this drying causes checking of the wood and renders the logs useless for many purposes.

4. ***Trapping and destroying the insects.*** In many European countries it is common practice to trap tree-boring insects by providing quantities of girdled or felled trees. After attacking and entering the trap logs or trees, the borers are killed by burning or debarking the host material. Trapping methods have not been used much in North America, but at the present time more attention is being directed toward this approach.

5 and 6. ***Insect exclusion or physical destruction.*** Exclusion of insects by using tightly constructed buildings and screening all openings seldom would be practical, even for storing the more valuable susceptible wood products. The costs would be high and the maintenance of tight storage would be very difficult. As everyone knows, however, it is a most useful method for the protection of people and food from insect attacks. Likewise, the use of hand and mechanical methods of insect destruction must be limited to very small isolated infestations. Often, in places such as forest nurseries or on small ornamental trees, a few insects can be more easily destroyed by hand when first seen than by taking the trouble to prepare and apply a spray. Even in young stands of forests, infested tree parts (leaders infested with boring weevil larvae) sometimes can be gathered by hand more economically than by using other applied methods.

BIOLOGICAL CONTROL

The type of applied control in which parasites and predators are either collected in an area where they are abundant or propagated under confinement and then released in another place where they are needed is known as biological control. Sometimes introductions of the parasite must be made annually in order to insure continued effectiveness, but such intensive practice under forest conditions generally is not practicable. To many people this biological method is very appealing, but its use is not a panacea. In some cases, however, spectacularly effective results have been achieved. Several agricultural crop pests have been adequately controlled when these insects occurred in certain ecologically isolated areas, such as insular and irrigated desert regions; but the results against forest pests have not been outstandingly successful. In one instance, however, an epidemic of the European spruce sawfly was effectively controlled by a virus disease introduced—apparently inadvertently—into eastern Canada. The greatest effort to control a forest pest by biological means has been directed against the gipsy moth. Even though many species of parasites and predators were introduced and became established, the gipsy moth is still a serious problem.

Biological control is a most obvious and logical method of attacking pests introduced from foreign countries; therefore, it has been frequently used by governmental agencies. Collections are made of the various parasites and predators known to attack the pest in its native home, and then these are introduced in the new areas where the pests have become established. When this method is effective, the population of the pest is reduced to a level similar to that of the location where both the pest and entomophageous insect(s) came naturally. Thereafter the population of the introduced insect pest should respond in a way similar to that of a native insect, with the combined action of climate and the entomophagous organisms keeping the insect at a sufficiently low population level so that economic damage seldom occurs.

Whenever insect parasites and predators are to be introduced, they must be studied carefully in order to be sure that they themselves do not also become a problem by parasitizing other valuable native parasites (hyperparasitism) or, in some cases, by causing destructive multiple parasitism.

A parasite or predator selected for use must have several attributes. These are: (1) high reproductive powers, (2) ability to utilize many hosts, (3) good host-finding ability, (4) life cycle synchronized with that of host to be controlled, and (5) must not be a serious hyperparasite. In addition, sometimes the habit of parasitizing hosts already parasitized (multiple parasitism) may be objectionable. When planning this type of control, a sequence of parasitic species should be sought to attack each stage of the insect. Usually the combined effect of many parasite species attacking a single stage is greater than when the most effective one is used alone.

Biological control methods are very complex; therefore, probably the only concern most foresters would have with these methods would be to release or distribute parasites and predators. Later they may be asked to collect samples of the hosts for the purpose of determining whether the entomophages have been successfully established.

REGULATORY CONTROL

Various public governing bodies often help to control insects and diseases by regulatory means. The insect control laws enacted and enforced by authorized agencies may have one or more of several objectives as follows: (1) to prevent the introduction and/or spread of injurious pests, (2) to enforce the application of pest control when people or properties are endangered by the inaction or carelessness of others, (3) to combat pests directly on public lands or when the problems involve mixed ownerships or are too great for private owners to handle by themselves, and (4) to prevent adulteration and misbranding of insecticides, fungicides, and other poisons.

In the first group are those laws that regulate the movement of pests, animals, and plant materials. These are known as *plant quarantine laws.* Provisions of these laws contain prohibitions regarding the importation of certain pests and all plants that may harbor these pests. To enforce these prohibitions, inspection stations are maintained and manned at all ports or places of entry through which people and freight pass. All infested or infected materials intercepted are either destroyed or treated before being released for entry. This method of inspecting at ports of entry is used chiefly to prevent the introduction of pests from foreign continents, although some domestic state quarantines are enforced in this way. Another method is to inspect plant materials at the sources of production. If the plant stock is found uninfested and healthy, the producer is allowed to move his stock freely anywhere throughout the state and often into various other states. Reciprocal agreements between any two states make a single inspection at one place by one agency applicable for both. Movement of these outgoing shipments is facilitated by requiring the producer to attach official inspection tags to all packages. Failure to pass inspection prohibits movement of the infested materials.

The first effective federal quarantine law was passed in 1912, and since then many others have been promulgated. At the present time all federal plant quarantines are enforced by the Bureau of Entomology and Plant Quarantine. Although most of the serious forest insect pests in North America are native, some have been introduced from foreign continents. These include several defoliators such as the notorious gipsy moth, the satin moth, the European spruce sawfly, the elm leaf beetle, and some borers.

A federal policy regarding the use of legislative means to aid in combating forest insects and diseases has developed slowly. It was initiated in 1928 when the McNary-McSweeney Forest Research Act directed attention to the need for research by enacting a program to guide the appropriation committees during the following ten-year period. Since then the most important act dealing with the control of forest insects and diseases is the Forest Pest Control Act of 1947. According to the provisions of this act, the United States Government can act alone or cooperate with states, territories, or with private timber owners to control destructive forest insects or diseases. The federal money appropriated can be used for surveys to detect and appraise problems, to determine practical control methods, and to plan and direct needed control measures. The Secretary of the Agriculture is responsible for determining the amount of cooperation to be required from the other agencies or private parties involved. The first of these cooperative arrangements was made in connection with spruce budworm control operations in Oregon and Washington. For this project the federal government contributed 25 per cent of the cost of spraying on private lands, whereas the state governments and the private owners each paid half of the remaining 75 per cent. On federal lands the United States paid the full cost, whereas on state lands the states were solely responsible. In the future, however, this pattern of contributions may vary somewhat from current practice.

SELECTED REFERENCES

Beal, J. A., and L. M. Hutchins. 1955. The role of the forest service in control of insects and diseases. *J. Forestry* 53:129–32.

Graham, S. A. 1951. Developing forests resistant to insect injury. *Sci. Monthly* 73:235–244.

———. 1956. Forest insects and the law of natural compensations. *Can. Entomologist* 88:45–55.

Morris, R. F. 1951. The importance of insect control in a forest management program. *Can. Entomologist* 83:176–181.

Morris, R. F., and C. A. Miller. 1954. The development of life tables for the spruce budworm. *Can. J. Zool.* 32:283–301.

Prebble, M. L. 1955. Entomology in relation to forest protection in Canada. *Forestry Chronicle* 31:314–323.

Turner, K. B. 1952. The relation of mortality of balsam fir caused by the spruce budworm. *Can. Dept. Agri. Publ.* 875:107 pp.

Sweetman, H. L. 1958. *The principles of biological control.* Wm. C. Brown Co. Dubuque, Ia. 560 pp.

1959. *J. Forestry* 57:243–289.

CHAPTER 6

Insecticides

Insecticides are substances highly poisonous to insects, which are applied as dusts, mists, fumigants, dips, or by pressure treatments. Foresters now need more knowledge regarding these materials because insecticides are being used in increasing amounts for the protection of forests and forest products.

TOXICOLOGY

Insecticides gain entrance in one or more of three ways. These methods are (1) directly through the body wall, (2) by way of the digestive tract, or (3) via the respiratory system.

The physiology of insecticide action is complex and only imperfectly understood, but it appears that insecticides function in several ways. They may act as (1) physical poisons, (2) general protoplasmic poisons, (3) respiratory enzyme poisons, and (4) nerve poisons.

Physical poisons kill by suffocating the insects or by causing dehydration. Dehydration results when the thin outer waterproof layer of the cuticle is abraded or when the insecticide actively absorbs water from the affected insect. Protoplasmic poisons kill by coagulating the cell contents of any tissues contacted. Respiratory enzyme poisons inactivate catalysts and thereby block cellular respiration. And lastly, the nerve poisons act in either of two ways: (1) by injuring the nerves so that they can no longer transmit impulses (cause paralysis) or (2) by continuously stimulating the affected nerves so that uncontrolled activity results. In the latter case, death probably occurs when the metabolic wastes accumulate in the body faster than they can be excreted (auto-intoxication).

INSECT RESISTANCE TO POISONS

The resistance insects develop to poisons is not a newly discovered phenomenon, although it has become more widely known only recently, following the development of the very effective synthetic insecticides. This resistance develops when insects are exposed for many successive generations to amounts of a poison that kill only part of the population and leave the most resistant individuals to survive and reproduce. Thus, by repeating this selective process over a long period of time, varieties or races evolve which are very resistant to the particular insecticide being used. This resistance, therefore, can develop only when the same poisons are applied repeatedly at frequent intervals over a long period of time, as is commonly the practice for protecting many types of crops. Even though large dosages may be used, uneven distribution and deterioration of the poisons between successive applications helps to explain why a few of the insects survive. Insecticide resistance is not likely to develop in populations of most forest insect pests, because these poisons are generally used only at infrequent intervals.

INSECTICIDE FORMULATION

The formulation of an insecticide refers to the physical form and characteristics of an insecticide preparation as it is manufactured for use by the consumer. Good insecticides may be useless in the pure or technical form if they are too expensive, too poisonous for animals other than insects, too gummy and lumpy to be distributed evenly, or have other undesirable characteristics. Such poisons usually must be combined with other materials in order to overcome these difficulties. The various components of a formulated insecticide commonly include at least the first two of the following: (1) insecticide chemicals, (2) diluents (carriers), (3) stabilizers (emulsifiers or deflocculating agents), (4) wetting and spreading agents, (5) adhesives, (6) synergists.

Diluents (carriers) include materials such as kerosene and fuel oil for sprays and various clays, lime, and other flocculent, flowable solids for dusts. Other diluents commonly used are water and air, but these are usually supplied by the consumer. In addition to the two main ingredients (diluents and insecticides) other accessory materials such as those listed previously sometimes must be used in order to achieve desired results. If the pure insecticides are oily or are dissolved in oil, an emulsifier must be added if the oil is to be dispersed and carried in water. Likewise, when solid insecticides are to be suspended in water a deflocculating material often must be combined with the insecticide so that the resulting powder can be wetted by and suspended in the water. Sometimes, when water is the main carrier, the foliage being sprayed may be so waxy that the spray, instead of wetting the surface, merely runs off. Such circumstances require the use of wetting-spreading agents. Adhesives may also be needed to reduce washing by rains following spraying. Sometimes materials are added which have little or no insecticidal value by themselves but greatly increase the killing efficiency of another substance. Such materials are named *synergists.*

The types of formulations commonly used are: (1) true solutions, (2) wettable powders, (3) emulsions, (4) dusts, (5) aerosols and mists, and (6) fumigants. In a true solution the insecticide is dissolved in the carrier so that all particles are reduced to molecular size. Kerosene, fuel oil, and water often are used as solvents. Wettable powder formulations are water insoluble flour-like materials which can be suspended readily in water. Often the insecticide compounds, if used alone, are too heavy or lumpy to form a suspension readily; therefore, light, flocculent, readily wettable materials (deflocculating agents) are milled with the insecticides to produce the wettable powders. The insecticide concentration in most of the newer wettable powder formulations varies between 15 and 80 per cent.

Like wettable powder formulations, emulsions consist of three phases. In emulsions, water is the continuous phase, oil is the dispersed phase, and the emulsifier is the stabilizing phase. The oil and water phases are almost insoluble in one another, and even mixtures (emulsions) of these two materials cannot be formed readily without the addition of a stabilizer. In an insecticide emulsion formulation the poison chemical is dissolved in the oil and the oil solution then emulsified in the water.

Some emulsifiers consist of asymmetrical molecules that have some affinity for both the main emulsion phases. It appears that opposite ends of these emulsifier molecules are attracted to each of the oil and water phases, so that they become regularly oriented around the oil droplets and coat each of them. This emulsifier coating reduces the interfacial tension between the main phases so that the resulting mixture becomes more stable. Soaps and other detergents are examples of emulsifiers. Often, however, special types of these are more efficient and, therefore, are used when large quantities of emulsion concentrates are prepared. Emulsion concentrates commonly contain 25 to 40 per cent insecticide. They are sometimes called miscible oil formulations.

Dust formulations are similar to the wettable powders in physical appearance, but they are usually more flocculent. The insecticide concentration, however, is usually much lower than in wettable powders (2 to 12 per cent). Dusts may or may not be water wettable. As the name "dust" indicates, the dry powder is blown out as a cloud of dust, which settles on the plants or object being treated. Dusts have not been used against forest insect pests in North America, although they are used for this purpose in Europe.

Aerosols and mist formulations are usually concentrated (10 per cent) oil solutions of insecticides which are prepared especially to be dispersed as very small droplets (diameters of 2 to 50 micron) by special equipment. Mists have been found to be most useful for the control of shade tree defoliators.

Fumigants are poisons that are gases under atmospheric conditions; therefore, they are most useful when they can be confined with the insects in places such as enclosed structures or in soil. Some insecticides may act as both a fumigant and as a contact poison. Benzene hexachloride, for example, acts in both these ways.

CLASSIFICATION OF INSECTICIDES

INORGANIC INSECTICIDES

Insecticides that contain elements such as arsenic, fluorine, mercury, copper, or zinc are general protoplasmic poisons. In order to act as stomach poisons these compounds must be insoluble in water, whereas those that are soluble are contact poisons.

The arsenicals are white powders, commonly dyed pink to make them readily distinguishable from food materials having a similar appearance. The water insoluble arsenates of lead and calcium are stomach poisons. Both are rather safe to use on living plants, but, of the two, lead arsenate is the safer. It is formulated as a wettable powder and used at the rate of 2 to 4 pounds per 100 gallons of spray. Calcium arsenate has been used most as a dust. Neither of these arsenates is used much at the present time against forest pests, but in the past lead arsenate was widely employed against leaf-chewing shade tree defoliators. Since white arsenic and the various arsenites are very soluble in water, they penetrate and kill living plant tissue as well as insects. This characteristic restricts their use to such things as baits or wood preservatives. Even then, however, these materials are so poisonous to higher animals that other insecticides usually are preferred.

Fluorine compounds are also white powders, but these are dyed blue. They are general protoplasmic stomach poisons. Water insoluble compounds of fluorine, such as cryolite, have been used mainly for the control of certain vegetable-infesting insects (blister, flea, and leaf beetles) and have not been used much against forest pests. Cryolite is used at the rate of 10 pounds per 100 gallons of spray. Water soluble sodium fluoride is used in some wood preservative mixtures and is also the insecticide in older types of roach poisons. Newer insecticides, however, such as chlordane are better roach poisons.

Sulfur is a well-known yellow-colored element. It acts as a respiratory poison by combining with organic materials to form the toxic compound, hydrogen sulfide. For example, sulfur sometimes is used as a dusting powder on the skin and clothes to prevent attacks by chiggers. The main objection to its use for this purpose is that the user becomes somewhat malodorous because of the H_2S generated. The compound lime sulfur is effective against some scale insects, many overwintering eggs, and some mites.

Copper sulfate, zinc chloride, and mercuric chloride are white or bluish crystalline powders. These water-soluble compounds are general protoplasmic stomach poisons used to some extent as wood preservatives. Because they are water soluble, however, they leach from the treated wood rather quickly when such wood is exposed in wet ground. One way this water solubility limitation can be improved is to use a double diffusion method consisting of soaking the wood first in a solution of either zinc chloride or copper sulphate and then in a second solution of sodium chromate. Inside the wood the two chemicals react to form a toxic water insoluble zinc or copper chromate.

When any of the previously mentioned water soluble inorganic salts or mixtures of these are used as wood preservatives, they are commonly applied as 2 to 5 per cent solutions, so that retention of about ½ pound of the salt is obtained per cubic foot of wood.

BOTANICAL INSECTICIDES

The best known of the plant organic insecticides are rotenone, nicotine, and pyrethrum. These are all contact nerve poisons that cause paralysis. They may be formulated as wettable powders, emulsions, or dusts. In addition, pyrethrum is often formulated as an aerosol.

The alkaloid *rotenone* is obtained from the roots of several species of tropical plants. It is commonly formulated as wettable powders, emulsion concentrates, and as dusts. Insecticides containing rotenone are especially useful for the control of insects that infest plants having aerial edible fruits, because this poison is detoxified rapidly when exposed to the sun. It is a slow-acting paralytic type of contact nerve poison that has not been used much against forest pests. The slow toxic action probably is partly due to the slow penetration of the cuticle. It is commonly used as a 1 per cent dust or as 1/10 pound of rotenone in a wettable powder or emulsion diluted with 100 gallons of water.

Nicotine is extracted from tobacco and then combined chemically with a sulphate to produce nicotine sulphate. The nicotine sulphate is dissolved in water made alkaline, usually by adding some soap, and used as a spray. The alkalinity of the soap causes the acid stable nicotine compound to decompose and release the free nicotine alkaloid. Dust formulations are also available. These consist of nicotine sulphate sprayed on and mixed with a nonalkaline dust carrier. Nicotine is used for the control of soft-bodied sucking insects but has only limited use in forestry practice, for example in forest nurseries. It is a fast acting contact and nerve poison that acts by blocking nerve impulse transmission in the ganglia. One pint of the 40 per cent nicotine sulphate formulation added to 100 gallons of water makes an effective spray mixture.

Pyrethrum, the insecticide obtained from certain chrysanthemum flowers, is usually formulated as a solution in deodorized kerosene. This poison causes a fast paralytic action, is nontoxic to humans, and lacks a strong odor. These characteristics make this contact nerve poison most useful as a household spray. The pyrethrin content of these sprays is usually about 1/10 per cent, but frequently the pyrethrins are combined with slower-acting but more effective poisons such as DDT. Synergists also commonly are incorporated with pyrethrins to increase their effectiveness by 3 to 5 or more times. A synthetic pyrethrin-like insecticide, *allethrin,* is also available.

MINERAL ORGANIC INSECTICIDES

Crude petroleum oils are mixtures of many (mostly aliphatic) hydrocarbon compounds. They can be separated by distillation into various fractions of increasing viscosities such as petroleum ether, gasoline, fuel oils, lubricating oils, and the solid residues such as paraffin or asphalt tars. Kerosene and other fuel oils are somewhat insecticidally active when used alone, but these are most useful as carriers for other insecticides. For example, the main carrier used for aerial forest spraying is domestic heating fuel oil, also known as fuel oil No. 2 or diesel fuel. The heavier lubricating oils, however, are more useful as insecticides per se. These lubricating oil sprays kill the insects either by suffocation or by acting as general protoplasmic poisons. Asphyxiation results when the oil forms an air-excluding coating over the treated insects. These spray oils are usually formulated as emulsion concentrates, which the user dilutes with water to produce a ½ to 4 per cent oil emulsions. Oil emulsions are used chiefly against scale insects or as ovicides. On deciduous trees late winter or early spring applications are more effective, because during these times heavier, less purified oils can be used without causing tree injury. Even during this dormant season, however, the temperature should never drop below freezing for a period of 24 hours following application. Trees such as sugar maple, walnuts, and beeches are injured by oil applied at any time.

Two characteristics of these oils which are used to indicate toxicity are viscosity and degree of saturation. The greater the viscosity (the heavier the oil) the greater the toxicity to both plants and insects. Oils used during the winter or dormant season can be heavier, hence are more toxic than summer oils. This viscosity characteristic is responsible for the smothering action of the oil. The degree of saturation refers to the number of double or triple bonds connecting adjacent carbons within the hydrocarbon molecules. The greater the number of these multiple bonds within the molecules the more the amount of unsaturation and the greater the toxicity of the oils to both plants and insects. For insecticide oils this unsaturation characteristic is commonly indicated by a U.R. number (unsulfonatable residue). When hydrocarbon oils are treated with sulphuric acid the two combine at the places of unsaturation forming oil sulphate compounds. If this unsaturation is sufficiently great, the sulphate compound precipitates and leaves the remaining liquid oil as the residue. Dormant oils have U.R. numbers of 65 to more than 75 per cent, whereas oils that can be used during the summer must have U.R. numbers greater than 90 per cent.

Coal tar creosote is the most important and most extensively used wood preservative. It is a dark brown liquid obtained as the middle fraction during the distillation of coal tar. The many advantages of this material are that it is an easily applied, economical, general protoplasmic contact poison with a high toxicity for all wood-destroying organisms. It is rather insoluble in water and has a low volatility. Creosote is not the perfect wood preservative, however, for it has several disadvantages. It has an unpleasant penetrating odor, which precludes its use in dwellings or where food must be stored. The treated wood cannot be satisfactorily painted for the dark color of the creosote bleeds through and the caustic nature of the substance may cause burns to the skin of workmen and others who come in contact with the liquid or freshly treated wood.

Coal tar creosotes are mixtures of various poisonous chemicals such as phenols, cresols, organic acids, naphthalenes, and anthracenes. It is too difficult to routinely determine the chemical composition; therefore, it is commonly bought and sold on specifications based on source and physical characteristics. The latter include water content, solubility in benzol, and specific gravity of the various distillation fractions.

For many years coal tar creosote has been diluted with various petroleum oils for the purpose of reducing its cost, but when these mixtures are used, the coal tar creosote should never be less than 50 per cent by volume. As might be expected these mixtures are less toxic than the pure coal tar creosote, but even so, they are sufficiently effective to give a high degree of protection. Coal tar creosote is a general protoplasmic poison effective against all forms of life. It is usually applied by forcing the preservative into the wood by using pressure technique so that various retentions of creosote are obtained. These vary from 6 to 25 pounds of creosote per cubic foot of wood.

Anthracene oils (carbolineums) are distillates also obtained from coal. They have higher boiling points than coal tar creosote; consequently, they have higher specific gravities. Sometimes it may be advantageous to use these more viscous preservatives for open-tank treatments involving heating, since losses due to evaporation are less. These preservatives are also general protoplasmic contact poisons that are sold under various trade names.

Wood tar creosotes have not been used much because they have not been produced in sufficiently large quantities and have not had uniform quality to interest users.

SYNTHETIC ORGANIC INSECTICIDES

Many of the synthetic insecticides are chlorinated hydrocarbons, and of these DDT is the best known. The common name, DDT, is taken from the three underlined letters in the abbreviated chemical name, dichlorodiphenyltrichloroethane. It is a white, amorphous powder with a fruity odor, is very soluble in aromatic oils, has only limited solubility in aliphatic oils, and is almost insoluble in water. There are many isomers of this compound, but only the para para compound is of much insecticidal importance. It was developed in Switzerland about the time World War II started.

DDT has been used extensively for the control of forest defoliators, and millions of acres have been treated with DDT by aerial spraying. For these applications it is usually formulated as a solution consisting of one pound of DDT dissolved in fuel oil together with a lesser quantity of an auxiliary aromatic solvent, so that each gallon of spray solution contains one pound of the poison. This spray is usually applied at the rate of one gallon, hence one pound of DDT, per acre.

The amount of auxiliary solvent used depends on the efficiency of the solvent and the temperature. Commonly a methyl naphthalene auxiliary solvent is used at the rate of one quart per pound of DDT when the spray is used or stored during warm weather; but if it must be exposed to subfreezing temperatures, the amount of auxiliary solvent must be increased to 1.25 quarts or more per pound of DDT. For large control projects it probably is most practical to buy the DDT formulated and delivered ready for use, whereas for smaller projects, purchase of DDT in concentrated solutions may be cheaper. Such concentrated solutions can be mixed easily with the correct amount of diesel oil and be ready for use immediately. If the ingredients are purchased separately, special mixing equipment must be available and the mixing often requires much time.

When DDT is applied from the ground to control shade tree pests, wettable powders, emulsions, and mist formulations are commonly used. The wettable powders usually contain 50 per cent DDT, but higher DDT concentrations sometimes are available. Emulsion concentrates commonly contain about 25 to 40 per cent DDT. Either of these formulations is used by diluting with water so that each 100 gallons of spray solution contains 1 pound of actual DDT. Insecticidal mists are applied to trees with special blower equipment in which about 10 per cent concentrations of DDT oil solution are used. Mist-blower application is the most practical method for treating the foliage of shade trees to kill defoliating insects when many trees are involved.

DDT is also commonly formulated and used as 2 to 5 per cent dusts, but these dusts are used mainly for garden or truck crop insect control and not for treating forests or shade trees. Like many of the synthetic compounds, DDT is a contact nerve poison. More specifically it is a slow-acting poison which induces continuous stimulation of the sensory nerves and in this way causes continuous uncontrolled muscular activity. Accumulation of metabolic wastes from this continued activity probably is the immediate cause of death.

There are various other insecticides related to DDT, such as TDE (DDD), methoxychlor, and the difluro analogue of DDT, but these have not been used extensively against forest or shade tree insects.

Benzene hexachloride (BHC and lindane) (a hexachlorocyclohexane) is a buff-colored powder with a characteristic musty odor. The use of this chemical as an insecticide developed simultaneously in England and France during the early 1940's. There are many isomers of this compound, but only the gamma isomer is insecticidally important. The technical grades commonly contain 12 to 36 per cent of the gamma isomer, but it can be highly purified and is then named *lindane.* BHC is cheaper than lindane; therefore, it is usually used for the control of insects such as bark and ambrosia beetles. BHC is generally formulated as an oil solution for use against forest pests, but other available formulations include wettable powders and dusts. The latter two are used mostly for the control of various agricultural pests. Gamma concentrations of ¼ to 1 per cent by weight of the finished spray are used for controlling the various tree- and wood-boring insects. It is generally formulated as an oil solution and should be applied heavily enough so that all surfaces are thoroughly wet. This wetting will take about one gallon per 100 square feet of bark area. The mode of action of benzene hexachloride is similar to that of DDT in that it is a slow-acting contact nerve poison which causes continuous stimulation of the sensory nerves with resulting hyperactivity. For foliage sprays the gamma concentration usually used is ¼ pound per 100 gallons of spray.

Chlordane, heptachlor, aldrin, and dieldrin are bicyclic chlorinated hydrocarbon (cyclodienes) insecticides that were developed in the United States. These are aromatic insecticides which are commonly formulated as emulsion concentrates, wettable powders, oil solutions, and dusts. Several of these are especially effective for controlling soil-infesting insects in forest nurseries. They are also most useful for killing ants, roaches, grasshoppers, and termites. Toxicologically, these cyclodienes are slow-acting contact nerve poisons that produce symptoms similar to those caused by DDT. Foliage sprays usually are formulated so that they are used at the rate of one pound of actual chlordane per acre, or ¼ as much for aldrin, dieldrin, or heptachlor.

Toxaphene, another cyclodiene, is made by chlorinating terpenes obtained from the resins of old pine stumps. It is a thick, resinous material which has solubility characteristics similar to the other chlorinated hydrocarbons; therefore, it commonly is formulated as oil solutions, emulsions, wettable powders, and dusts. The toxicity of toxaphene is similar to DDT, but it has been little used in forest insect control because it is very toxic to fish. One pound of the actual insecticide per 100 gallons of spray is the dosage usually used.

The chlorinated benzene, orthodichlorobenzene, is an aromatic liquid, which is usually formulated as a solution consisting of 1 part of the insecticide to 6 parts of No. 2 fuel oil. Although in the past it has been extensively used for the control of certain western bark beetles that infest thin-barked trees, it is now being replaced by ethylene dibromide.

Pentachlorophenol is a buff-colored fibrous powder that is commonly formulated as a 40 per cent oil concentrate. This is diluted with fuel oil or other oil solvents and used as a 5 per cent solution. It is a general protoplasmic contact poison which is effective against both insects and fungi; consequently, it is an excellent wood preservative. Application methods consist of dipping, spraying, or pressure treatments of seasoned wood products. One advantage it has over coal tar creosote is that the treated wood can be painted more satisfactorily. Retentions usually desired are ⅓ to ⅔ pounds of the poison per cubic foot of wood.

Organic phosphate insecticides such as TEPP, parathion, EPN, and malathion are brown liquids often having a strong garlic-like odor. Many are very toxic to mammals and therefore very dangerous for humans to use. Of those in general use, malathion is the safest, being about as toxic to humans as is DDT. These insecticides are commonly formulated as 15 to 25 per cent wettable powders, or as water soluble concentrates for TEPP, or as arosols. The dosages used are usually ⅛ to ½ pound actual insecticide per 100 gallons of spray. These materials are especially useful for killing sucking insects and mites but have not been used much for controlling forest pests. These organic phosphates cause hyperactivity due to inactivation of an enzyme involved in nerve transmission at nerve junctions. During recent years most of the newly developed insecticides belong to this type.

REPELLENTS

Materials that repel insects are of much interest to foresters, because they provide protection against the biting insects, ticks, and mites that attack man. Only a few of the many chemicals tested for this purpose have proved useful. The most outstanding of these discovered to date is the ortho isomer of *N*, *N*-diethyltoluamide, formulated as a 50 per cent solution in alcohol. Other effective compounds are dimethyl phthalate (40 to 60 per cent), dimethyl carbate (30 to 50 per cent), ethyl hexanediol (30 to 50 per cent), and indalone (20 per cent). Two or three of the aforementioned materials are usually combined, because such mixtures are more effective against a greater variety of pests. Dimethyl phthalate generally is one of these components. None of these substances is completely effective against ticks, but indalone, diethyltoluamide, and dimethyl carbate have some value for this purpose. Two other materials, *N*-propylacetanilide and *N*-isopropylacetanilide, are more effective than the others but are expensive and not easily obtained.

The repellents may be applied as a spray or thinly but thoroughly smeared on all parts of the body to be protected. Of course the eyes and tender mucus parts should be avoided. One disadvantage of most of these repellents is that they dissolve paints and many plastics. Nylon, cotton, and wood are not affected, but rayon may be. Ethyl hexanediol and diethyltoluamide are the least injurious in this respect.

For use against chiggers the repellents should be applied to the clothing, especially that covering the lower part of the body. When a person is to be constantly exposed to heavy chigger infestations it probably would be best to impregnate all of the outer clothing and the socks with benzyl benzoate. Five tablespoons of this material should be emulsified in 3 pints of water using any commercial emulsifier or 2 tablespoons of common laundry soap. The clothing should be dipped, wrung lightly, and dried thoroughly before use. The treatment will remain effective after two launderings. Overdosing should be avoided, and this chemical should not be used on the skin.

Powdered sulfur liberally applied to skin and clothing that come in contact with the vegetation is also a cheap, good method to prevent chigger attacks. Its disadvantages are that the skin of some people is sensitive to sulfur, and the users become somewhat malodorous due to the generation of hydrogen sulfide.

HANDLING INSECTICIDES

Most insecticides are not extremely poisonous to man but nearly all can cause injury if they are handled carelessly. The user should be thoroughly familiar with the poisons and exercise adequate care. Vapors and dusts should never be inhaled. No insecticides should be allowed to come in contact with the skin or mucus tissues of the mouth, nose, or eyes, for many of these poisons can be absorbed directly through skin and other tissues. Whenever any insecticide is spilled on the skin it should be washed off immediately with soap and water, and the clothes should be changed. Goggles and suitable respirators should always be worn when vapors, concentrated sprays, or dusts are being handled. If any insecticide is accidentally swallowed, a doctor should be called and vomiting induced immediately. Vomiting can be facilitated if the patient drinks a glass of warm water containing one tablespoon of table salt. If suitable information regarding treatment is not available locally, the U. S. Public Health Service should be called. The branches of this organization most familiar with insecticide poisoning are located in Savannah, Georgia and in Wenatchee, Washington.

LABELING

The user of insecticides must rely on the statements presented on container labels. Labels can be depended upon, since the accuracy of these statements is enforced by state and federal labeling and registration laws. The latest federal legislation dealing with this problem is the

Insecticide, Fungicide, and Rodenticide Act of 1947. This law controls all economic poisons moved in interstate commerce. In addition, most states also have similar laws which cover poisonous products sold within the state.

In order for insecticides, and the other agricultural poisons, to comply with this federal act they must be registered with and approved by the U. S. Department of Agriculture and must not be misbranded. Acceptable labeling includes the following: (1) name and address of the manufacturer or person for whom manufactured, (2) brand name or trade mark, (3) net contents, (4) active ingredients and warning statement to prevent injury to animals and plants for which use of the material is not intended (the antidote must be included), (5) adequate directions for use and a listing of the insects the poison is effective in killing. Of course, all statements must be correct.

Even though much useful information is included on the labels of insecticide packages, research by various organizations is continuously producing better materials and devising better methods of application. If a moderate to large forest insect control project is under consideration, the latest information regarding all aspects of carrying out such a project should be obtained from the state regulatory agency, the state experiment station or the Forest Insect Division, Forest Service, U. S. Department of Agriculture, Washington, D. C.

SELECTED REFERENCES

Brown, A. W. A. 1951. *Insect control by chemicals.* John Wiley and Sons, N. Y. 817 pp.

Metcalf, R. L. 1955. *Organic insecticides.* Interscience Publishing Co., N. Y. 392 pp.

Shepard, H. H. 1951. *The chemistry and action of insecticides.* McGraw-Hill, N. Y. 504 pp.

Shepard, H. H. (editor). 1958. *Methods of testing chemicals on insects.* Vol. 1. Burgess Publishing Co., Minneapolis, Minn. 356 pp.

Smith, C. N., I. H. Gilbert, and H. K. Gouck. 1957. Use of insect repellents. *U.S.D.A., ARS* 33-26 (revised).

U.S.D.A. 1957. Complete summary of pesticidal chemicals. *Agri. Chem.* 12:81.

CHAPTER 7

Methods of Applying Insecticides

Insecticides can be applied to trees or wood products to be protected in a number of ways, depending on the insecticide, the formulation, and the equipment available. These general methods include spraying, fogging or aerosoling, misting, dipping and pressure treatments, dusting, and fumigating.

Sprays can be applied by means of hydraulic pump sprayers either from the ground or from the air. When only a few shrubs or small trees are to be sprayed, hand-operated sprayers such as compressed air or small household aspirator type sprayers can be used. For spraying large forested areas, however, the only practical method is to use airplanes. Aerial forest dusting and spraying have been experimented with at various times since 1925, but this method became practical only after the extremely effective insecticide, DDT, was developed. At the present about a million acres of forests are sprayed each year in the United States.

AERIAL SPRAYING

There are several advantages of aerial spraying. The sprays can be applied rapidly to large forested areas, inaccessible areas can be treated, and the cost is sufficiently reasonable to make the operations practical. With favorable weather, one of the commonly used smaller 450 horsepower engine planes can spray 5,000 to 7,000 acres per week. The total costs average between 1 and 3 dollars per acre, with the larger areas of 100,000 acres or more approaching the lower figure. There are, however, certain disadvantages of aerial spraying. The most important of these is that the weather conditions must be favorable. The spray can-

Fig. 7.1. Aerial forest spraying. *Courtesy U.S.D.A., Forest Service.*

not be applied during rains or when the foliage is still dripping wet. Fortunately, however, a small amount of moisture on the foliage, such as dew, does not interfere seriously. Air movement also must be light, for when the wind is above 8 miles per hour much of the spray drifts away. Thermal updrafts, as indicated by bumpy flying, also interfere with the deposition of the spray, for this weather condition carries the light spray cloud upward. Finally, the low flying required is more dangerous than ordinary flying. Nevertheless, in spite of these limitations, aerial forest spraying has been used extensively for the control of defoliators and certain other forest pest insects.

The most popular planes for forest spraying have been biwinged planes equipped with 450 horsepower engines. With these planes a spray load of 125 gallons can be carried safely, and if the ferry distance between the airfield and the treating area is not too great, two or more trips can be flown per hour of good flying weather. Smaller planes can also be used, but they are not as efficient because the spray load is smaller. For large, relatively flat areas larger planes with load capacities up to 1,000 gallons often can be used advantageously. At the present time, however, the smaller planes have greater applicability; therefore, most of the new spraying aircraft being developed are of this size. In certain cases, where great maneuverability is required and limited landing space is available, helicopters may be used advantageously, but for general use they are too expensive to buy and operate.

The spraying equipment that must be fitted into or on a plane includes a spray tank, a pump, a discharge nozzle system, and certain accessory parts such as pipes and valves. The tank is commonly placed in the front cockpit and is equipped with suitable strainers, air vents, and quick-dump valves. The spray pump is usually attached to the forward part of the plane immediately in back of the propeller and engine. This pump is usually powered by a small fan, driven by the main air stream of the plane. The spray discharge equipment may be of various designs, but probably the one most used at the present for forest spraying consists of a pipe boom fitted with nozzles. The boom consists of a discharge pipe attached along the length of the lower surface of the underwing, with the nozzles either evenly spaced or concentrated along the inner half of each boom. When too many outlets are placed near the wing tips, two peaks in spray deposition occur without producing a

Fig. 7.2. Spray pump and boom attached to a plane. *Courtesy U.S.D.A. Forest Service.*

greater swath width. This ununiform deposition occurs because the airstream vortexes developing off the wing tips carry and deposit an excessive proportion of the spray.

After a plane has been equipped for spraying it must be calibrated to discharge the desired quantity of spray per unit of area covered. This calibration is accomplished by varying the nozzle size, nozzle number, and/or the operating pressure. The discharge rate is determined by timing the period of discharge for a known quantity of the type of spray to be used, whereas the swath width can be estimated roughly by multiplying the boom span by 4. More accurate determinations, however, can be made by conducting flight tests. Such a test consists of placing small (4 by 4 inches) aluminum plates or oil-dyed cards 10 feet apart along a line and then flying the plane across this line while discharging the spray. After the plane has passed, a short period of time is allowed for the spray cloud to settle before the plates are collected and the amount of deposit determined. These quantity determinations are made by comparing the spray deposit on the test plates with photographic or other standards made with known quantities of spray. Usually the effective swath width is considered to be the area where ¼ gallon or more of the spray is deposited per acre. For the 450 horsepower biplane mentioned previously, the effective swath width is about 132 feet.

Airplane forest-spraying operations usually must be carried out during a relatively short period of time; consequently, everything must be ready when spraying is to start. The actual spraying is usually done by commercial operators, but the other aspects are usually handled by foresters and entomologists. These duties include procuring the insecticide, obtaining the use of suitable air fields, subdividing the area into treating blocks, and supervising the airfield crew and the forest insect checking crew.

The spray planes must be approved by the Civil Aeronautic Authority (C.A.A.), and waivers must be obtained for low flying. All spray pilots must have commercial licenses, and in addition there are certain state laws that must be observed.

The insecticide is usually purchased completely formulated according to specifications. Part, and sometimes most, of the spray must be stored at the airfields, because the application period is usually very short and the roads often impassable in areas where spring spraying must be done.

Airfields already in operation are used when available, but sometimes temporary fields must be constructed. For the smaller, most-used planes the fields must be at least 800 feet long and be sufficiently smooth so that an automobile can be driven safely on them at a speed of 40 miles per hour.

Spraying units or treating blocks between 500 and 1,000 acres in size are most easily treated. They are usually delineated by topographical and other features that can be readily seen from the air. These include ridges, streams, roads, and changes in forest types. Various man-made markers such as flags and balloons have not proven very practical, except when reproduction-size stands are being treated or when numerous roads are available. The presence of these roads makes it possible for the crew members to move quickly from one place to another and fly the balloons to mark the successive swaths as they are flown. Aerial photographs and topographic maps are used by the pilots to identify the areas where the spray is to be applied. The spraying is usually conducted from daybreak until 8 or 9 A.M., but sometimes it may be continued longer if the weather conditions remain favorable. On flat or rolling land the pilot flies in straight lines forth and back over the area and usually spaces the successive swath by judgement. In rough or mountainous areas safety precautions require that the flight lines must be along contours or down the slopes. Spraying over lakes, rivers, and streams is avoided whenever possible, because fish and other aquatic animal life are very susceptible to the DDT insecticide usually used. Aerial spraying is dangerous work primarily because of the low altitudes (50 to 150 feet) at which the flying must be done. Consequently, all possible precautionary safety measures must be taken. Plans must be made to conduct possible rescue operations, and in wild, inaccessible areas, special equipment such as a helicopter may have to be engaged on a stand by basis for such emergencies.

The forest insect checkers are located in the areas being treated. Their duties are varied. During a two or three week period before and after spraying, they collect samples of the living insect population for the purpose of determining the effectiveness of the treatment.

One of the field men also determines whether the weather conditions are favorable for spraying and then transmits this information to the spray pilots by means of radio or flag signals. Other members of the field crew are responsible for placing cards treated with oil soluble dye in various stand openings. The purpose of these cards is to determine the uniformity of spray coverage.

Effect on wildlife. Aerial insecticide applications on large areas cause much concern regarding the effect such poisons may have on other forms of life. The various studies made on this problem indicate that much remains to be learned; but it appears that DDT (1 pound per acre), which has been extensively used in aerial forest spraying, causes little injury to most vertebrate wildlife. When misapplied, however, many fish, reptiles, and other cold-blooded animals may be killed. Birds can tolerate as much as 2 pounds of DDT per acre and mammals as much as 5 or more pounds per acre, but fish frequently suffer severe mortality following applications of more than ¼ pound per acre. Often the spray pilots can avoid the larger bodies of water, and the foliage cover over many of the smaller streams screens out much of the spray. Nevertheless, the killing of fish and other cold-blooded animals in sprayed areas can be serious. The general reduction of insect life can also have an indirect effect on the insect-eating animals such as birds and fish. The dosage of one pound of DDT per acre does not seriously affect the general terrestrial arthropod fauna for longer than a week, but it often requires as much as 2 months to a year or more following treatment before the aquatic insect population returns to the original level.

Soil accumulations of insecticides probably never will be serious in forests, because spraying is a type of control used only infrequently. Seldom will the applications have to be repeated oftener than every 10 to 20 years, and often one application for each crop of trees will be sufficient.

"GROUND" SPRAYING

Although hydraulic spraying equipment operated from the ground has been used a great deal in the past for control of shade tree insects, at present much of this work can probably be done faster and at less cost with mist blowers. These hydraulic sprayers have never been practical for general forest spraying, but they are useful when complete coverage is required. The relatively large quantity of diluent that must be used, however, makes their operation rather slow and expensive.

The equipment consists of tanks to hold the spray, suitable pumps, and engines together with accessory parts such as air chambers, pressure relief valves, hoses, and nozzles. The pumps are usually either of the piston or the diaphragm types with ball-type valves.

When logs are sprayed, the quantity of spray needed to thoroughly wet the bark is about one gallon per 100 square feet of bark area. This dosage requires about 3¼ gallons per 100 board feet for logs averaging 18 inches in diameter. The smallest of the power sprayers, such as those with a rated output of three or four gallon per minute and powered by 1½ to 2 horsepower engines, has been found suitable for most log spraying. However, for treating big log decks it might be more practical to use larger equipment.

COMPRESSED AIR SPRAYERS

These are small, hand-operated units that have only limited use for certain types of spraying. They meet a need in forest nurseries or for spot infestations in plantations. The output is small and much labor is required. The equipment consists of an elongated cylindrical airtight tank, about 7 inches in diameter and 16 to 24 inches long, with a built-in hand pump to compress the air present above the spray. Some of these are adapted for using compressed carbon dioxide to supply the pressure. This compressed air or carbon dioxide pushes the liquid spray mixture out through the hose and nozzle to form a spray. Solutions, emulsions, and wettable powders can be dispersed with this type of sprayer.

MIST BLOWERS

A mist blower consists of an engine-powered fan that produces a blast of air. A concentrated solution of the insecticide is injected into this air stream as it leaves the cylindrical discharge pipe where the liquid is broken up to produce a mist. The larger machines, powered with 25 to 50 horsepower engines, develop the output of 8,000 to 15,000 or more cubic feet of air per minute at 100 to 125 miles per hour required to successfully treat the tops of trees 80 to 100 feet tall.

Misting machines have been used extensively in recent years, especially for the application of insecticides to control shade tree defoliators. They have the advantage of using air as the main carrier; therefore, heavy loads of diluents such as water do not have to be handled. Mist blowers must discharge only 1 to 2 per cent as much spray gallonage as that used by ordinary hydraulic spraying in order to achieve equal effectiveness. In addition, other aspects of insecticide application with mist blowers is that misting is easier and faster, so that the same amount of treating can be done in one fourth the time and at a cost of only one third as much as when hydraulic sprayers are used. The disadvantages of mist blowers are that they often cover less completely, and wind greatly affects application.

WOOD PRESERVATION

Methods used for treating wood with poisons are also of interest to forest entomologists, because part of the action of wood preservatives is insecticidal. To be effective, these materials must persist for long periods of time in a state highly toxic to both wood-boring insects and wood-destroying fungi. Various application methods can be used, but certain of these are more effective than others. Brushing preservative on the surface of the wood is the least effective method, and frequently it is not worth the effort. Dipping is slightly better and prolonged soaking (24 hours) is much better. The best method, however, is impregnation. The less common of these consist of injecting water soluble salts into the sap stream of living trees or of using gravity to force the pre-

Fig. 7.3. A mist blower in operation. *Courtesy The John Bean Division, Food Machinery and Chemical Corp.*

servative liquids into the wood. For the first of these two methods the liquids are introduced into living trees by cutting a single girdle well into the sapwood around the base of each tree. A waterproof sheet of material is then draped around this girdled part and fastened so as to form a bag-like container. This bag holds the preservative solution in contact with the cut so that the liquid slowly enters the tree, ascends, and permeates the sapwood. Only living trees can be treated, because the transpirational pull of the crowns is essential to move the liquid upward. Of course, as soon as the toxic material reaches the leaves they die; therefore, the wood is impregnated but the trees are killed. Even though this method is effective, it has been used very little. Wood of trees so treated has remained sound 15 years or longer.

Forcing preservatives through small logs by means of gravity is accomplished by fitting a rubber or plastic tube around one end of the tree bole to be treated. By elevating the tube and tilting the post the preservative liquid placed in the tube will enter and permeates the wood by means of gravity.

The hot-cold bath treatment is one of the better and simpler methods for treating seasoned wood when pressure equipment is not available. The seasoned wood is immersed in the hot (225° F for creosote) preservative liquid and kept there for varying periods of time, depending on the size of the wood. After completing the required time in the heating bath, the wood is immediately immersed in a cold bath of the same preservative material. As the wood and the enclosed cellular air cools, the air volume decreases and draws the liquid into the cells. Actually the liquid is pushed into the wood by atmospheric pressure. Air cannot re-enter the wood because the latter is covered with the liquid preservative.

The injection of preservatives into wood with pressure requires special equipment consisting of large, strong, steel treating cylinders, fitted with pumps and gages, plus preservative storage tanks. Some of the larger of these cylinders are 8 or 9 feet in diameter and 100 to 150 feet long. The pumps are used for introducing and removing the preservatives and for adding air. Various procedures are used for operating the treating cylinders, but these can be classified as either the *full-cell* process or the *empty-cell* process. As the terms imply, the difference between the two methods is in the amount of preservative retained by the wood. In the full-cell process a vacuum is first drawn on the charged cylinder for the purpose of removing part of the air from the wood. The preservative liquid is then introduced and air pressure applied. The result of using these two operations is that more of the preservative liquid enters and is retained by the wood. With the empty-cell method, on the other hand, air is not removed before the preservative is introduced; consequently, the retained air causes much of the preservative to be expelled soon after the applied pressure is removed.

AEROSOLS

Aerosols are extremely fine mists that can be produced by various generating machines or simply by releasing certain combinations of materials into the air. In the latter method the insecticides are combined with low-boiling carriers or propellents (dichlorodifluoromethane and methyl chloride are most commonly used) that are maintained as liquids by containing them under pressure in strong containers. The aerosols are generated when the mixtures are released into the air, where the propellants immediately vaporize and leave the insecticides as clouds of minute particles (diameter 1 to 10 microns) floating in the air. Small to large sizes of these "bombs" are available for treating homes, greenhouses, and other enclosed spaces.

Some of the machine aerosol generators operate by vaporizing and blowing the insecticide into the air where it condenses immediately to produce the fog. They have been used in a limited way around resorts, camps, and homes, but this method can not compete with aerial spraying for treating large forested areas.

FUMIGANTS

Gaseous insecticides can be applied in several ways. One method is to promote chemical reactions that yield the poison gases. For example, calcium cyanide and water yield hydrogen cyanide and calcium hydroxide. Some fumigants can be liquified by compression and then held in strong containers until used. Others vaporize so slowly that they can be handled as liquids or even solids under ordinary atmospheric pressures and temperatures. Several of these high-boiling fumigants, including orthodichlorobenzene, ethylene dibromide, and BHC, have been used against bark beetles.

SELECTED REFERENCES

Anonymous. 1953. Methods of applying wood preservatives. U.S.D.A. *Forest Products Lab. Rept.* D154.

Balch, R. E., F. E. Webb, and J. J. Fettes. 1955, 1956. The use of aircraft in forest insect control. *Forestry Abstracts* 16:453–465, 17:3–9, 149–159.

Brann, J. L., Jr. 1956. Apparatus for application of insecticides. *Ann. Rev. Entomol.* 1:241–260.

George, J. L. 1957. *The pesticide problem.* The Conservation Foundation, N. Y. (This publication has a rather complete bibliography.)

Hoffmann, C. H., H. K. Townes, H. H. Swift, and R. I. Sailer. 1949. Field studies on the effects of airplane applications of DDT on forest invertebrates. *Ecol. Monographs* 19:1–46.

Linduska, J. P., and E. W. Surber. 1948. Effects of DDT and other insecticides on fish and wildlife. *U. S. Fish and Wildlife Serv. Circ.* 15.

Potts, S. F. 1958. *Concentrated spray equipment.* Dorland Books, Caldwell, N. J. 598 pp.

Weick, F. E., and G. A. Roth. 1957. Aerial application of insecticides. *Ann. Rev. Entomol.* 2:297–318.

Yuill, J. S., and C. B. Eaton. 1951. Airplane spraying for forest pest control. U.S.D.A., *Bur. Entomol. Plant Quarantine* E-823.

1949. Forest spraying and some effects of DDT. *Ontario, Canada Dept. Lands and Forests, Div. Research Biol. Bull.* 2:174 pp.

1959. *J. Forestry* 57 (4) 243–289.

CHAPTER 8

Forest Insect Surveys

Various methods are used for detecting and appraising the seriousness of forest insect pests. Some of these require special techniques and the full time of the survey crews, whereas under other conditions surveys can be conducted in conjunction with the performance of other forest work. One way to assure good detection is to require that every woods forester also be a forest pest survey man. To be most effective in performing these insect detection duties, foresters should be familiar with the local tree pests, and they should always be alert to any unusual appearance of the trees. Since these skills develop with experience, the better field men soon are able to detect when something unusual is happening, even when the trees are viewed from a distance.

Field men often neglect to report minor insect troubles because they do not have time to write the long reports commonly required. Sometimes as many as twenty-five or more items of information are requested for these reports. Undoubtedly much of this information is desirable, but from the viewpoint of the busy field man, the work of completing such long forms is onerous. On the other hand, if only a minimum amount of information is required, the reporting forms can be short enough to be printed on post cards. These can be easily carried and used in the field and thereby stimulate greater participation by the field foresters. The information requested should include the following: (1) location, date, and observer, (2) host species and approximate size or age class, (3) rough estimate of type and extent of damage and (4) the insect causing the trouble, if known.

Of course, it is not necessary to report the occurrence of every insect found feeding on trees, but insect damage should be reported when the insect species is an important pest or when the insect or damage becomes prevalent. Whenever additional data or insects and tree damage specimens are desired by the interested specialists, they can be obtained by making personal inspections or, in some cases, by having the field foresters do the collecting.

Insect specimens should not be sent through the mail in ordinary paper envelopes, since the insects invariably become mashed and when in this condition they are most difficult to identify. A preferable method, especially for insects that are soft bodied, is to send them in small vials or bottles containing any available preservative such as rubbing alcohol, vinegar, concentrated salt water, or even kerosene. Hard-bodied insects can be sent in small boxes loosely packed with cotton or tissue paper to prevent them from being shaken about. Samples of damage should always accompany the insect specimens.

In the various ways indicated, field foresters can perform a most important function by helping to detect forest insect pest activity. The degree of coverage obtained in this way cannot be duplicated as economically with other methods utilizing full-time pest survey crews. This does not mean that insect survey crews are neither needed nor important, but the work of these specialists can be greatly facilitated if a primary detection system is conducted by the larger group of general foresters. The specialized survey men can then concentrate their efforts where they are most needed.

The objectives of the more intensive forest insect surveys, commonly conducted by trained men, are to detect, appraise, and estimate whether more insect-caused damage is likely to occur in the near future.

AERIAL SURVEYS

Many of the most serious insect-caused tree injuries result in loss of foliage or change in color of crowns from the normal green to yellow and then brown. These tree crown changes usually can be seen from a distance; therefore, surveys made from airplanes often are most useful for detecting and evaluating certain types of insect damage, especially when extensive areas are involved. Another advantage of aerial observation is that insect-caused tree injury frequently appears first in the tops of the crowns. The disadvantages are that some types of tree injury cannot be detected from the air, and no causes of tree injury can be positively identified when the damage is viewed from a distance. These limitations often make it necessary to check the injured trees more closely from the ground for the purposes of determining the actual cause of the injury and determining the condition and abundance of the causal organism.

For aerial surveys the smaller slower-flying (60 to 90 miles per hour) planes are best, and of these, the high-winged monoplanes provide the best ground visibility. The crew usually consists of a pilot and one or two observers.

The area surveyed is covered in a systematic manner by flying forth and back along straight imaginary lines at a height from which the injury can be detected readily. In mountainous country the flight pattern often must be altered so that the flight lines are along the land contours. The altitudes flown usually vary between 200 and 5,000 feet above the ground, with the flight lines spaced 1 to 10 or more miles apart. Various methods may be used to record the data collected. One of these consists of sketching on maps or aerial photographs the boundaries of the areas involved according to various degrees of injury. For this type of survey, flying usually is at elevations of between 1,000 and 5,000 feet. Often it is advantageous to have foresters familiar with the areas being surveyed act as observers because, in many cases, they are better able to locate landmarks and because they can map more quickly and more accurately than others less familiar with the areas.

A second aerial survey method consists of actually counting trees with discolored crowns. For this work more accurate counts can be obtained when lower flight altitudes of 200 to 500 feet are used. Often when large tracts are involved, only part of each is sampled; that is, tree counts are restricted to strip plots over which the plane is flown. These plot data are then used to estimate conditions for the whole area, according to general sampling procedures. Sometimes certain devices and instruments are used to help the observers determine strip width. One of these instruments limits the observer's field of view to that of the area from which he should make the injured tree count. Instruments for recording the data have also proven useful. One of these, the *operation recorder,* conveys a strip of paper past a series of inked pens that trace separate lines on the paper. These pens, which can be activated individually by observers to produce jogs in the lines, register pertinent information. Each of the 9 or 10 pens usually is used to record only one specific item of information, but of course, the use of various combinations of the pens can greatly increase the descriptive potential of the device. The approximate location of the damage recorded on the paper can be determined, because the speed at which the paper feeds through the machine is synchronized with the speed of the plane. Topographic landmarks also may be registered on the paper strip at the instant the plane passes over them.

Since atmospheric conditions greatly influence visibility, they determine when aerial surveys can be made. Often it is impractical to work when clouds cover more than half the sky, when the haze is too dense, or when more than a small amount of smoke is present.

At the present time, there is a great deal of interest in the use of photographic methods for making aerial insect injury surveys. Of the various types of film tried, only color photography sufficiently differentiates the visible differences between injured and uninjured trees. One limitation of this method is that large quantities of high cost film must be used; nevertheless, the method appears promising.

"GROUND" SURVEYS

Forest insect surveys must also be made from the ground in order to supplement data obtained from the air and because, in some cases, aerial surveys are not suitable. The people who conduct these ground surveys must be familiar with such habits of the insect as the season of the year when each stage is present and the exact habitat occupied by each. Sometimes it is also necessary to know the daily activity rythm of the insect. Usually the most intensive surveys should be made during the season when the insect stage most easily observed is present. Even though the results of these intensive surveys may indicate the need for insecticidal treatment, time of application frequently must be delayed in order to synchronize control with the presence of the insect stage most vulnerable to the poison. As a result, insecticide applications may have to follow the main appraisal survey by six or more months. During this time interval various factors may cause great declines in insect populations; consequently, some additional checking usually must be done immediately preceding the control application. At these times the insects may be most difficult to observe, because they are either inconspicuous or well concealed. Even so, these additional observations often must be made regardless of the difficulties involved.

SAMPLING METHODS

Whenever possible the insect population should be expressed in absolute units such as the number of insects per plant or per branch of a certain size or per unit area of bark or of soil. Since smaller trees can be observed readily, any insects present can often be hand picked, the more active ones can be caught in nets, or they can simply be counted on the trees.

Sample collecting is a more difficult problem on taller trees, and special collecting equipment often must be used. On these trees the foliage-feeding insects can be sampled by cutting the infested branches with a pruning hook attached to the top of a pole 20 to 30 feet long. The jarring and general disturbance of cutting the branches usually causes some of the insects to drop off. These falling insects have to be caught if they are to be counted. For this purpose nets are commonly attached just below the cutting head, so that both the dropping insects and the cut branches fall into the net.

Other methods must be used for sampling insects in other habitats. For example, the bark must be removed to expose inner bark borers, and sifting devices are useful for sampling soil-inhabiting insects. Often, however, tree injury is easier to observe than the insects themselves; consequently, injury measurement may be the most satisfactory method for sampling certain species.

Some sampling methods are based on relative rather than absolute counts, in which the data are obtained by trapping, by collecting, or by making time counts of the insects sampled. Insects that react strongly to specific attractants, can be concentrated easily and caught in traps. The number of insects so caught are proportional to their abundance. As is well known, many, but not all, nocturnal species are attracted to light. The more effective odor attractants are those associated with food or those produced by the females for the purpose of attracting the males. These sexual attractants often are obtained in quantity from virgin female moths by extracting the effective chemical from macerated abdomens. Simply caging the live virgin female moths is also an effective method.

Some foliage-inhabiting insects can be caught on cloth trays after being dislodged from the trees by jarring. Of course, only smaller trees can be shaken easily, but sometimes hitting the boles of larger trees with an ax is an effective method for dislodging the insects. Another useful method, especially for the more active insects that infest reproduction or smaller saplings, is to make time counts. The observer simply collects or counts the insects observed during a specified period of time by scanning the foliage and collecting or counting all the insects seen. By moving slowly but continuously, each particular tree or part thereof is observed only once. If possible, the time periods used should be long enough (but never longer than about 5 minutes) so that 10 to 30 insects are counted per observation. Sweeping the crowns of small trees with insect nets also may be used for sampling the more active insects. One difficulty with this method is that the nets easily catch in the branches and throw out the insects already caught.

Collecting the excrement (frass) produced by insects feeding on the foliage of tall trees is another indirect way for measuring some insect populations. This method is applicable only for those insects that excrete solid frass and do not enweb the foliage or branches in which the frass can collect. Identification of many insect species can also be accomplished by means of only the frass (Hodson and Brooks, 1956).

Often it is desirable to be able to predict events such as when insect activity will begin in the spring or the progress of insect development.

For this purpose *Hopkins' Bioclimatic law* may be applied. The law states that any phenological event in the spring occurs 4 days later for each 1 degree latitude northward, 5 degrees longitude eastward, or 400 feet upward. Therefore, if the stage of insect or plant development is known for one locality, the stage of development for the same species in other areas can be estimated.

STATISTICAL ASPECTS

Although a thorough discussion of sampling statistics can not be considered here, a few pertinent points on the subject are presented in the following discussion.

Counts of the insects made for each sample are used for estimating the total insect population present in the areas studied. The most common statistic—the average or arithmetic mean—is calculated from the individual sample values and indicates the average insect population per tree, per branch, or per acre.

To obtain the most accurate estimate of the mean the observations must not be biased, and they must be representative of the whole area being sampled. Bias is the tendency to err persistently in one direction—usually unknowingly—when making the observations. Freedom from bias often can be achieved best by using mechanical methods for selecting the plots or units from which the sample counts are to be made. Representative sampling requires that the observations be collected equally from all parts of the areas. This statement suggests that the sampling should be conducted in a systematic way whereby each observation is spaced at a constant definite distance from the adjacent sampling spots.

Another statistic, which is most useful for interpreting sampling data is the standard error of the mean. This statistic, together with the mean, is an estimate of the variation that can be expected in the mean if the sampling were to be repeated in the same way in the same population but not necessarily using the same sampling spots. Two thirds of the times the true mean will be within the range of the sampled mean plus or minus the calculated standard error of the mean.

The standard error of the mean can be calculated from a series of observations by means of the equation

$$s = \sqrt{\frac{S(X^2) - \frac{(SX)^2}{N}}{N(N-1)}}$$

where s = standard errorr of the mean, S = sum of, X = value of each individual observation, and N = number of observations. If N is fairly large—more than 50—an estimation of the range within which the true mean should occur 95 per cent of the time is derived by multiplying the standard error by 2, whereas the range for the true mean 99 per cent of the time is computed by multiplying by 2.7.

The best estimate of the standard error is obtained when the observations are collected in a random manner, which means that every unit of the population has an equal chance of being selected as a sample observation. Systematic methods of sampling usually result in better estimates of the mean, but they do not necessarily yield the best estimate of the standard error. Conversely, the opposite may be true for a randomly collected sample. When there is little variation between different parts of the area, however, randomly and systematically collected samples yield similar results.

Representation and randomization may be somewhat reconciled by dividing the area into blocks and then collecting two or more randomly collected sample observations from each block. The individual observations might be either tree injury or insect counts on plot or strip areas or on individual trees or parts thereof. A sampling plan can be devised from information consisting of the area size to be sampled, the time required to collect each sample, and the help and time available; but the degree of confidence that can be placed in any estimate will depend on the standard error of the mean. For large areas that are extensively surveyed the standard error commonly obtained often has been as high as 50 per cent of the mean value. The importance or significance of such large errors of the estimate depend on what use is to be made of the survey data. Of course, the smaller the standard error the more precise will be the population estimates. A higher degree of precision is more desirable when planning for control operations than for simple detection surveys. In any case, the significance of the standard error should be understood by those engaged in survey work.

Sequential sampling. This is a special sampling technique being developed at the present for use in forest insect surveys. In sequential sampling, for each area sampled a series of randomly collected observations is made in sequence, and the resulting numerical values are added. The total sum obtained after each observation is then used to determine whether additional sampling is needed. The objective of this method is to classify populations into broad categories such as small, medium, and large, for precise numerical estimate of actual population sizes are not obtained. The advantage of this procedure as compared with the more conventional types of sampling is that when populations fall definitely within certain limits, the number of observations needed to obtain reliable conclusions is reduced. Borderline cases require more intensive sampling to obtain the same degree of confidence regarding the reliability of the insect population estimates. In this manner, the number of observations needed to sample each area may vary, but in all cases the question of whether enough observations have been made can be quickly determined in the field as the data are being collected. The method appears to be especially useful for deciding whether or not artificial control should be applied.

If only two categories—that is, control or no control—are to be used, the information which must be taken to the field can be readily tabulated in three columns as shown in Table 8.1.

TABLE 8.1 SEQUENTIAL SAMPLING TABLE FOR SAMPLING A HYPOTHETICAL INSECT POPULATION

	Total Number of Insects Counted	
Observation	No Control Needed	Control Needed
1	—	18
2	4	26
3	11	33
4	18	40
5	26	48
6	33	55
7	41	63
8	48	70

The values of the observations made in each sample area are added together as soon as they are obtained, so that they can be compared readily with the tabulated values. Sampling is continued in each area until the total sum obtained is either above the highest or below the lowest of the two values listed for that particular number of observations. For example, in Table 8.1, if the sum of three observations results in counting 29 insects, one or more additional observations must be made. If, after the next observation, the sum totals 42, the population in that area is classified as needing control, and no further sampling is needed. Thus, after four observations have been made, any sum obtained between 18 and 40 indicates the need for additional sampling, whereas populations yielding sums that total less than 18 or more than 40 are classified.

In order to use this sampling method, the first thing to be done is to establish the limits of the various categories—that is, to prepare a table like Table 8.1. If a decision is to be made as to whether or not control is needed, it must be based on the size of the insect population (per twig, per square foot, or any other unit most applicable) that must exist before applied control is needed. The second thing that must be done is to discover the way in which the insect population is distributed. This distribution can be determined only from previously collected sampling data.

The various possible distributions are the binomial, the negative binomial, the Poisson, and the normal. The binomial distribution prevails when each sample is recorded simply as "insects present" or "insects absent." The negative binomial equation frequently fits the distribution of many forest insect populations. With this type of distribution, the presence of one individual increases the chances of others occurring in the immediate vicinity. For example, the distribution usually occurs when the eggs are laid close together in masses, or when the insects have gregarious feeding or attacking habits. The variance (standard error squared) for this type of distribution is always significantly greater than the mean. When insect population densities are low, the distribution may be of the Poisson type (the variance is equal to the mean). For the last type of distribution (normal) the variance is independent of the mean.

The calculations necessary for computing the required tables or graphs varies for these different types of distributions; therefore, it is evident that whenever this method is to be used, a great deal has to be known about the distribution of the insect being sampled. The methods used for making these graphs and tables are presented in papers by Stark (1952), Ives (1954), and Waters (1955).

SELECTED REFERENCES

Heller, R. C., J. L. Bean, and J. W. Marsh. 1952. Aerial survey of spruce budworm damage in Maine in 1950. *J. Forestry* 50:8–11.

Heller, R. C., J. F. Coyne, and J. L. Bean. 1955. Airplanes increase effectiveness of southern pine beetle surveys. *J. Forestry* 53:483–487.

Hodson, A. C., and M. A. Brooks. 1956. The frass of certain defoliators of forest trees in north central United States and Canada. *Can. Entomologist* 88:62–68.

Ives, W. G. H. 1954. Sequential sampling of insect populations. *Forestry Chronicle* 30:287–291.

McGugan, B. M. 1958. The Canadian forest insect survey. *Tenth Intern. Congr. Entomol., Proc.* (1956) 4:219–232.

Morris, R. F. 1955. The development of sampling techniques for forest insect defoliators. *Can. J. Zool.* 33:225–294.

Orr, L. W. 1954. The role of surveys in forest insect control. *J. Forestry* 52:250–252.

Plant Pest Control Division (Weekly). Cooperative economic insect report. *U.S.D.A., ARS.*

Schumacher, F. X., and R. A. Chapman. 1948. Sampling methods in forestry and range management. *Duke Univ. School of Forestry Bull.* 7 (revised) 222 pp.

Stark, R. W. 1952. Sequential sampling of the lodgepole needle miner. *Forest Chronicle* 28:57–60.

Wadley, F. M. 1952. Elementary sampling principles in entomology. *U.S.D.A. Bur. Entomol. Plant Quarantine* ET–302.

Washburn, R. I. 1956. Electronic recorder as an aid in aerial insect surveys. *U.S.D.A. Forest Serv., Intermountain Forest and Range Exp. Sta. Res. Note* 31.

Waters, W. E. 1955. Sequential sampling in forest insect surveys. *Forest Sci.* 1:68–79.

Wear, J. F., and P. G. Lauterback. 1955. Color photographs useful in evaluating mortality of Douglas-fir. *Proceed. Soc. Am. Foresters* 169–171.

SECTION *II*

The More Important Forest, Shade Tree, and Wood Product Insect Pest Species of North America

THE REMAINING PART OF THIS BOOK DEALS WITH THE MORE IMPORTANT and commonly observed insect pests that infest trees and wood. The main classification category used is type of tree damage. These are as follows:

TYPES OF TREE DAMAGE CAUSED BY INSECTS

1. DEFOLIATING. All or part of the foliage is eaten by insects that have mouthparts adapted for chewing. Some insects mine inside the leaves, others only eat the outer leaf tissues (skeletonizers), but most consume the leaves or parts thereof with a high degree of thoroughness. See Chapter 9.

2. INNER-BARK BORING. The nutritious succulent phloem and cambium tissues are destroyed by many insects that chew their way through

the inner bark. In trees so damaged, the open or frass-packed tunnels are most evident. See number 6 below and Chapter 10.

3. Wood Boring. Holes of various sizes and shapes are bored in the wood of the main tree boles, the larger branches, and in dried wood by many species of insects and by a few other kinds of animals that chew their way through these tissues. See Chapter 11.

4. Sapsucking. Spotting, discolorations, malformations and/or general devitalization of the foliage, twigs, or other plant parts are caused by insects and mites that have mouthparts adapted for piercing and extracting plant sap. Solid parts of the plants are never consumed. See Chapter 12.

5. Oviposition. A type of damage consisting of slit-like punctures in twigs. These punctures are made by the female insects when they lay their eggs. See Chapter 13.

6. Bud and Twig Boring. See Chapter 13.

7. Bark Eating. The thin bark on seedlings and the twigs of older trees are chewed off and eaten by certain insects. See Chapter 13.

8. Root Feeding. Roots damaged in various ways as described previously in numbers 2, 3, 4, and 7. See Chapter 14.

9. Destruction of Cones and Seed. This type of damage is always done by borers that excavate the seed or chew holes through the cones, nuts, or other types of fruit. See Chapter 15.

GENERAL REFERENCES FOR SECTION II

Beal, J. A. 1952. Forest insects in the Southeast. *Duke Univ. School of Forestry Bull.* 14:168 pp.

Craighead, F. C. 1950. Insect enemies of eastern forests. *U.S.D.A. Misc. Publ.* 657: 679 pp.

Doane, R. W., E. C. Van Dyke, W. J. Chamberlin, and H. E. Burke. 1936. *Forest insects.* McGraw-Hill, New York. 463 pp.

Essig, E. O. 1958. *Insects and mites of western North America* (revised). Macmillan, New York. 1050 pp.

Felt, E. P. 1905–1906. Insects affecting park and woodland trees. *N. Y. State Museum Memoir* 8; 2 vol. 877 pp.

Graham, S. A. 1952. Forest entomology. 3rd edition, McGraw-Hill, New York. 351 pp.

Herrick, G. W. 1935. Insect enemies of shade trees. Comstock Publishing Co., Ithaca, New York. 417 pp.

Houser, J. S. 1918. Destructive insects affecting Ohio shade and forest trees. *Ohio Agri. Exp. Sta. Bull.* 332:487 pp.

Keen, F. P. 1952. Insect enemies of western forests. *U.S.D.A. Misc. Publ.* 273 (revised). 280 pp.

Packard, A. S. 1890. Insects injurious to forest and shade trees. 5th report of *U. S. Entomol. Comm.* 955 pp.

U.S.D.A. Yearbook, 1952. Insects. U. S. Government Printing Office, Washington, D. C. 779 pp. and 71 plates.

CHAPTER 9

Defoliating Insects

INJURY

Insect defoliation damage can be recognized readily by the absence of foliage or from uneaten leaf parts such as petioles, larger veins, or other remaining pieces of the leaf blades or needles. Many small insects and the very young of larger species often feed by eating only the softer parts and leaving the vascular or skeletal network. These insects are called leaf skeletonizers. Leaf miners, on the other hand, bore inside and eat the tissues between the upper and lower epidermal walls. The latter injury consists of serpentine or blotchy leaf areas that are blister-like and whitish colored. Most of the larger defoliators, however, consume most leaf tissues including both the inner and outer parts.

The injurious effects of defoliation are obvious to those who know that the chlorophyll-bearing leaves synthesize the food needed by the plants for sustenance and growth. When defoliation is sufficiently severe, continuous, or frequently repeated, death of the whole plant occurs, whereas partial or less frequent defoliations cause reduced growth and branch dying.

The degree to which a tree is injured depends on the extent of defoliation, the species, the season of year, and the frequency of successive defoliations. Evergreens usually suffer more severely than deciduous trees. Conifers such as jack and Scotch pines are killed by a single complete defoliation, irrespective of when it occurs during the year. One summer defoliation of only the new foliage also causes heavy tree mortality, but complete loss of only the old needles, even when repeated annually several times, usually does not kill pines. Spruces and firs are even more likely to be injured by defoliation than are the pines cited previously, whereas some southern pines (slash and longleaf) are not

Fig. 9.1 Types of defoliation damage. Upper left: almost complete leaf destruction by the orange-striped oak worm; upper right: holes made by a leaf beetle; lower left: leaf mining by the locust leaf miner; lower right: skeletonizing by sawfly slugs.

effected as much. Deciduous conifers (larches) and hardwoods, on the other hand, can be completely defoliated annually for many years and still survive. Any severe defoliation regardless of tree species always greatly reduces the production of wood, especially when the damage occurs early in the growing season. Successive annual defoliations also cause branch dying. The more severe injury that results from spring defoliation occurs because the food reserves are low immediately after the leaves have developed. At the end of the growing season the foliage has replenished the food supply, and the time is approaching when the leaves are to be dropped from the trees; therefore, defoliations at this time have little effect on deciduous trees.

CLASSIFICATION OF DEFOLIATORS

Species of insect defoliators belong to five orders. These are the moths and butterflies (Lepidoptera), the sawflies and leaf cutting ants (Hymenoptera), the leaf beetles and chafers (Coleoptera), the grasshoppers and walking sticks (Orthoptera), and fly leaf miners (Diptera). Only the larvae (caterpillars) of the Lepidoptera, Hymenoptera, and maggots of Diptera are leaf feeders, whereas both the immature and adult stages of Coleoptera and Orthoptera can cause feeding damage.

It is usually necessary to examine some stage of the insect, or have some part such as a molted skin or pupal case, in order to determine the order to which the insect causing the damage belongs. Some Lepidoptera caterpillars spin silk over the twigs and foliage on which they feed; therefore, when leaf-chewing damage occurs and this silk is present, the cause most likely is a lepidopterous feeder. Identification of the order to which an insect defoliator belongs can be accomplished by referring to Table 9.1.

TABLE 9.1 INSECT ORDERS THAT CONTAIN LEAF CHEWERS*

1.	Two or more pairs of short, fleshy, unjointed abdominal legs (prolegs) present, and/or paired groups of minute hooks (crochets) present on underside of some abdominal segments	2
1.	Prolegs absent or present only on the last abdominal segment; crochets absent	3
2.	Crochets present; usually with 5 pairs of prolegs, but sometimes only 2 or 3 pairs (loopers), and sometimes more (page 99)	(moth and butterfly caterpillars) LEPIDOPTERA
2.	Crochets absent; usually 6 to 8 pairs prolegs present (page 174)	(sawfly caterpillars) HYMENOPTERA
3.	Wings, wing pads, wing cases, or wing covers (elytra) present	4
3.	Wingless	9
4.	Elytra present (page 191)	(beetles) COLEOPTERA

4. Elytra absent 5
5. Wing pads or cases extend backward and somewhat downward toward the underside of body; insect relatively inactive, usually being able to move only its abdomen; appendages encased in membranes so that the joints are indistinct . . . . (pupae) 6
5. Wings fully developed, or if only wing pads are present they extend straight back; posterior legs enlarged and adapted for jumping (page 198) . . . . (grasshoppers) ORTHOPTERA
6. Wing cases, mouthparts, and other appendages tightly appressed or glued to the body so insects appear to lack appendages; body enlarged and blunt anteriorly and tapering behind; insect often enclosed in a cocoon (page 99) . . . . (moth and butterfly pupae) LEPIDOPTERA
6. Wings and other appendages distinct, not tightly glued to the body 7
7. Only one pair wing cases present; often inside a segmented hardened larval skin (puparium) (page 198) . . . . (fly pupae) DIPTERA
7. Two pairs wing cases present 8
8. Prothorax large (page 191) . . . . (beetle pupae) COLEOPTERA
8. Prothorax small and interlocked with mesothorax; each insect usually enclosed in a small tan to brown papery cocoon (page 174) . . . (sawfly pupae) HYMENOPTERA
9. Thoracic legs present 11
9. Thoracic legs absent 10
10. Head distinct, globular, and colored darker than body (page 195) . . . . (weevil larvae) COLEOPTERA
10. Head indistinct; anterior end pointed and may have a pair of dark colored mouth hooks at tip or inside tip; always leaf miners (page 198) . . . . (maggots) DIPTERA
11. Body constricted between the thorax and abdomen with 1 or 2 small scale-like or node-like segments forming a petiole between the two body regions (page189) . . . . (ants) HYMENOPTERA
11. Thorax and abdomen broadly joined; insects not ant-like; either slender, elongate and stick-like, or fleshy and grub-like 12
12. Body slender and stick-like; antennae and legs long and well developed (page 197) . . . . (walking sticks) ORTHOPTERA
12. Body soft, and fleshy, not stick-like; antennae short and inconspicuous; 3 pairs short, jointed thoracic legs 13
13. Head small and partly to deeply retracted into body; body may be naked or hairy, sometimes with poisonous spines or setae; sucker-like discs may be present on underside of abdomen (page 164) . . . . (slug caterpillars) LEPIDOPTERA
13. Head not retracted into body; body not hairy, but may be spined; poisonous hairs or spines never present 14
14. Two pairs of breathing pores (spiracles) on thorax; a pair of small appendages below anus; never more than 1 pair simple eyes (ocelli); thoracic legs never elbowed (page 183) . . . . (web-spinning sawflies) HYMENOPTERA
14. Only 1 pair of spiracles on thorax; 0 to several pairs ocelli present; legs usually elbowed at 2 or more joints (page 191) . . . . (beetle larvae) COLEOPTERA

* Only the stages that cause the damage or remain with the damage are included.

SELECTED REFERENCES ON THE EFFECTS OF DEFOLIATION

Craighead, F. C. 1940. Some effects of artificial defoliation on pine and larch. *J. Forestry* 38:885–888.

Duncan, D. P., and A. C. Hodson. 1958. Influence of the forest tent caterpillar upon the aspen forests of Minnesota. *Forest Sci.* 4:71–93.

Mott, D. G., et al. 1957. Radial growth in forest trees and effects of insect defoliation. *Forest Sci.* 3:286–304.

Rose, A. H. 1958. The effect of defoliation on foliage production and radial growth of quaking aspen. *Forest Sci.* 4:335–342.

LEPIDOPTEROUS DEFOLIATORS

The larvae of many species of moths and butterflies eat the foliage of trees, but only a few are major forest pests. Many, however, become locally abundant as pests of shade trees and then they are commonly observed.

INJURY. The leaf-eating damage is caused only by the caterpillar stage. Most of these feed by consuming nearly all of the foliage tissues, but some are skeletonizers when very young or throughout their larval life, and a few are miners. The adult moths and butterflies are unable to eat solid foods. Some have tube-like beaks through which they can ingest nectar, but many have nonfunctional mouthparts. The various leaf-chewing caterpillars can be separated into a number of groups as presented in Table 9.2.

TABLE 9.2 THE MORE IMPORTANT GROUPS OF FOLIAGE EATING LEPIDOPTEROUS CATERPILLARS

1. Caterpillars bore inside the leaves or needles (page 173) . . . . NEEDLE AND LEAF MINERS
1. Caterpillars do not mine inside the foliage 2
2. Caterpillars construct and live inside individual movable cases made of silk and parts of the foliage (page 170) . . . . CASE AND BAG MAKERS
2. Caterpillars do not make and inhabit individual movable cases or bags, but some make case-like structures of foliage which remain attached to the trees by the leaf petioles 3
3. Caterpillars have only 2 or 3 pairs of fleshy abdominal legs (prolegs)* (page 124) . . . . LOOPERS
3. Caterpillars have 5 pairs of prolegs 4
4. Small naked larvae that enweb individual shoots with a light covering of silk, roll or fold individual leaves, tie adjacent leaves together, or construct small tubes of needles (page 103) . . . . NEEDLE TIERS AND LEAF ROLLERS
4. Caterpillars either live exposed, or they make small dense webs or large loose webs 5

5. Caterpillars make and live within conspicuous silken tents or webs (page 111) . . . . TENT CATERPILLARS AND WEBWORMS
5. Caterpillars live exposed on the foliage 6
6. Caterpillars armed with spines, enlarged tubercles, or lobe-like extensions of body wall (page 152) . . . . SPINY CATERPILLARS
6. Hairy or naked caterpillars which are not armed with distinct spines or enlarged tubercles 7
7. Caterpillar body naked, only sparsely covered with hairs (setae); without spines or tubercles (page 165) . . . . NAKED CATERPILLARS
7. Caterpillars hairy, setae may be either short or long 8
8. Some hairs in dense tufts which are either short and brush-like or long and slender (pencil-tufts) (page 132) . . . . TUSSOCK CATERPILLARS
8. Setal tufts not dense; they may be arranged in groups arising from low warts, but in these places the body wall can be seen through the hairs (page 140) . . . . HAIRY CATERPILLARS

* Slugs have no prolegs. See page 164.

INSECT APPEARANCE. *Larvae* (caterpillars): body elongate, cylindrical, soft, segmented, and fleshy (worm-like), variously colored; head globular with chewing jaws; antennae small, bristle-like; 1 to 6 pairs simple eyes; 3 pairs of small, peg-like, jointed thoracic legs; usually 2 to 5 pairs of thick, short, unjointed fleshy legs (prolegs) on abdominal segments 3 to 6 and on the last (10th); prolegs with rows or circles of minute hooks (crochets). *Pupae:* tan to dark brown; body with anterior end enlarged and blunt; posterior end distinctly segmented and somewhat tapered to a point; all appendages are tightly appressed to body so they are not obvious; are found inside silken cocoons, in webbing, or exposed on foliage, twigs, and bark, and in ground litter or soil. *Adults:* well known moths or butterflies; bodies usually hairy; wings membranous, covered with scale-like setae that dislodge readily when handled and appear as dust; these scales produce wing color patterns; males often have feathery antennae, whereas on females antennae are slender and often thread-like; mouthparts, either rudimentary or consist of a tube-like beak held coiled like a watch spring. *Eggs:* variously shaped and colored depending on species; often laid in groups or masses on foliage or on twigs and bark, when winter is passed in this stage; egg masses often covered with a hardened froth-like secretion (*spumaline*).

LIFE CYCLE. The incubation period for the eggs and the duration of pupation usually last about 1½ to 2 weeks, unless the winter is passed in either of these stages. There are usually 5 or 6 larval instars.

The seasonal developmental patterns are various, but three general types are most common. The early season defoliators include those species which generally live through the winter either in the egg stage or as young larvae. They hatch or become active at about the time the new foliage appears and complete larval development 6 to 8 weeks later. Pupation usually occurs inside cocoons or webbing above the ground, among the remaining leaves, or on the bark of the branches and trunk. This stage lasts 1½ to 2 weeks. When the adult moths emerge they mate, and the females immediately lay the eggs for the next generation. Generally these moths do not feed; consequently, they live and perform the reproductive function within the very short period of time of one to a few days at the most. Usually there is only one generation per year. Many of the most serious forest pest species have this general type of seasonal development.

A second type of life cycle is similar to the early season feeders but differs in that the caterpillars develop more slowly and mature later in the summer. Species with this type of development feed on the foliage of conifers.

The third type is the late season feeders. These usually winter as pupae either in the leaf litter or in cells in the soil, and transform to moths during late spring or early summer. A few, however, hibernate over winter as adults. The moths mate, and the females soon lay eggs that hatch in about 2 weeks. The larvae complete their development in a period of a month or two and then drop to the ground to pupate. These late season feeders may have 1 or several generations per year depending on the species. The caterpillars are most abundant during late summer and fall.

It should be understood that the aforementioned statements are general seasonal development patterns and that exceptions occur.

CONTROL. Lepidopterous leaf chewers are subject to various natural controls. Probably the most common predators are birds and small mammals, whereas the most commonly observed internal parasites are insects such as the ichneumonid wasps and the tachinid flies. At times polyhedral virus diseases also are very effective as control agents, and unfavorable weather conditions sometimes cause great reductions in the populations. Cold, wet weather is especially unfavorable for many of the early spring feeding caterpillars.

Many insecticides are effective killers of lepidopterous leaf chewers, but, of these, DDT has proven so effective for most forest and shade tree species that others have not been used much. For the few species that are more susceptible to other insecticides, however, the specific material and details of application are presented at the time the insects in question are discussed.

Aerial application is the only practical method for treating large forest areas. DDT is formulated as a solution in petroleum solvents so that

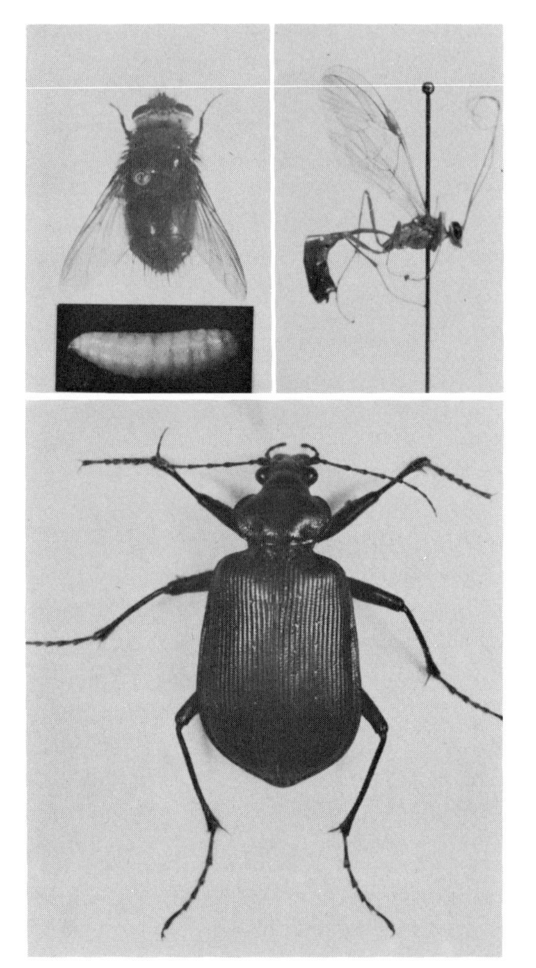

Fig. 9.2. Typical parasites and a predator of defoliating caterpillars. Upper left: parasitic tachinid (laraevorid) fly and maggot (X1¾); upper right: adult parasitic ichneumonid wasp (X1½); lower: predator ground beetle, *Calosoma scrutator* Fab. (X1½).

each gallon of spray solution contains one pound of DDT. The solvents are diesel oil (fuel oil No. 2) and auxiliary aromatic hydrocarbons—usually methyl naphthalenes. Usually the dosage is one gallon of spray, hence one pound of DDT, per acre.

When aerial spraying cannot be used, mist-blower application probably is the most practical method for treating large numbers of shade trees. The formulation commonly used consists of a concentrated 10 per cent DDT oil solution. Only a small quantity must be applied.

Power driven hydraulic sprayers or even hand-operated sprayers also can be used effectively but less efficiently than mist blowers. For these, one pound of actual DDT formulated either as a wettable powder or an emulsion is used per 100 gallons of water spray. This amount is equivalent to about 2 level tablespoons of 50 per cent powder per gallon of spray. If the leaves are very waxy so that the spray doesn't adhere well, a detergent such as soap should be added in sufficient amounts so that the spray will wet the foliage. These sprays should be applied lightly but completely, so that little runs off.

The best time to apply an insecticide is when the caterpillars are small, preferably before they are half grown, because three fourths or more of the total food eaten is consumed during the last feeding instar. Often, however, the damage is not noticed until after most of the injury has occurred. Spraying at this time only helps to reduce the population of the following generation. Additional information on all these methods of application is found in Chapter 6.

NEEDLE TIERS AND LEAF ROLLING CATERPILLARS

Lepidoptera, Tortricoidea

The larvae of needle tiers lightly enweb the shoots with silk and then feed on the needles within these webs, whereas the leaf rollers either form a tube or fold each leaf attacked or else fasten together two or more adjacent leaves. The naked caterpillars of the latter species feed on the leaf surfaces (skeletonize the leaves) while living within the enclosed leaf spaces. Some of the most serious forest pests are needle tiers, but very few leaf rollers cause serious injury to either forest or shade trees. Neither the naked caterpillars nor the drably colored, bell-shaped moths are distinctive enough to be identified easily. Nevertheless, some of the common species of the more important needle tiers can be identified by using Table 9.3. One species of leaf roller is discussed on page 111.

TABLE 9.3 COMMON SPECIES OF NEEDLE TIERS*

1.	Infest pines	2
1.	Infest conifers other than pines	4
2.	Each larva forms a tube by webbing together 5 to 20 needles so that they are arranged parallel and adjacent to one another, tubes remain attached to tree; head greenish brown with dark patch on sides (page 110)	PINE TUBE MOTH
2.	Larvae do not make and live in tubes made of needles	3
3.	Infest monterey pine; caterpillars a dirty white color	ORANGE TORTRIX
3.	Infest other species of pine; caterpillars light to dark brown (page 108)	PINE BUDWORMS
4.	Head tan to dark brown or black	5
4.	Head yellow or pale green	8
5.	Body with longitudinal stripes	*Dioryctria* spp. and *Zeiraphera* spp.
5.	Body not striped	6

6. Body greenish yellow (page 109) . . . . BLACK-HEADED BUDWORM
6. Body light to dark brown 7
7. Comb-like structure above anus (anal comb) with 7 or 8 teeth (page 104) . . . . SPRUCE BUDWORM
7. Anal comb with 6 teeth, inner pair sickle shaped . . . . *Recurvaria sp.*
8. Body green . . . . *Archip* spp. and *Tortrix* spp.
8. Body not green; no anal comb (page 110) . . . . BUD MOTHS

* Refer also to Table 9.4 and the section dealing with leaf miners.

SPRUCE BUDWORM

Choristoneura fumiferana* (Clem.), Lepidoptera, Tortricidae; Generic synonyms are *Tortrix, Harmologa, Archips,* and *Cacoecia

The spruce budworm has been the most injurious pest of the spruce-fir forests of northeastern North America, and recently it also has been causing increasing amounts of trouble in the West. During the tremendous 1910 to 1920 epidemic in eastern Canada, the amount of timber killed probably was more than half as much as was utilized from all Canadian forests.

INJURY. The larvae usually eat only the bases of the needles on the current, young, succulent foliage, but when this foliage is exhausted they also eat the older needles. The uneaten, severed needle parts are held on the twigs for several weeks by the silk webbing. These soon turn brown, which makes the trees appear fire scorched. Later in the summer the webbing disintegrates, and the browned needles drop. Spruce and fir suffer severely by defoliation, but inasmuch as budworms usually eat only the currently produced foliage, several successive years of heavy feeding must occur before serious tree mortality results. The trees seldom survive when more than two thirds of the total foliage has been removed during the several years of an outbreak. Stand mortalities during outbreaks commonly range from 25 to 75 per cent, and the increment on the surviving trees may be only 25 per cent that of normal growth for a period of six or more years. Often, the trees weakened by these repeated partial defoliations die as the result of subsequent attacks by various inner-bark boring insects such as bark beetles.

HOSTS. Firs, spruces, and Douglas-fir. The most favored species are balsam fir in the East and Douglas-fir in the West. Black spruce is seldom severely injured. Growth on this species starts so late in the spring that suitable food is not available for the young larvae.

RANGE. Wherever the hosts grow in North America with the possible exception of the southern Appalachians.

INSECT APPEARANCE. *Caterpillars:* typical; ¾ to ⅞ inch long when full grown; head dark, mottled brown; body dark brown with lighter area along each side, with conspicuous yellowish setae bearing warts; hairs sparse so insects appear naked; pronotal shield tan with scattered irregular darker spots; thoracic legs dark brown; 5 pairs prolegs armed with circles of minute hooks (crochets) of intermixed lengths; comb-like structure above anus (anal comb) has 7 or 8 teeth. *Pupae:* about ⅝ inch long; typical shape; not enclosed in cocoons; occur among the enwebbed needles. *Moths:* small; wing spread ¾ to 1 inch; wings broad; forewings indistinctly mottled gray to tan with no good distinguishing characteristics, abruptly widened at base so resting moths appear bell-shaped when viewed from above; hind-wings a uniform gray; antennae thread-like. *Eggs:* greenish, flattened, oval, and overlap one another like fish scales; arranged in 2 rows along needles in masses of 10 to 20; empty egg shells appear white.

LIFE CYCLE. The moths appear from late June through July depending on weather conditions. They are most active during the evenings when, during epidemic years, the mass flights may appear as snow whirling around the tree crowns. After mating, each female immediately lays 100 to 200 eggs in small masses of 10 to 20 on each needle. One to 1½ weeks later the eggs hatch. The resulting larvae immediately find protected places, such as beneath bark scales, where they enclose themselves in individual light silken webs (hibernacula). Here they hibernate

FIG. 9.3. Spruce budworm (*Choristoneura fumiferana*). Insects on right (X2½): twig on left showing enwebbed pupae (X2).

over the winter. Early the following spring the tiny caterpillars emerge and either mine the needles of the old foliage or feed in the expanding, staminate flowers of fir. As soon as the new foliage appears, the needle-mining larvae start to feed on the fresh succulent needles. Later, when the flowers dry, the flower-feeding caterpillars also move to and start feeding on the new foliage. Feeding then continues on the new foliage until the larvae mature. Pupation occurs next and requires 1½ to 2 weeks. The pupae are formed among the enwebbed needles. There is only one generation per year.

SILVICULTURAL CONTROL. In the northeastern spruce-fir regions certain conditions prevail which suggest that lasting, widespread protection against budworm losses can be obtained by managing these forests in certain prescribed ways. This idea is not new, and although it has not yet proven to be workable, many who have studied the problem think that the method has possibilities. The main requirements are to reduce the quantity of the favored host, balsam fir, and reduce the vulnerability to injury by keeping both the spruces and firs growing thriftly. Mature and overmature balsam stands are more susceptible to damage because they are not growing vigorously and because they consistently produce heavy crops of flowers. The staminate flowers of firs are a preferred type of food; and the heavy-bearing trees produce less foliage; therefore, defoliation usually is more severe on these trees.

From the previously mentioned facts it appears that the eastern spruce-fir forests can be made budworm resistant by regulating the stand composition. Sometimes this can be done, but often it is not practical because the balsam fir is too abundant and comprises the major part of the stands over large areas. In addition, balsam fir is a desirable tree because it reproduces readily, grows rapidly, and the wood yields a good pulp; therefore, often it is desirable to favor and grow this species. It should be grown on a short rotation of not more than about 40 years, because the older trees are the heavy flower producers.

The major present problem is to reduce the budworm susceptibility of large areas containing much mature and overmature timber. A preferred method is to selectively cut the stands as lightly as possible so as to reduce the hazard quickly. If possible, all the fir larger than 7 inches DBH should be cut, plus the overmature spruce (14^{+} inches DBH). The amount of fir permissible in budworm-resistant mixed stands has not been definitely determined, but it has been suggested that 2 cords of balsam is the maximum. Sometimes, however, cutting according to these standards removes too much of the stands. This result is undesirable because both fir and spruces are susceptible to wind throw; therefore, no more than about one third of any stand should be removed at one time. Old stands should be either clear cut, if adequate reproduction is present, or a shelterwood system should be used to obtain the desired reproduction. After the first cutting cycle the stands should be operated on a selection system in which the spruce is favored by cutting the fir when still young. Large blocks of one age class should also be avoided.

When budworm control by silvicultural methods is to be initiated, the first thing to be determined is the degree of budworm susceptibility of the various stands. The criteria used for evaluating stand susceptibility usually include quantity of fir and spruce, age of stand, and species of spruce. After the stands have been classified, a cutting plan is prepared so that the most vulnerable areas are logged first and then followed by those less susceptible until the whole forest has been treated.

In the Douglas-fir forests of the West, conditions are different; therefore, the methods described above are not useable.

INSECTICIDAL CONTROL. When forests are being severely attacked there is not sufficient time to remove the susceptible balsam before serious damage occurs; therefore, direct control methods must be used. At the present, the only practical direct control method is aerial spraying with DDT.

Preparations for spraying require both aerial and ground surveys. Aerial surveys must be made the previous summer at the time the browned foliage, resulting from the past season's defoliation, is on the trees. This survey is to determine where the heavy past season defoliation occurred and should indicate where the infestations probably will again be heavy the following year. The ground surveys are used to determine whether there will be heavy tree mortality if the stands are subjected to another year of heavy defoliation and whether any significant declines in the budworm population occurred during the winter and early spring. The latter insect counts usually are obtained by collecting branch samples from the crowns at heights of 20 to 25 feet, as described in Chapter 8. An average of ten or more caterpillars per 15-inch branch indicates a heavy population.

The actual spraying consists of aerial applications of DDT applied at the rate of 1 pound per acre during the last few weeks of the caterpillar stage. At the present time even lower dosages are being tried. The details of this aerial application of sprays have been discussed on page 101 and in Chapters 6 and 7. During recent years millions of acres have been sprayed for the control of the spruce budworm. In the United States nearly all of this spraying has been in the West, whereas in Canada it has been mostly in the East.

Balch, R. E. 1946. *Can. Dept. Agri. Publ.* 1035.
Blais, J. R. 1952. *Can. J. Zool.* 30: 1–29.
Brown, R. C. 1944. *U. S. Dept. Agri. Leaflet* 242.

Brown, R. C. 1952. *U. S. Dept. Agri. Yearbook.* pp. 683–694.
Brown, R. C., H. J. MacAloney, and P. B. Dowden. 1949. *U. S. Dept. Agri. Yearbook.* 1949. pp. 423–427.
Graham, S. A., and L. W. Orr. 1940. *Minn. Agri. Exp. Sta. Tech. Bull.* 142.
Heller, R. C., J. L. Bean, and J. W. Marsh. 1952. *J. Forestry* 50:8–11.
McLintock, T. F. 1955. (*U. S. Forest Serv.*) *Northeast Forest Exp. Sta.* Paper No. 75.
Morris, R. F., and R. L. Bishop. 1951. *Forestry Chronicle* 27:1–8.
Swaine, J. M., F. C. Craighead, and I. W. Bailey. 1924. *Can. Dept. Agri. Tech. Bull.* 37 (N.S.).
Turner, K. B. 1953. *Canadian Dept. Agri. Publ.* 875.
Webb, F. E. 1955. *Forestry Chronicle* 31:343–352.
Westveld, M. 1954. *J. Forestry* 52:11–24.
Refer also to the references cited at the end of Chapters 4 and 5.

PINE BUDWORMS

Choristoneura spp., Lepidoptera, Tortricidae

The pine budworms have not been as destructive as the spruce budworm; nevertheless, the jack pine budworm (*C. pinus* Free.) has caused serious damage in the Great Lakes region. The most injurious of the many western species is the sugar pine budworm [*C. lambertianae* (Busck)].

INJURY. The characteristics of defoliation are similar to those caused by the spruce budworm; consequently, several successive years of defoliation causes a weakening of the trees so that they are eventually attacked and killed by inner bark borers.

HOSTS. Pines. Preferred hosts of various species are jack and Scotch pines in the East and lodgepole and sugar pines in the West.

RANGE. Northern and western pine regions of North America.

INSECT APPEARANCE. All stages are very similar to those of the spruce budworm. The moths differ slightly in that they are slightly smaller (9⁄16- to 15⁄16-inch wingspread) and the forewings are always reddish tan and have a more distinct mottling.

LIFE CYCLE. Similar to that for the spruce budworm except that the jack pine budworm moths emerge about 2 weeks later than *C. fumiferana.* This variation in the time of moth appearance for the two species effectively isolates them so that they rarely interbreed even when both occur in the same area.

CONTROL. Of the various pine budworms only the jack pine species has caused serious outbreaks. For this species the most practical type of applied control is based on silvicultural methods. This insect is even more dependent on pollen than is the spruce budworm; consequently, outbreaks seldom develop in the absence of an abundance of staminate flowers. Management methods should stress the removal of the heavy flower-producing trees. These are the mature and overmature trees, the large crowned wolf trees and sometimes the suppressed trees.

Insecticidal control with aerial applications of DDT also can be used when the values warrant this type of treatment. Procedures would be the same as those suggested for the spruce budworm.

Freeman, T. N. 1953. *Can. Entomol.* 85:121–127.
Graham, S. A. 1935. *Univ. of Mich. School Forest Conserv. Bull.* 6.
Hodson, A. C., and P. J. Zehngraff. 1946. *J. Forestry* 44:198–200.
MacAloney, H. J., and A. T. Drooz. 1956. *U. S. Dept. Agri. For. Pest Leaflet* 7.

BLACK-HEADED BUDWORM

Acleris variana (Fern.) Lepidoptera, Tortricidae

This insect is a serious defoliator of northern coniferous forests, and several severe outbreaks have occurred to hemlock stands on the Olympic Peninsula and on Vancouver Island. It has also been epidemic in the northern Rockies and in the Northeast. Frequently, both the black-headed budworm and the spruce budworm become abundant at the same time. The black-headed budworm is not as serious a forest pest as the spruce budworm.

INJURY. The damage caused by the black-headed budworm is indistinguishable from that caused by the spruce budworm; therefore, in order to differentiate the two, insect specimens must be available.

HOSTS. Hemlock, firs, Douglas-fir and spruce. Preferred species are hemlock and silver fir in the West and balsam fir, hemlock, and spruce in the East.

RANGE. Wherever the hosts grow in Canada and northern United States.

INSECT APPEARANCE. *Caterpillars:* typical; about ½ inch long when full grown; head shiny black during first 4 instars, tan or orange in fifth instar; body naked, yellowish to bright green, not striped; 5 pairs prolegs. *Moths:* small, wingspread ⅝ to ¾ inch; similar in shape to spruce budworm; color varies from gray or brownish with variable brown, black, orange, and white markings. *Eggs:* yellow; oval and flattened about 1/25 inch long; on undersides of needles.

LIFE CYCLE. Winter is passed as eggs on the needles. These hatch in the spring at the time the foliage buds are swelling. The tiny caterpillars first mine inside the expanding buds, but later move to the new shoots where they enweb and feed on the foliage. The larval stage lasts about a month and a half. Pupation occurs among the enwebbed needles and lasts about 3 weeks. The moths usually are present during July to August, depending on the weather conditions in the locality where they occur. Each female lays 100 to 200 eggs. There is only one generation per year.

CONTROL. This insect has not been sufficiently serious to require the development of either insecticidal or silvicultural controls. Nevertheless, it does have the potentiality of becoming a serious forest pest. Virus and fungus diseases commonly are very prevalent and undoubtedly help to hold this species in check. Should conditions ever warrant insecticidal treatment, however, aerial spraying with DDT at the same dosage as used for the spruce budworm probably would be effective.

Balch, R. W. 1932. *Can. Dept. Agri. Entomol. Branch Spec. Circ.*

PINE TUBE MOTH

Argyrotaenia pinatubana (Kearf.), Lepidoptera, Tortricidae

INJURY. The larvae construct, live in, and partially eat cylindrical tubes made of needles. Each tube consists of 5 to 20 parallel needles, which are fastened together and lined with silk. Only one larva inhabits each. When a tube has been shortened to about 1 inch by feeding, the larva abandons it and makes a new one. HOSTS AND RANGE. Lodgepole pine, white bark pine, and eastern white pine, throughout the ranges of the hosts. INSECT APPEARANCE. *Caterpillars:* typical; ½ inch long when full grown; body naked; greenish yellow to dark green; head greenish brown with a dark spot on each side; 5 pairs prolegs. *Moths:* small, wingspread ¼ to ⅝ inch; forewings mottled brownish gray to reddish with lighter cross lines; hindwings uniform light gray; shape similar to the spruce budworm. LIFE CYCLE. Winter is passed as pupae either inside the tubes (East) or in the ground litter (West). In the East moth emergence occurs in the spring at about the time the new growth starts, but in the West emergence may be after midsummer. The eggs hatch after about a week and the resulting minute larvae are needle miners for the next 2 or 3 weeks. After this needle mining period they become tube builders. They probably have two generations per year in the East and only one in the West. CONTROL. Applied control has never been needed for this insect, but should it ever be needed, DDT applied by any of the usual methods would be effective. See page 101.

Hartzell, A. 1919. *J. Econ. Entomol.* 12:233–237.

BUD MOTHS

Zeiraphera spp. Lepidoptera, Olethreutidae

INJURY. The spruce bud moth has become abundant several times along the Maine coast and the larch bud moth has defoliated thousands of acres of larch and fir in eastern Washington. The foliage damage is similar to that caused by the spruce budworm. HOSTS. *Z. ratzeburgiana* Sax.—spruces, firs, and Douglas-fir; *Z. griseana* (Hbn.)—larch, and firs. RANGE. Throughout the regions where the hosts grow. These two species were introduced from Europe. INSECT APPEARANCE. *Caterpillars:* typical; small, about ⅜ inch long when full grown; body naked, pale brownish or grayish green, never bright green; head and pronotum yellowish; anal comb absent; 5 pairs prolegs. *Moths:* small, wingspread about ½ inch; color variable gray to light brown; forewings with darker diagonal markings. **LIFE CYCLE.** Similar to that for the spruce budworm except that winter probably is passed in the egg stage. **CONTROL.** Applied methods have never been used against these species, but should the need ever arise the insecticidal methods as used for the spruce budworm probably would be effective.

LARGE ASPEN TORTRIX

***Archips conflictana* (Wlk.) Lepidoptera, Tortricidae**

INJURY. This is one of the numerous leaf rollers that sometimes becomes very abundant, but it can not be classified as a major forest pest. The larvae cause leaf rolling and skeletonizing. **HOSTS.** *Populus* spp. Aspen is preferred but birch, willow, chokecherry, and alder also may be infested. **RANGE.** Throughout northern North America and south in the Rockies to Utah. **INSECT APPEARANCE.** *Caterpillars:* typical; about ¾ inch long when full grown; body naked, dark green to almost black with pronotum brown to black; 5 pairs prolegs. *Moths:* small, wingspread 1 to 1¼ inches; forewings dull, light gray and diffusely mottled; hindwings uniform gray; shape similar to the spruce budworm. **LIFE CYCLE.** Similar to that for the spruce budworm. The flattened, pale green eggs are laid in large masses (up to several hundred) on the upper leaf surfaces. The young larvae feed by skeletonizing the leaves for about a week and then spin hibernacula in which they overwinter. Early in the spring the larvae become active and cause the typical leaf rolling. Usually only one larva lives in each rolled leaf. Pupation also occurs inside the rolled leaves. There is only one generation per year. **CONTROL.** Natural control agents usually hold this insect in check; therefore, insecticidal control has not been used. DDT probably would be effective if the values involved ever warrant the use of applied control. See page 101.

Prentice, R. M. 1955. *Can. Entomol.* 87:461–473.

TENT CATERPILLARS AND WEBWORMS

Lepidoptera, Several Families

The caterpillars of many species of moths construct webs or tent-like structures in which they remain during part or most of their larval life. The various species can be identified in the field on the basis of web appearance, time of year the caterpillars are present, the type of damage caused and the appearance of the caterpillars. See Table 9.4.

TABLE 9.4 COMMON LEPIDOPTERA CATERPILLARS THAT CONSTRUCT SILK TENTS OR ENWEB THE LEAVES AND BRANCHES*

1.	Infest conifers	2
1.	Infest broad-leaved deciduous trees	6
2.	Caterpillars rather naked, being clothed with only a few scattered setae	3
2.	Caterpillars more or less hairy; occur only in the West (page 137)	SILVER SPOTTED HALISIDOTA

3. Silken webs almost completely filled with masses of excrement pellets and dried needles; hosts spruces and pines 4
3. Webs are not packed with excrement although some frass always present . . . 5
4. Infest pines (page 122) . . . . PINE WEBWORM
4. Infest spruces (page 124) . . . . SPRUCE WEBWORM
5. Infest only *Juniperus* spp. (page 123) . . . . JUNIPER WEBWORMS
5. Infest spruces, firs, pines, and hemlock . . . . Table 9.3
6. Naked caterpillars with only 2 or 3 pairs of fleshy abdominal legs (prolegs) . . . . LOOPERS, Table 9.5
6. Caterpillars with 5 pairs of prolegs 7
7. Caterpillars with a small projection (gland) on mid-back of the 6th and/or 7th abdominal segments; small, tough, gray webs on ends of twigs that insects inhabit from August to May (page 145) . . . . BROWN-TAILED MOTH
7. Caterpillars without glands on mid-back of the 6th and/or 7th abdominal segments 8
8. Yellow-striped brown caterpillars with a two-tipped black tubercle on back of each of the 1st and 8th abdominal segments; tents consist of a few leaves fastened together; hosts poplar and willows (page 119) . . . . . POPLAR TENT CATERPILLAR
8. Caterpillars without tubercles on the back of the 1st and 8th abdominal segments 9
9. Caterpillars naked being clothed with only a few scattered setae 10
9. Caterpillars more or less hairy 13
10. Infest mimosa (*Albizia*), honey locust, or tree of heaven (*Ailanthus*) 11
10. Feed on foliage of other tree species 12
11. Infest mimosa and honey locust; larvae gray to brown with 5 white stripes along body (page 121) . . . . MIMOSA WEBWORM
11. Infest tree of heaven . . . . [*Atteva aurea* (Fitch)] AILANTHUS WEBWORM
12. Caterpillars yellow to green with pronotum black (page 121) . . . . UGLY NEST CATERPILLARS
12. Caterpillars tan to brown usually with stripes along body (page 121) . . . . *Tetralopha* spp.
13. Many long whitish to reddish setae arise in groups from low black or orange wart-like spots; caterpillars yellow, tan, or gray; large loose webs enclosing skeletonized foliage; these constructed and occupied during the summer and fall (page 117) . . . . FALL WEBWORM
13. Setae scattered over body and do not arise in groups from low wart-like spots; caterpillars marked with blue, red, or yellow; tents constructed and occupied during spring of year and consist of many sheet-like layers of silk (page 112) . . . . TENT CATERPILLARS

* Also refer to Table 9.3.

TENT CATERPILLARS

Malacosoma spp., Lepidoptera, Lasiocampidae

1. Eastern tent caterpillar [*M. americanum* (F.)]
2. California tent caterpillar [*M. californicum* (Pack.)]

3. Blue-sided tent caterpillar [*M. constricta* Stretch]
4. Great Basin tent caterpillar [*M. fragile* Stretch]
5. Western tent caterpillar [*M. pluviale* (Dyar)]
6. Forest tent caterpillar [*M. disstria* Hbn.]

The last species listed is the most important, but because it differs from the others in several respects it is discussed separately.

INJURY. Tent caterpillars are not very injurious forest pests even though they often cause much defoliation. Trees so affected seldom die even after many annual defoliations. The main tree injury consists of reduced growth and sometimes branch dying. The caterpillars feed by consuming most of the leaf substance except the larger veins and petioles. All species except *M. disstria* construct large characteristic tent-like silken webs that are most conspicuous during the first 5 to 6 weeks of spring plant growth. The larger tents often are a foot in diameter and 2 or more feet long. They consist of numerous layers of dense sheets of silk webbing and contain much excrement and numerous molted skins.

INSECT APPEARANCE, HOSTS, AND RANGE. *Caterpillars:* 2 to 2½ inches long when full grown; young larvae (first 2 instars) dark, older larvae brightly colored with blue, orange, and rust; body hairy, but setae never in tufts, nor do hairs arise from warts; 5 pairs prolegs each with a half circle of crochets of 2 intermixed lengths. *Moths:* moderately robust; wingspread 1 to 1¾ inches; color tan to reddish tan; front wings each have 2 thin parallel, lighter or darker cross lines, or a broad, darker, transverse band; color of hindwings a uniform tan. *Pupae:* typical; reddish brown; inside well-made cocoons placed in protected places such as bark crevices or in rolled leaves. *Eggs:* in masses of 100 to 300, usually the bands encircle the twigs; covered with a hardened, porous, glistening material that soon becomes dull and often partially wears off exposing the eggs. The egg masses of the eastern tent caterpillar in parts of the northern Great Lakes region may differ in that they are oval masses located on the tree trunks near the ground.

1. EASTERN TENT CATERPILLAR. *Caterpillars:* dark colored with a white stripe along the mid-back line bordered on both sides with orange; a small, oval, blue spot within a larger black spot on sides of each segment. *Hosts:* species of *Prunus* preferred, but apple and other woody *Rosaceae* species sometimes attacked. *Range:* throughout the East and the Rocky Mountain regions.

2. CALIFORNIA TENT CATERPILLAR. *Caterpillars:* reddish brown with a narrow blue stripe along each side. *Hosts:* oaks, madrone, ash, poplar, willow, and many other hardwood species. *Range:* California.

3. BLUE-SIDED TENT CATERPILLAR. *Caterpillars:* body orange-brown

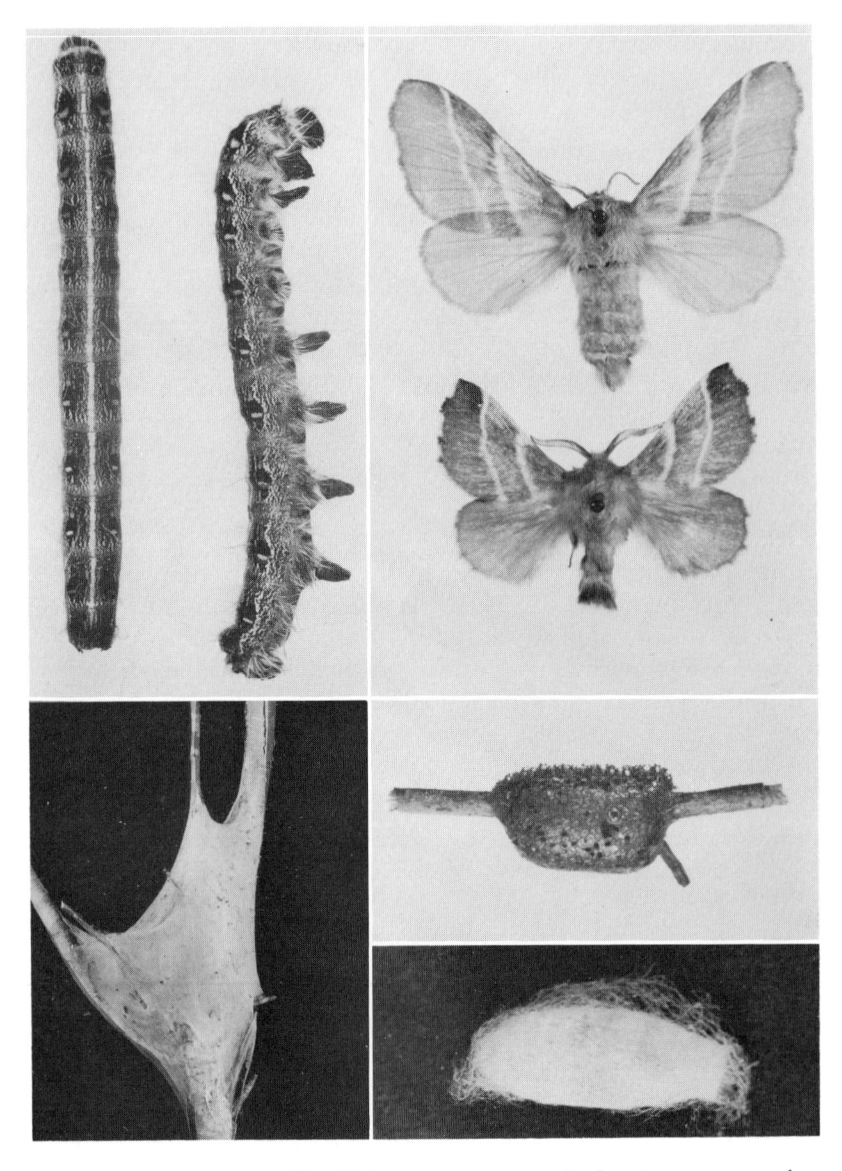

FIG. 9.4. Eastern tent caterpillar (*Malacosoma americanum*). Insects, egg mass and cocoon (X1½): lower left: tent made by the caterpillars (X¹⁄₁₀). Moths—female above, male below. Note the light line along the back of the caterpillar.

with blue sides and blue dots along upper part of each side. *Hosts:* oaks and other woody plants. *Range:* Oregon, California, and Arizona.

4. GREAT BASIN TENT CATERPILLAR. *Caterpillars:* dark colored; head blue gray; a broad, pale blue stripe bordered with fine orange lines along the mid-back line; 2 blue spots on each segment; whitish hairs on sides. *Hosts:* bitter brush, aspen, oaks, poplar, and many other hardwood species. *Range:* throughout the region between the Rocky Mountains and the Cascade–Sierra Nevada ranges (Great Basin region).

5. WESTERN TENT CATERPILLAR. *Caterpillars:* brown with a row of elongate blue spots along the back; on each side of these blue spots on each segment are two orange spots; pale, diffused, orange lines and spots on sides. *Range:* west of the Cascades in the Pacific Northwest and also in the East in the region north of the Great Lakes. *Hosts:* alder is preferred, but cherry, apple, and other hardwood species are also infested.

6. FOREST TENT CATERPILLAR. See below.

LIFE CYCLE. The caterpillars hatch when the leaves start to unfold in the spring and complete their larval development about 6 weeks later. They remain inside the tents most of the time except when feeding, which is usually during evenings, nights, mornings, or on cloudy days. The pupal stage lasts 1½ to 2 weeks. Moths emergence occurs early in July in the North or at higher elevations, but in warmer climates the moth may appear as early as May. A small caterpillar soon forms within each egg, where it remains until hatching time the following spring. There is only 1 generation per year.

CONTROL. Natural controls usually prevent tent caterpillars from causing much damage. Adverse weather conditions, a wilt disease, and insect parasites appear to be most important in this respect. Applied methods have not been used to protect forests from the attacks of these defoliators. Nevertheless, in order to prevent the loss of woody increment or to protect ornamental trees, DDT applied in the usual dosages by any practical method is most effective for killing the caterpillars. See page 101.

Essig, E. O. 1958. *Insects and mites of western North America.* Macmillan Co., N. Y. P. 695–697.
Wadley, F. M. 1947. *U.S.D.A. Leaflet* 161.

FOREST TENT CATERPILLAR

Malacosoma disstria Hbn., Lepidoptera, Lasiocampidae

During outbreaks, which occur at intervals of 10 to 15 years, this insect defoliates forests over millions of acres. The deciduous hosts seldom are killed even by many successive annual defoliations; therefore, branch killing and loss of woody growth is the main type of damage. An indirect undesirable effect is caused by the presence of large numbers of caterpillars that discourage vacationers from using the recreational lake areas of the North. See Fig. 4.1.

INJURY. Forest tent caterpillars are unlike other *Malacosoma* spp. in that they do not construct tents. Instead they make silk mats on the larger branches and on the main tree boles upon which the older larvae rest when not feeding. The feeding habits of *M. disstria* are similar to those for the other tent caterpillars in that they consume most of the foliage and waste little.

HOSTS. Aspen, hard maple, gums, and oaks are preferred. Other hosts are birch, basswood, cherry, elm, alder, willow, and hazel. Red maple is never eaten.

RANGE. Throughout North America wherever the hosts grow. The larger outbreaks have occurred in the Great Lakes region.

INSECT APPEARANCE. *Caterpillars:* see the general description for *Malacosoma* spp. on page 113. Body color bluish to brownish with keyhole or somewhat diamond-shaped white spots on the middle of the back

FIG. 9.5. Forest tent caterpillar (*Malacosoma disstria*) (X1⅓). Moths—female above, male below. Note the characteristic markings on the back of the caterpillar.

of each segment; 2 thin, broken, yellow lines extend along each side. *Pupae:* typical, reddish brown in a loose cocoon commonly formed inside rolled leaves. *Moths:* wingspread 1 to 1¾ inches; both pairs colored tan; front wings each have either two thin darker parallel lines or only a single broad dark band crossing the middle. *Eggs:* similar to those for other tent caterpillars.

LIFE CYCLE. Similar to that for the other tent caterpillars. See page 115.

CONTROL. The more important natural agencies that eventually have controlled big outbreaks in the past are lack of food, unfavorable weather conditions, and parasites. Foliage destruction by freezing accompanied by cold rains shortly after the eggs have hatched have been most effective in destroying many insects. The most abundant pupal parasite in the North is a large gray flesh fly, *Sarcophaga aldrichi* Parker. This viviparous fly deposits small maggots directly on the cocoons. These maggots work their way through the silk and bore into and kill the pupae. Parasitized cocoons can be recognized readily by the stained appearance caused by bleeding of the pupae through the maggot entrance holes.

Recreational areas such as cottage sites and camp grounds commonly are sprayed with DDT to prevent defoliation and to eliminate the caterpillars. Other forest areas generally have not been treated, because the defoliated trees are seldom killed. In the future, however, it may become practical to treat infested areas solely for the purpose of preventing the loss of woody growth. Aerial spraying with DDT at the usual rate of one pound per acre is most effective, but other application methods can be used if only a few trees are involved. See page 101. The spray should be applied about a week after the trees have leaved out. At this time the caterpillars are about ½ inch long. Spraying later is also effective, but the leaf destruction is progressively greater. To protect small areas from invading caterpillars the spraying should include a barrier strip 200 to 400 feet wide.

Batzer, H. O., and W. E. Waters. 1956. *U.S.D.A. Forest Pest Leaflet* 9.
Hodson, A. C. 1941. *Univ. Minn. Agri. Exp. Sta. Tech. Bull.* 148.

FALL WEBWORM

Hyphantria cunea (Drury), Lepidoptera, Arctiidae

The conspicuous webs of this species are often abundant, but the insect is not a major forest pest.

INJURY. The leaf skeletonizing, gregarious caterpillars live in large (2 to 3 feet or longer), loose, irregular, silk webs, which cover many leaves, twigs, and smaller branches and contain various amounts of excrement, molted skins, and leaves. The webs and inhabitants are present from about midsummer until fall. The larvae remain inside the webs all the time. When more food is needed the webs are enlarged to enclose more leaves.

HOSTS AND RANGE. Almost all deciduous hardwood species. In the South, pecan, sourwood, and persimmon are preferred, whereas in other regions boxelder, poplar, willow, and ashes are most frequently infested.

RANGE. Throughout most of North America.

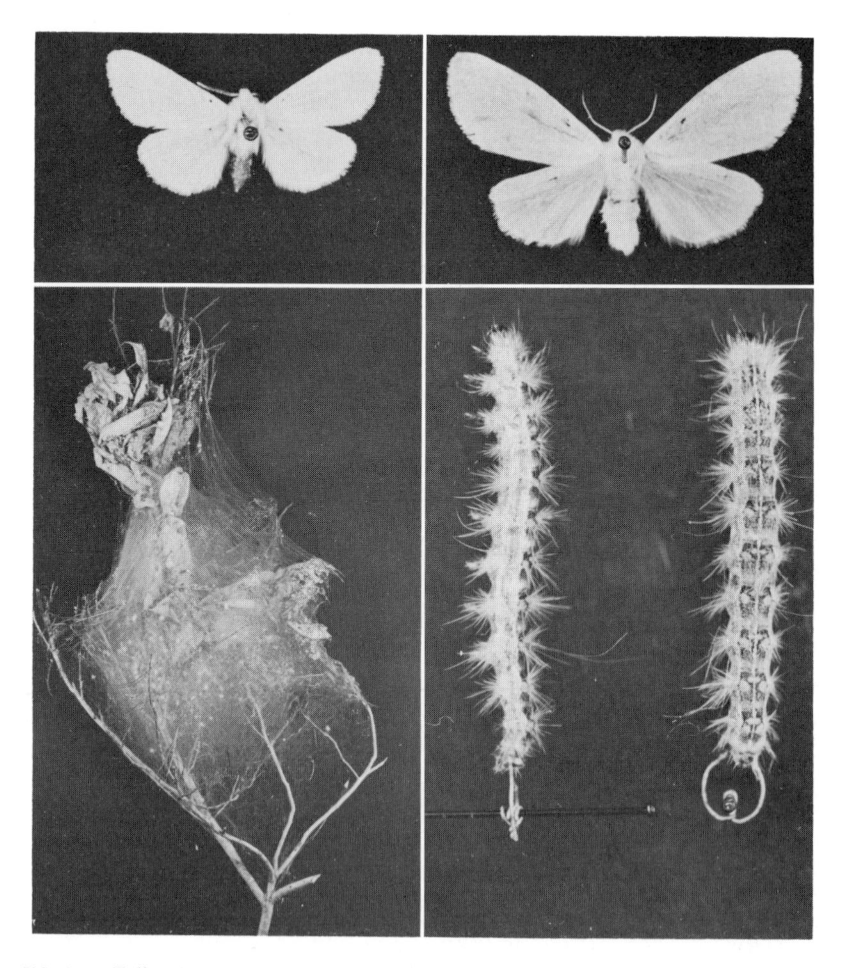

FIG. 9.6. Fall webworm (*Hyphantria cunea*). Insects (X1¾): lower left: silk web enclosing skeletonized leaves made by the caterpillars (X1/12). Moths—female right, male left.

INSECT APPEARANCE. *Caterpillars:* typical; about 1 to 1¼ inches long when full grown; body colored variously from pale yellow to gray or brown, darker individuals appear to have a dark band with a thin yellowish line along the back; body covered with long (⅜ inch) silky hairs that arise in groups from black, orange, or yellow warts; 12 of these setae bearing tubercles occur in a transverse row on each segment; 5 pairs prolegs. *Pupae:* typical; colored brown; inside a gray cocoon consisting of silk, hairs, and other debris. *Moths:* medium sized with stout bodies, wingspread ⅞ to 1½ inches; color pure white, or wings may be marked with one to many small brown or black spots. *Eggs:* laid in masses of up to 400 or 500; usually deposited on undersides of leaves; globular with fine surface markings; color yellow or light green, but turn gray before hatching.

LIFE CYCLE. Winter is passed as pupae located in the surface layers of the soil. The moths are present throughout the summer and early fall. In the middle latitudes of the United States the moths emerging from the overwintering pupae are most common during May, whereas in other areas emergence may be either earlier or later. After mating the females lay their eggs in large masses on the undersides of the leaves. The caterpillars hatch within 1 to 1½ weeks and immediately construct and live in the loose tent-like webs already described. When ready to pupate, the larvae leave the webs and pupate in protected places such as bark crevices or, for the last generation of the year, in the soil. In the South there are two or more generations per year, but in the North there is only one.

CONTROL. Applied control has never been used against this insect in forests. For protecting orchard or shade trees, spraying with DDT in the usual dosages and formulations is very effective. For detailed instructions see page 101. When only a few webs are involved, however, it may be easier to simply remove and burn them.

POPLAR TENT CATERPILLARS

***Ichthyura inclusa* Hbn., Lepidoptera, Notodontidae (There are also several other related species)**

INJURY. This leaf feeder makes small silk-lined, nest-like tents consisting of several leaves webbed together in which the larvae rest when not feeding. Although they consume the foliage rather completely and leave only the petioles and larger veins, the leaves comprising their nests are not eaten. They cause little damage but the nests may be unsightly on ornamental trees.

HOSTS. Poplars and willows.

RANGE. Throughout eastern United States west to the Rockies.

INSECT APPEARANCE. *Caterpillars:* 1¾ to 2 inches long when full grown; body tan to brown with 4 yellow stripes along back and 3 lighter stripes along each

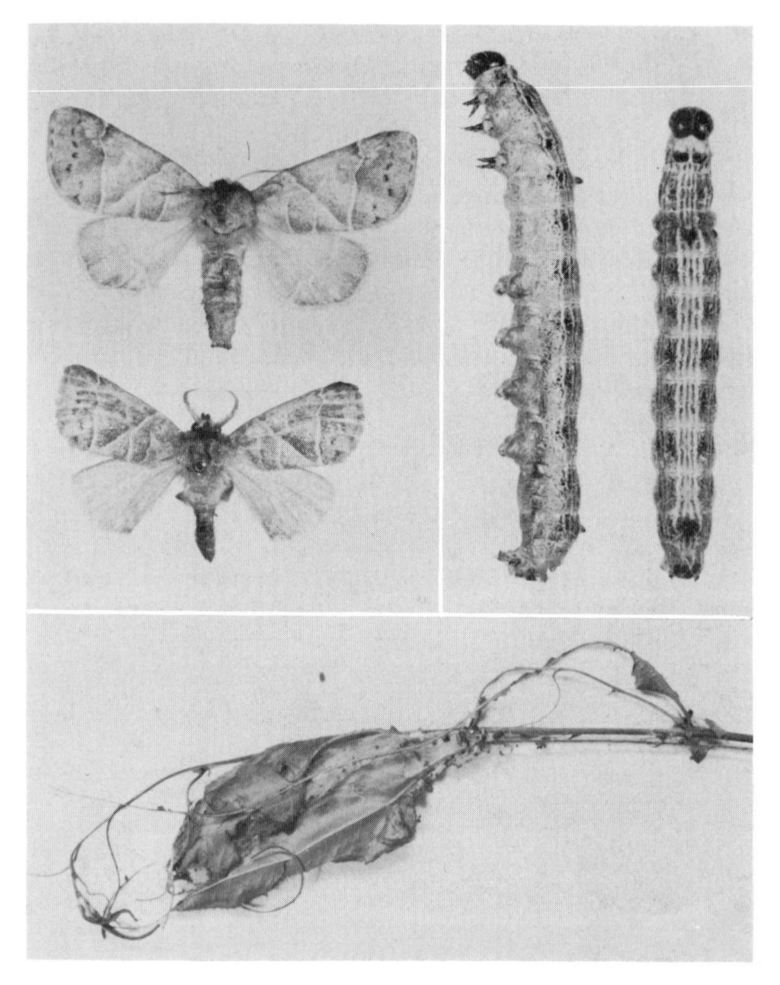

FIG. 9.7. Poplar tent caterpillar (*Ichthyura inclusa*). Insects (X1¾): below: tent made by the caterpillars (X½). Moths—female above, male below. Note the tubercles on the back of the first and eight segments of the caterpillars.

side; head black with scattered white hairs; pronotum with 2 black spots; back of 1st and 8th abdominal segments each with a black 2-knobbed wart; thoracic legs and posterior abdominal tergite black; with 5 pairs prolegs. **Pupae:** typical; reddish brown; inside loose cocoons rolled in leaves or, for the overwintering generation, in the soil. **Moths:** medium size, 1 to 1¼ inch wingspread, body stout; color grayish tan; forewings short and broad with yellowish spot on front margin near each tip, 4 thin light lines cross each forewing, the 2 middle ones frequently meet or cross to form a "V" or "X," outer edge of forewings with 2 rows of dark dots.

LIFE CYCLE. Winter is passed as pupae in the ground. The first generation moths emerge during the spring, whereas those of the second generation appear in July and August. The eggs are laid in clusters on the undersides of the leaves. Pupation for the first generation occurs inside rolled leaves. Throughout most of the range of the insect there are two generations per year.

CONTROL. Insecticidal control is seldom needed even on shade and ornamental trees. If ever needed, however, DDT would be effective. See page 101 for the details regarding dosages and application.

UGLY NEST CATERPILLARS

Archips cerasivorana (Fitch) and *A. fervidana* (Clem.), Lepidoptera, Tortricidae

INJURY. These defoliators make irregular webs enclosing and fastening together several to many leaves. Even though these webs are rather conspicuous during the spring and early summer, these insects are of insignificant importance as forest pests.

HOSTS. *A. cerasivorana* infests *Prunus* spp., whereas *A. fervidana* feeds on various species of oaks.

RANGE. Eastern United States to Minnesota and Missouri and southeastern Canada.

INSECT APPEARANCE. *Caterpillars:* ¾ to ⅞ inch long when full grown; body naked; tan to gray-green (*A.f.*) or yellowish (*A.c.*) with head and pronotal shield dark brown to black. *Moths:* wingspread ¾ to ⅞ inch; *A.f.* with forewing brown to reddish brown, each with two darker brown patches on the front edge and a smaller spot behind; hindwings reddish gray. *A.c.* with forewings dull orange irregularly speckled with brown with 3 brown patches, two of which are near middle of the front wing margin; hindwings orange.

LIFE CYCLE. The gregarious larvae live and feed inside the webs during the early part of the season. In June or July they pupate within the webbing, and the moths emerge during the following 2 months. The winter is passed as eggs, which probably are placed on the twigs.

CONTROL. Insecticidal control has never been needed in forests. To protect ornamental trees, however, the usual DDT emulsions or wettable powder sprays should be effective. See page 101.

TETRALOPHA WEBWORMS

Tetralopha spp., Lepidoptera, Pyralidae (Epipaschiidae)

INJURY. The larvae of these species enweb and feed on the leaves but have never caused much injury.

HOSTS. Oaks, maple, poplar, and beech.

RANGE, INSECT APPEARANCE, LIFE CYCLE, AND CONTROL. Similar to that for the pine webworm. See page 122.

MIMOSA WEBWORM

Homadaula albizziae Clarke, Lepidoptera, Glyphipterygidae

INJURY. This is an introduced, webforming, defoliator pest that is a pest chiefly of its ornamental hosts.

HOSTS. Mimosa or silk tree (*Albizzia julibrissin*) and the honey locust.

RANGE. Maryland, Virginia, and North Carolina, but it is constantly spreading to adjacent areas.

INSECT APPEARANCE. *Caterpillars:* about ⅝ inch long when full grown; head and pronotal shield brownish, marked with dark brown or black irregular longitudinal bands; body gray to brown with 5 longitudinal white stripes; abdomen sometimes pinkish and mottled with white patches; anal tergite dark brown mottled with white; tubercles and spiracles dark; thoracic legs dark brown to black, joints white ringed. *Moths:* gray with silvery luster, forewings marked with about 20 conspicuous black spots; hindwings brownish gray. *Eggs:* minute, oval, pearly gray soon changing to pink.

LIFE CYCLE. The moths appear during early June. After mating each female deposits 60 to 70 eggs on the flowers and leaves of the host trees. These hatch within a few days and the resulting larvae immediately start enwebbing and feeding on the foliage, flowers, and seed. When the larvae are full grown many descend to the ground on silken threads where they pupate in sheltered spots such as in bark crevices or in the litter. Others of the first brood remain and pupate within the webs. There are 2 to 3 generations per year. Winter is passed in the pupal stage.

CONTROL. Lead arsenate or DDT applied at the usual rates are both effective insecticides for controlling this insect on ornamentals. It is not a forest pest.

Clarke, J. F. G. 1943. *Proc. U. S. Nat. Museum* 93(3162) 205–208.
Wester, H. V., and R. A. St. George. 1947. *J. Econ. Entomol.* 40:546–553.

PINE WEBWORM

Tetralopha robustella Zell., Lepidoptera, Pyralidae (Epipaschiidae)

This insect frequently infests seedling pines but large, even mature, trees may also be infested. Recently, it has become rather prevalent throughout the South in newly established plantings, and sometimes seedlings are killed or damaged. Trees older than a year or two, however, are seldom, if ever, severely injured.

INJURY. Infested trees can be readily identified by the presence of irregular to globular compact masses of enwebbed caterpillar excrement surrounding the twigs and enclosing the basal portions of the needles. These masses are commonly 1½ to 2 inches long and an inch or more in diameter, although larger masses are sometimes seen.

HOSTS. Pines.

RANGE. Throughout eastern United States west to Minnesota.

INSECT APPEARANCE. *Caterpillars:* about ¾ inch long when full grown; tan or light brown with two darker longitudinal stripes along each side, lower one faint; head and pronotum also tan with irregular brown line on posterior half of head; *Pupae:* typical; robust, about ½ inch long; reddish; in soil. *Moths:* small, wingspread of ¾ to 1 inch; dark colored with head thorax and wings medium gray, forewings with basal third purple black and outer half darker gray; raised tufts of scales on forewings.

LIFE CYCLE. The biology of this species is imperfectly known. Winter and spring are passed in the soil as pupae and the moths emerge from late June until August. Most likely the eggs are laid on the needles. There appears to be only one generation per year.

CONTROL. Applied control has not been used against this insect under forest conditions. In nurseries and other valuable plantings DDT applied as a wetta-

FIG. 9.8. Pine webworm (*Tetralopha robustella*). Caterpillars (X2); webs made by the caterpillars contain compact masses of excrement (frass) (X½).

ble powder or emulsion at the rate of one pound of DDT per 100 gallons of spray has been effective in keeping the trees free of this pest. See page 101.

JUNIPER WEBWORMS

***Dichomeris marginella* (F.), Lepidoptera, Gelechiidae and *Phalonia rutilana* (Hbn.), Lepidoptera, Tortricidae (Phaloniidae)**

INJURY. These two species, which enweb and feed on the foliage of junipers, are not very injurious, but are pests chiefly of ornamentals. HOSTS. Junipers. RANGE. Both species were introduced from Europe into the Northeast. From here they have spread so that now they occur west to Michigan and Indiana. *Dichomeris* has also spread south to North Carolina. INSECT APPEARANCE. *Caterpillars:* small, less than ½ inch long when full grown; colored brown; with 5 pairs prolegs. *Dichomeris* larvae have 3 longitudinal dark brown stripes. *Moths: Dichomeris*—small ⅝ inch wingspread; pale pinkish brown moths; forewings margined with white; *Phalonia*—small, ⅜ to ½ inch wingspread; bright yellow with 4 broad red bands across forewings; hindwings and abdomen gray.

LIFE CYCLE. The life histories of these species are imperfectly known, but it appears that there may be 2 generations per year. The moths for the first generation commonly emerge during June. **CONTROL.** Applied control with insecticides has never been needed in forested areas. Ornamental trees probably can be protected with a spray of DDT. See page 101.

SPRUCE WEBWORM

***Epizeuxis aemula* (Hbn.) Lepidoptera, Phalaenidae (Noctuidae)**

INJURY. The larvae live and feed on the foliage forming webbed masses of dried needles and excrement. It is not a serious forest pest. **HOSTS.** Spruces. **RANGE.** Wherever hosts grow throughout eastern North America. **INSECT APPEARANCE.** *Caterpillars:* about ⅝ inch long when full grown; body brown; with a darker stripe along back, spiracles and tubercles black. *Moths:* wingspread about ⅞ inch; forewings gray crossed by 3 wavy or zigzag black lines, plus other partial or indistinct lines; hindwings also crossed by 4 diffused wavy dark lines. **LIFE CYCLE.** The moths emerge during the early summer, and the females lay eggs that do not hatch until sometime later; consequently, the young caterpillars are present only for a few weeks late in the summer. They then hibernate over the winter in the enwebbed nests and resume feeding the following spring. There is only 1 generation per year. **CONTROL.** Insecticidal control has never been needed for this insect but, if ever needed, DDT should be effective. For application methods and dosages refer to page 101.

LOOPER CATERPILLARS

Lepidoptera, Geometridiae

Looper caterpillars have only 2 or 3 pairs of prolegs instead of the usual 5 pairs, and these are located on the posterior abdominal segments. These characteristics cause the larvae to crawl with a characteristic looping motion. The mid-portions of their bodies loop upward at regular intervals as they move along. Other common descriptive names are inchworms, measuring worms, spanworms, and geometers. There are also a few species of Phalaenidae that have fewer than 5 pairs of prolegs, but none of these infests trees. Often the naked looper caterpillars are not easily seen because they are drably colored and become immobile whenever disturbed. The adults are slender-bodied, delicate, but broad-winged moths that are drably colored tan to gray with fine irregular lines crossing the wings. The characteristics of the commoner species are presented in Tables 9.5 and 9.6 and in the descriptions that follow.

HEMLOCK LOOPERS

***Lambdina fiscellaria* varieties *fiscellaria* (Guen.) and *lugubrosa* (Hulst.), Lepidoptera, Geometridae**

Brief descriptions of several other hemlock-feeding geometrids are presented in Table 9.6 and in the text.

TABLE 9.5 COMMON SPECIES OF LOOPER DEFOLIATORS THAT INFEST BROAD-LEAVED TREES

1. Larvae present during spring through June 2
1. Larvae present from late June through the summer 15
2. Larvae construct and live within webs formed by rolling together leaves or by fastening together several leaves 3
2. Larvae do not construct webs 6
3. Body color dark brown or black on back 5
3. Body color yellow, green, or pink with lighter yellow or green stripes along sides and/or back 4
4. Western species; larvae large, up to 1¾ inches long when full grown; color variable yellow, green, or pink with yellow or green stripes along sides and back, black markings over body (page 131) . . . . OMNIVEROUS LOOPER
4. Present only in eastern Canada (Nova Scotia); larvae about ¾ inch long when full grown; color bright green with paler stripes along sides (page 131) . . . . WINTER MOTH
5. Head dark reddish brown, body blackish with 4 broken lines along body, yellow underneath; less than 1 inch when full grown; host only cherry (page 131) . . . . CHERRY SCALLOP SHELL MOTH
5. Head black, body dark brown to black with slender black stripe along each side which are formed by a series of small black dots, white to red spots below these lines; about 1 inch long when full grown; hosts birch, willow, and other hardwoods (page 131) . . . . SPEAR MARKED BLACK MOTH
6. Larvae with a pair of long slender filaments on each of 2nd and 3rd abdominal segments, a pair of warts on each of the 1st and 8th abdominal segments; about ¾ inch long when full grown (page 131) . . . . FILAMENT BEARER
6. Larvae not as above 7
7. Head and anal segments bright red, body brown to black; about 1½ inches long when full grown (page 131) . . . . ELM SPAN WORM
7. Larvae not colored as above 8
8. Larvae more than 1 inch long when full grown 9
8. Larvae about 1 inch or less in length when full grown 11
9. Body bright yellow with 10 wavy black lines along back with outer 2 being more distinct than others; eastern North America west to Rockies, about 1½ inches long when full grown (page 131) . . . . LINDEN LOOPER
9. Larvae without lines along back 10
10. Back of larvae brown, sides yellow and yellow to greenish underneath; about 1½ inches long when full grown; a European species which has been introduced into the Pacific Northwest (page 131) . . . . MOTTLED UMBER MOTH
10. Color variable, yellow green to gray, head and body flecked with black; about 1¼ inches long when full grown; found only in Pacific Northwest (page 131) . . . . OAK LOOPER
11. Larvae with 3 pairs of fleshy abdominal legs (prolegs) but pair on 5th abdominal segment may be small 12
11. Larvae with only 2 pairs of prolegs 13
12. Larvae pinkish gray varied with darker gray or black and yellow; present only in the Pacific Coast States (page 131) . . . . WALNUT SPANWORM

12. Larval color variable, light to dark green, gray, brown to almost black with whitish stripes along sides (page 129) . . . . FALL CANKERWORM
13. Larvae usually bright green with narrow light stripes along each side 14
13. Larval color variable, usually gray or brownish but sometimes may be yellowish green, head dirty white mottled with brown; body lines numerous, irregular and broken, but may be absent on some individuals (page 129) . . . . SPRING CANKERWORM
14. Many larvae are leaf rollers and feed within the shelters formed; present only in Nova Scotia (page 131) . . . . WINTER MOTH
14. Larvae are not leaf rollers (page 130) . . . . BRUCE'S SPANWORM
15. Larvae make and live within webs formed by fastening together a number of leaves 3
15. Larvae do not construct webs 16
16. Head indented along center so as to form a two-lobed head 17
16. Head not bilobed 18
17. Pair of lateral tubercles on 5th abdominal segment and another pair of converging granulated tubercles on the 8th abdominal segment; about 2 inches long when full grown (page 131) . . . . CLEFT-HEADED SPANWORM
17. Reddish swollen areas back of abdominal segments 2 to 5 and on underside of 3rd; about 2½ inches long when full grown (page 131) . . . . NOTCHED-WING GEOMETER
18. Body yellowish with lines along each side consisting of short lines and spots which suggests a chain; about 2 inches long when full grown; very slender (page 131) . . . . CHAIN-SPOTTED GEOMETER
18. Color variable, yellow green to gray, head and body flecked with black; about 1¼ inches long when full grown; found only in the Pacific Northwest (page 131) . . . . OAK LOOPER

TABLE 9.6 COMMON SPECIES OF LOOPERS THAT INFEST CONIFERS

1. Usually infest pines 2
1. Usually infest conifers other than pines 5
2. Infest only Monterey pine in California (page 131) . . . . MONTEREY PINE LOOPER
2. Infest other species of pines 3
3. With light stripes along back or sides 4
3. No light stripes along back or sides, body greenish to brown flecked with black; about 1½ inches long when full grown (page 131) . . . . *Lambdina athasaria pellucidaria* (G. and R.)
4. Light longitudinal stripes on each side of back with the finer lines between, body light green with brownish tinger above; about ⅞ inch long when full grown (page 132) . . . . *Semiothisa granitata* (Guen.)
4. Three light lines along back with the mid-line being broadest, body dark green; about 1 inch long when full grown (page 132) . . . . *Eufidonia notataria* (wlkr.)
5. Prominent swellings on sides of second abdominal segment, body green tinged with brown, head broader than thorax and bilobed; about ⅞ inch long when full grown (page 132) . . . . *Anacamptodes ephyraria* (Wlkr.)

5. No swellings on sides of second abdominal segment 6
6. Larvae with light or dark longitudinal stripes 7
6. Larvae lacking longitudinal stripes (page 124) . . . . HEMLOCK LOOPERS
7. Stripes light in color 8
7. Sides with wavy brown lines sometimes interrupted by dashes of white, 5 broken wavy lines beneath, body yellowish indistinctly marked with brown white and yellow; about 1¼ inches long when full grown (page 132) . . . . *Lambdina athasaria athasaria* (Wlkr.)
8. Light longitudinal stripes on each side of back with two finer lines between. See number 4 above; about ⅞ inch long when full grown (page 132) . . . . *Semiothisa granitata* (Guen.)
8. With 4 equally spaced thin light lines along sides and back (page 132) . . . . NEW MEXICO FIR LOOPER

The hemlock loopers are most destructive forest pests. There have been several major epidemics in the coastal areas of the Pacific Northwest in which the western variety (*lugubrosa*) has killed an estimated ¾ million board feet of timber. During the late 1920's the eastern variety (*fiscellaria*) was prevalent in parts of the Lake States and the Northeastern States. The outbreaks usually last for about 3 years, and if the trees are not defoliated more than 50 per cent they usually recover. Sometimes even those defoliated as much as 75 per cent may recover if they are not attacked by inner bark borers.

INJURY. During severe epidemics most of the needles are consumed, and the older caterpillars even clip off small twigs. When the insects are less numerous, however, the needles are only partially eaten. The partly severed needles soon turn brown, and these make the heavily infested trees appear as if they were scorched by fire. The larvae also drop from the trees on silken threads, so that during epidemics these silken threads may make the infested woods look and feel like a big cobweb. The most rapid rate of defoliation occurs during late July and August.

HOSTS. Hemlocks are preferred but Douglas-fir, sitka spruce, western red cedar, and balsam fir may also be heavily attacked, especially when mixed with hemlock. Sometimes even broad-leaved trees and shrubs may be eaten when the other food is exhausted.

RANGE. The eastern variety (*fiscellaria*) occurs throughout the range of eastern hemlock from southern Canada south to Georgia in the mountains and west to Wisconsin, whereas the western variety (*lugubrosa*) occurs in the coastal region of British Columbia, Washington, and Oregon.

INSECT APPEARANCE. *Caterpillars:* about 1¼ to 1½ inches long when full grown; color varies from yellow-green to brown; both head and body flecked with black; with only 2 pairs of prolegs; *lugubrosa* has diamond-

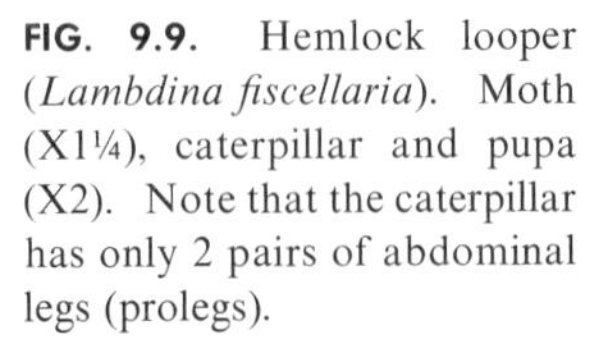

FIG. 9.9. Hemlock looper (*Lambdina fiscellaria*). Moth (X1¼), caterpillar and pupa (X2). Note that the caterpillar has only 2 pairs of abdominal legs (prolegs).

shaped markings along back. *Pupae:* about ½ inch long; mottled greenish brown; always naked (not in cocoons) in bark crevices or in litter. *Moths:* light tan slender moths with a wingspread of about 1½ inches; forewings each crossed with 2 wavy thin dark lines, whereas each hindwing is crossed by only one thin dark line. *Eggs:* about the size of a pinhead, color blue to brown.

LIFE CYCLE. Moth emergence occurs mostly during September. They mate immediately and each female lays about 100 eggs singly or in groups of 2 or 3 on the twigs, branches, and tree trunks. The eggs overwinter and then hatch during the late Spring (May and early June). Larval developments is usually complete by late August, and pupation lasts for about 3 weeks. There is only one generation per year.

CONTROL. The most important natural control agents are a virus disease that infects the caterpillars and heavy rains during the period of moth flight. Aerial applications of DDT at the rate of 1 pound per gallon of fuel oil spray per acre is effective. The best period for application is from about mid-June to mid-July.

Carroll, W. J. 1956. *Can. Entomol.* 88:587–599.

CANKERWORMS

Spring cankerworm [*Paleacrita vernata* (Peck)] and fall cankerworm [*Alsophila pometaria* (Harr.)], Lepidoptera, Geometridae

These two species of cankerworms have been pests of forests and shade trees since colonial times. They frequently completely defoliate forests over extensive areas, but because the epidemics usually last only for a few years and because these deciduous host trees seldom die as the result of the defoliations, these insects are not considered among our major forest pests. Nevertheless, cankerworms often are troublesome on shade trees, especially in cities and towns.

INJURY. The caterpillars frequently produce a characteristic feeding pattern that consists of irregular holes eaten through the leaves; but as the larvae reach maturity, especially during heavy infestations, most of the leaf substance is consumed so that only the mid-rib and larger veins are left.

HOSTS. Elm and basswood are preferred, but the foliage of a great many other species of deciduous trees is also eaten.

RANGE. Both species occur throughout much of eastern North America from southern Canada west to Colorado and also in California.

INSECT APPEARANCE. *Caterpillars:* about ¾ to 1 inch long when full grown; color variable from light green to dark green (fall species) to brown or black (spring species); some individuals of the latter have numerous irregular broken dark stripes along back, whereas the fall species has a darker stripe along the mid-back line and other faint whitish lines along back; spring cankerworm has only 2 pairs of prolegs, but fall species has a small nonfunctional 3rd pair on the 5th abdominal segment. *Moths:* females wingless, gray to black; about ½ inch long; males slender, delicate, brownish gray moths; wingspread ⅝ to 1⅕ inches; fall species sometimes with 2 irregular whitish bands crossing forewings, whereas spring moths sometimes have 3 irregular darker lines across forewings, but these lines may be indistinct; dorsum of first abdominal segment of both males and females of the spring cankerworm have 2 transverse rows of reddish spines. *Eggs:* in masses of about 100 for the fall species and about 50 for the spring moths, deposited on the twigs, limbs, and trunks of the host trees; although the spring cankerworm moths deposit eggs in irregular masses, those for the fall species lay their eggs in more regular groups.

LIFE CYCLE. These cankerworms are named "spring" and "fall" according to the time of the year when the moths appear and not the time when the caterpillars are present. Fall cankerworm moths emerge and deposit their eggs mostly during the late fall (October to December depending on location), whereas the spring cankerworm moths do not emerge until early spring. The eggs of both species, however, hatch in the spring at about the time the leaves start to unfold, and the resulting caterpillars complete their growth about 4 or 5 weeks later. They then drop to the soil and enter it to pupate. Here they remain until fall or the following spring, depending on the species.

CONTROL. Even though the defoliation caused by these insects may be rather spectacular, insecticidal control has not been used in forests. Eventually, however, it may be desirable to prevent the loss of woody growth. Then aerial applications of DDT at the rate of 1 pound per acre would be effective. For applying sprays to the foliage of shade trees, mist blowers, using about a 10 per cent DDT oil solution, have been found to be most practical. For hydraulic sprayers the dosage should be 1 pound of actual DDT (either the wettable powder or emulsion) per 100 gallons of spray. Powerful spraying equipment or mist blowers are needed to reach the taller trees; therefore, if such equipment is not available, spraying the trunks of the host trees with 5 per cent DDT emulsion may be the only available method. This trunk-spraying method kills the wingless female moths as they climb the trees to lay eggs; therefore, the time of application is critical. Sprays applied after the moths have climbed the trees are ineffective. Therefore, careful observations must be made during the fall and spring to determine when the moths start emerging. For this purpose bands of sticky materials can be used to encircle the trunks to catch the first emerging females. In the North this usually is about a week after the first major spring thaw and just after the first heavy fall frost for the two species respectively. A second spraying should be made about 10 days later.

Hodson, A. C. 1951. *Univ. Minn. Agri. Exten. Folder* 136.
Jones, T. H., and J. V. Schaffner. 1953. *U.S.D.A. Leaflet* 183.

LOOPERS OF MINOR IMPORTANCE THAT INFEST DECIDUOUS BROAD-LEAVED TREES*

Lepidoptera, Geometridae

The larval characteristics of the species listed below are presented in Table 9.5. Sometimes these species of loopers become abundant in localized areas, but these outbreaks are not as common as they are for the cankerworms.

1. CANKERWORMS. See page 129.

2. BRUCE'S SPANWORM [*Operophtera bruceata* (Hulst)]. *Range:* Canada and the northern United States west to the Rocky Mountains. *Life cycle:* similar to the fall cankerworm.

* Most are general feeders on numerous species of host trees.

3. WINTER MOTH [*O. brumata* (L.)]. **Range:** Western Nova Scotia. Probably introduced from Europe. **Life cycle:** similar to the fall cankerworm.

4. LINDEN LOOPER [*Erannis tiliaria* (Harr.)]. **Range:** Eastern United States and Canada west to the Rocky Mountains. **Life cycle:** similar to fall cankerworm.

5. MOTTLED UMBER MOTH (*E. defoliaria* Clerck). **Range:** a European species introduced into the Pacific Northwest. **Life cycle:** similar to the fall cankerworm.

6. WALNUT SPANWORM [*Coniodes plumogeraria* (Hulst.)]. **Range:** western United States. **Life cycle:** similar to spring cankerworm.

7. CHERRY SCALLOP SHELL MOTH [*Calocalpe undulata* (L.)]. **Range:** North America. **Life cycle:** winter as pupae in soil; moths appear from May to September; larvae present from June to October; 1 generation per year. **Host:** cherry.

8. SPEAR-MARKED BLACK MOTH [*Eulype hastata* (L.)]. **Range:** United States. **Life cycle:** winter as pupae in soil; moths appear from May to August; larvae present from June to September; may be a partial second generation per year.

9. CLEFT-HEADED SPANWORM (*Amphidasis cognataria* Guen.). **Range:** Atlantic States. **Life cycle:** winter as pupae in soil; moths appear from May to July; larvae present from July to August; 1 generation per year.

10. FILAMENT BEARER [*Nematocampa limbata* (Haw.)]. **Range:** eastern United States and Canada. **Life cycle:** winter probably is passed in the egg stage, larvae present from May to July; moths appear from June to August; 1 generation per year.

11. ELM SPANWORM [*Ennomos subsignarius* (Hbn.)]. **Range:** eastern United States and Canada west to the Rocky Mountains. **Life cycle:** winter is passed in the egg stage; larvae are present from time foliage appears until June; moths appear in July; 1 generation per year.

12. CHAIN-SPOTTED GEOMETER [*Cingilia catenaria* (Drury)]. **Range:** northeastern North America west to the plains. **Life cycle:** winter is passed in the egg stage; larvae present from June to August, moths appear late summer and early fall; 1 generation per year.

13. NOTCHED-WING GEOMETER [*Deuteronomos magnarius* (Guen.)]. **Range:** northern United States. **Life cycle:** winter is passed in the egg stage; larvae present from June to August; moths appear late summer and early fall; 1 generation per year.

14. OAK LOOPER [*Lambdina fiscellaria somniaria* (Hulst.)]. **Range:** Pacific Northwest. **Life cycle:** similar to hemlock looper. **Hosts:** chiefly oaks.

15. OMNIVEROUS LOOPER (*Sabulodes caberata Guenee*). **Range:** California. **Life cycle:** winter in all stages; several broods per year but that occurring in late summer is most abundant so moths and larvae are present throughout year.

LOOPERS OF MINOR IMPORTANCE THAT INFEST CONIFERS

Lepidoptera, Geometridae

The larval characteristics of the following species are presented in Table 9.6.

1. HEMLOCK LOOPERS. See page 124.

2. MONTEREY PINE LOOPER [*Nepytia umbrosaria* (Pack.)]. **Range:** California. **Life cycle:** no information available. **Hosts:** Monterey pine.

3. *Lambdina athasaria pellucidaria* (G. and R.). **Range:** Atlantic States. **Life cycle:** winter as pupae in litter; moths appear during May and June; larvae are present from June to September; 1 generation per year. **Hosts:** hard pines.

4. *Semiothisa granitata* (Guen.). **Range:** northeastern United States. **Life cycle:** winter as pupae in litter; moths appear in June; larvae present from June to September; 1 generation per year. **Hosts:** white pine, spruce, fir, and larch.

5. *Eufidonia notataria* (Wlkr.). **Range:** Atlantic States. **Life cycle:** winter as pupae in litter and soil; moths appear during May and June; larvae present from June to September; 1 generation per year. **Hosts:** white pine.

6. *Anacamptodes ephyraria* (Wlkr.). **Range:** northeastern United States. **Life cycle:** winter in egg stage; larvae are present May to July; moths appear in August; 1 generation per year. **Host:** hemlock.

7. *Lambdina athasaria athasaria* (Wlkr.). **Range:** northeastern United States west to Ohio. **Life cycle:** winter as pupae in litter; moths appear during May and June; larvae present from June to September; 1 generation per year. **Host:** hemlock preferred.

8. False Hemlock Looper [*Nepytia canosaria* (Wlkr.)]. **Range:** Canada and northern United States. **Life cycle:** probably winter in egg stage; larvae present from June to August; moths appear during August and September; 1 generation per year. **Hosts:** hemlock, spruce, fir, and larch.

9. *Nepytia phantasmaria* (Stkr.). **Range:** Pacific Northwest. **Life cycle:** probably winter in egg stage; larvae present from May to July; moths appear August and September; 1 generation per year. **Hosts:** hemlock, spruce, and Douglas-fir.

10. New Mexico Fir Looper [*Galenara consimilis* (Hein.)]. **Range:** southern Rockies. **Life cycle:** no information available. **Hosts:** Douglas-fir, true firs, and spruce.

TUSSOCK CATERPILLARS

Lepidoptera, Lymantriidae, Arctiidae, and Phalaenidae (Noctuidae)

These caterpillars are characterized by having some of the hairs (setae) arranged in dense tufts that are either short and brush-like or long and slender "pencil" tufts. The caterpillars of Lymantriidae—the most important family—are characterized by having a small spot or projection (gland) on the mid-back of each of the sixth and/or seventh abdominal segments. The general characteristics of the common species are presented in Table 9.7 and in the text.

TABLE 9.7 COMMON SPECIES OF TUSSOCK MOTH CATERPILLARS*

1.	Feed on the foliage of conifers	2
1.	Feed on the foliage of broad-leaved deciduous trees	4
2.	Small, circular, convex spot or small projection (gland) on mid-back of 6th and/or 7th abdominal segments	3
2.	Glands as described in 2 above not present; with brown to black and yellow tufts (page 137)	SILVER-SPOTTED TUSSOCK MOTHS
3.	Occurs in northeastern North America; hosts are chiefly pines and spruces (page 136)	PINE TUSSOCK MOTH
3.	Occurs in western North America; hosts are chiefly Douglas-fir and true firs (page 133)	FIR TUSSOCK MOTHS

4. Small, circular, convex spot or small projection (gland) on mid-back of 6th and/or 7th abdominal segments 5
4. Glands, as described in 4 above, not present 9
5. Two long slender "pencil" tufts present on prothorax 6
5. Only short, dense, brush tufts present; body light to dark brown with a tuft of white setae on both sides of each segment (page 145) . . . . BROWN-TAIL MOTH
6. Long slender "pencil" tufts also on both sides of 2nd abdominal segment (page 136) . . . . RUSTY TUSSOCK MOTH
6. Single long slender "pencil" tuft on back of 8th abdominal segment 7
7. Occurs in Pacific Coastal States; body gray, spotted with red, yellow, and blue (page 136) . . . . WESTERN TUSSOCK MOTHS
7. Occur in eastern North America west to the Rockies; head bright red or yellow . . 8
8. Broad black stripe along back (page 135) . . . . WHITE-MARKED TUSSOCK MOTH
8. Broad pale stripe along back; black spots behind 2nd and 3rd abdominal tufts (page 135) . . . . DEFINITE-MARKED TUSSOCK MOTH
9. Short brush tufts present; long slender "pencil" tufts on thorax and/or abdominal segments (page 138) . . . . TIGER TUSSOCK MOTHS
9. Short brush-like tufts absent; long slender "pencil" tufts present only on abdominal segments (page 139) . . . . DAGGER TUSSOCK MOTHS

* Some of the setae (hairs) are arranged in dense brush-like and/or long slender (pencil) tufts.

FIR TUSSOCK MOTHS

Douglas-fir tussock moth (*Hemerocampa pseudotsugata* McD.), and *H. oslari* (Barnes), Lepidoptera, Lymantriidae

The Douglas-fir tussock moth is a forest pest of major importance in the northern Rockies. This species was first discovered in 1818, and since then it has become epidemic several times with major outbreaks occurring during the late 1920's and the mid–1940's.

INJURY. The caterpillars are needle eaters which like many other important defoliators, first consume the foliage near the tree tops before moving to feed on the lower parts of the crowns. The young caterpillars feed on the buds and needle bases so that the remaining needle parts turn brown and, thereby, cause the crowns of heavily infested trees to appear scorched.

HOSTS. The chief host is Douglas-fir, but the foliage of true firs is also eaten.

RANGE. Northern Rockies from British Columbia south to Colorado and Utah. A closely related species, *H. oslari,* feeds on white fir in California and Colorado.

INSECT APPEARANCE. *Caterpillars:* about ¾ to 1 inch long when full grown; bodies light brown, heads shiny black; pair of long (¼-inch)

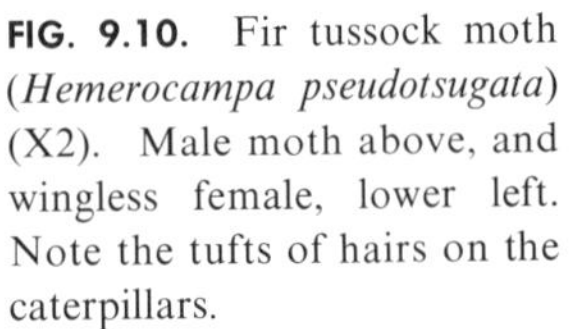

FIG. 9.10. Fir tussock moth (*Hemerocampa pseudotsugata*) (X2). Male moth above, and wingless female, lower left. Note the tufts of hairs on the caterpillars.

prothoracic horn-like pencil tufts; shorter, dense, tan to light-brown brush-like tufts (1/16 inch long) on each abdominal segments 1 through 4 and the 9th; somewhat broken narrow orange stripes along sides; underside body nearly naked. **Pupae:** in cocoons characterized by having many larval hairs intermixed with the silk; placed on various parts of trees and shrubs. **Moths:** females dull brownish gray and wingless, about 5/8 inch long; males with wingspread of 1 to 1¼ inches also colored brownish gray. **Eggs:** oval masses fastened to the cocoons and covered with spumaline—a dried frothy material.

LIFE CYCLE. This species winters as eggs. These hatch in the spring and mature by early August. The moths emerge during late August and after mating, the females immediately lay eggs. There is only 1 generation per year.

CONTROL. The most effective natural control is a virus disease. Aerial spraying with DDT at the rate of 1 pound per acre has been found to be effective. More than 400,000 acres were sprayed in 1947 in what was the first large scale airplane spraying control operation against a forest insect pest. See page 101 for details of application.

Roberts, P. H., and J. C. Evenden. 1949. *U.S.D.A. Yearbook.* Pp. 436–445.

WHITE-MARKED TUSSOCK MOTH

***Hemerocampa leucostigma* (J. E. Smith), Lepidoptera, Lymantriidae**

INJURY. This species is one of the more important shade tree pests but it is not a major forest pest. The small larvae are leaf skeletonizers but when larger they consume most of the leaf tissues except the larger veins.

HOSTS. Many species of deciduous trees and shrubs.

RANGE. Throughout eastern North America west to the Rockies.

INSECT APPEARANCE. *Caterpillars:* full grown about 1¼ inch long; head bright red, body white to yellowish with darker streak along back; pair long (⅜- to ½-inch slender "pencil" tufts on the prothorax and the 8th abdominal segment; short (⅛ inch) white tufts on mid-back of first 4 abdominal segments; red glands on mid-back of 6th and 7th abdominal segments. *Pupae:* in grayish cocoons containing many body hairs; on branches or twigs. *Moths:* females wingless with gray oval bodies; males with ashy gray wings with darker, wavy, somewhat indistinct bands

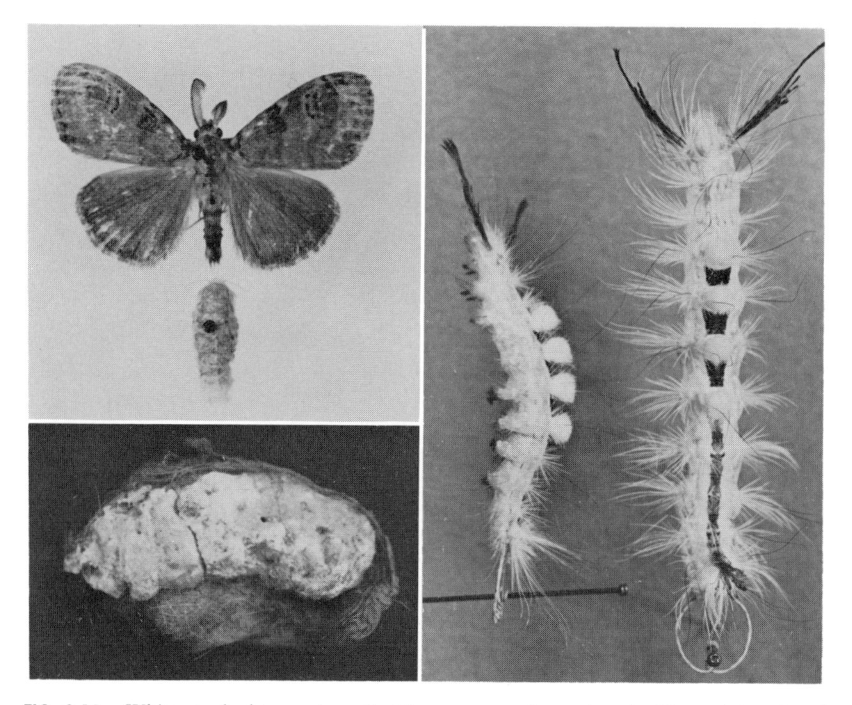

FIG. 9.11. White-marked tussock moth (*Hemerocampa leucostigma*). Upper left: winged male and wingless female moths (X1½); lower left: egg mass on an old cocoon (X2); lower right: caterpillars with white tufts of hairs (X2).

crossing forewings; wingspread about 1¼ inches. *Eggs:* in masses on the old cocoons; covered with a gray frothy material.

LIFE CYCLE. Winter is passed as eggs. These hatch early in the spring and the resulting caterpillars mature 5 to 6 weeks later. Pupation lasts for about 2 weeks. There are 1 to 3 generations per year depending on length of summer.

CONTROL. DDT applied as a mist or spray at the usual dosages is effective. See page 101 for details regarding dosages and application methods.

DEFINITE MARKED TUSSOCK MOTH

***Hemerocampa definita* (Pack.), Lepidoptera, Lymantriidae**

This insect is similar to *H. leucostigma* in most respects. The caterpillars differ in that they have a pale stripe along back and have a black spot behind each of the second and third abdominal tufts. The moths also differ in that they are clothed with golden brown hairs. This may cause one to confuse the egg masses of this species with that of the gipsy moth, for both pack body hairs over the egg masses.

WESTERN TUSSOCK MOTHS

***Hemerocampa vetusta* (Bdv.) and *H. gulosa* Hy. Edw., Lepidoptera, Lymantriidae**

Most of the characteristics of these species are similar to those for the *H. leucostigma,* except that they occur in the Pacific Coastal States and the insects are slightly different. INSECT APPEARANCE. *Caterpillars:* about 1 inch long when full grown; body gray, spotted with red, yellow, and blue; pair of black prothoracic "pencil" tufts and a similar single one at posterior end of body; 4 white, short, "brush" tufts on back of each of first 4 abdominal segments; these are black tipped on *H. gulosa.* *Moths:* females wingless, colored light gray; males brown and gray, wingspread about 1 inch.

RUSTY TUSSOCK MOTH

***Orgyia antiqua* (L.), Lepidoptera, Lymantriidae**

This species is chiefly a pest of shade trees. Most characteristics are similar to those for *H. leucostigma,* but they differ in range and insect appearance. RANGE. Northeastern United States and southeastern Canada. INSECT APPEARANCE. *Caterpillars:* about 1 inch long when full grown; body dark gray with head black; "pencil" tuft on each side of 2nd abdominal segment; other tufts on prothorax and abdomen like those on white-marked tussock moth; raised hair bearing spots reddish. *Moths:* females wingless with gray oval bodies; males resemble white-marked tussock moth but are rusty colored. *Eggs:* laid on cocoons but are not covered.

PINE TUSSOCK MOTH

***Olene plagiata* (Wlk.), Lepidoptera, Lymantriidae**

This species is a potentially dangerous forest pest, but in the past only a few small outbreaks have occurred in Wisconsin. HOSTS. Pines, spruces, and firs.

RANGE. Throughout northeastern North America. **INSECT APPEARANCE.** *Caterpillars:* head black; body gray or brown densely clothed with hairs. *Moths:* both sexes are winged; gray or brownish with no good distinguishing characteristics. **LIFE CYCLE.** The moths emerge during the summer and, after mating, the females lay their eggs in masses. These hatch in a couple of weeks and the resulting larvae feed for a few weeks and then hibernate in protected places such as bark crevices. They complete their development in the spring; therefore, there is only 1 generation per year. **CONTROL.** Aerial applications of DDT at the usual rates should be effective but, as yet, this has never been tried against this species.

SILVER SPOTTED TUSSOCK MOTHS

Halisidota argentata Pack., and related species, Lepidoptera, Arctiidae

INJURY. This species, a potentially dangerous pest, as yet has not caused serious forest defoliation. During the late summer the young larvae form silken webs that become filled with dead needles. These remain over the winter. They consume most of the needle tissues.

HOSTS. Douglas-fir is preferred, but true firs, Sitka spruce, and pines also may be infested.

RANGE. Western North America.

INSECT APPEARANCE. *Caterpillars:* about 1½ inches long when full grown; densely clothed with long brown to black and yellow brush-like tufts; poisonous barbed hairs present. *Moths:* wingspread 1¾ to 2 inches; forewings reddish brown with white spots; hindwings yellowish tan to whitish with a few brown spots near the outer margins.

LIFE CYCLE. The moths appear during July and August. After mating, each female lays several hundred eggs in small clusters on the twigs and needles of the host trees. The eggs hatch in about 2 weeks and the resulting larvae imme-

FIG. 9.12. Silver-spotted tiger moth (*Halisidota* spp.). Moth (X1⅓), caterpillar (X1¾).

diately start feeding on the needles. During this stage they are gregarious and form webs in which the partially grown caterpillars hibernate through the winter. The following spring they disperse and feed singly. Pupation occurs in dirty brown cocoons attached to the needles, twigs, and trunks of the host trees or among ground litter. There is only 1 generation per year.

CONTROL. Insecticidal controls have not been needed for this insect because, even though they are abundant at times, the natural occurring parasites are most efficient. Should applied control ever be needed, however, DDT applied in the usual ways probably would be effective. See page 101 for details of formulations, dosages, and application methods.

Silver, G. T. 1958. *Can. Entomol.* 90:65–80.

TIGER TUSSOCK MOTHS

Halisidota spp., Lepidoptera, Arctiidae

INJURY. These defoliators are not serious pests of either forest or shade trees but they commonly are encountered in small numbers. One species, *H. argentata,* differs in several respects from the other species considered here; therefore, it has been treated separately on page 137. The caterpillars feed by consuming most of the leaf tissues except the larger veins.

HOSTS. Many species of deciduous broad-leaved trees.

RANGE. Throughout eastern United States. The spotted tussock moth (*H. maculata*) also occurs throughout the West.

INSECT APPEARANCE. *Caterpillars:* all species are about 1¼ to 1½ inches long when full grown. *Hickory tussock moth* [*H. caryae* (Harr.)]—body grayish with black head; short "brush" tufts on back of each of 8 abdominal segments; pair of long "pencil" tufts on each of 1st and 7th abdominal segments. *Pale tussock moth* [*H. tessellaris* (J. E. Smith)]—head black; body blackish with gray to yellow compact body tufts, long brown to black "pencil" tufts on 2nd and 3rd thoracic segments with a whitish "pencil" tuft adjacent to each of the darker "pencils"; pair of dark "pencil" tufts on 8th abdominal segment. *Spotted tussock moth* [*H. maculata* (Harr.)]—row of short, mostly black, tufts along middle of back with those on 2nd and 3rd thoracic segments and the 8th abdominal segment longer and have some lighter colored hairs intermixed; general body tufts of black, whitish, and yellow hairs. *Sycamore tussock moth* (*H. harrisii* Walsh)—head yellowish brown; body yellowish with whitish to yellow hairs; pencil tufts are orange. *Moths:* wingspread 1½ to 2 inches. *H. caryae*—forewings light brown spotted with many silvery spots; hindwings yellowish and translucent. *H. tessellaris* and *H. harrisii*—wings pale yellow translucent with several faint broad irregular bands crossing forewings. *H. maculata*—pale yellow; forewings marked with brown forming irregular bands across wings; hindwings translucent.

LIFE CYCLE. The moths emerge during late spring or early summer. After mating, the females deposit egg masses on the undersides of leaves. The resulting larvae are rather gregarious and are present from June until October. They then pupate and pass the winter in gray, hairy cocoons. There is only 1 generation per year.

CONTROL. Under forest conditions insecticidal control will seldom if ever be needed. Should any of these insects become abundant on shade or ornamental trees, DDT should be effective when applied either as a mist or a spray.

FIG. 9.13. Pale tussock moth (*Halisidota tessellaris*) (X1¼). Note the tufts of setae on the caterpillar.

DAGGER TUSSOCK MOTHS

Acronicta spp. Lepidoptera, Phalaenidae (Noctuidae)

Species most commonly encountered are the cottonwood dagger moth (*A. lepusculina* Guen.), American dagger moth [*A. americana* (Harr.)], and the smeared dagger moth [*A. oblinita* (J. E. Smith)].

INJURY. These species are defoliators of forest and shade trees but seldom cause much damage. They consume most of the leaf tissues.

HOSTS. *A. lepusculina*—poplar and willow. The other two species are general feeders on the foliage of various deciduous broad-leaved trees.

RANGE. *A. lepusculina*—northern United States and southern Canada. The other two are eastern species.

INSECT APPEARANCE. *Caterpillars:* *A. lepusculina*—about 1½ inches long when full grown; body densely clothed with long, soft, yellow hairs; single, long, slender, black "pencil" tufts on back of each of 5 abdominal segments; head, pronotum, and thoracic legs are black; spiracles rimmed with black. *A. americana*—about 2 inches long when full grown; appearance similar to the previous species, except these have a pair of black "pencil" tufts on back of each of the 1st and 3rd abdominal segments and a single "pencil" tuft on back of 8th segment; broad black stripe on 9th and 10th abdominal segments and a broken black stripe along each side. *A. oblinita*—about 1½ inches long when full grown; black velvety body dotted with yellow band along each side which is notched around each spiracle leaving the white spiracles surrounded by black; body tubercles black to red with each bearing short, stiff, reddish hairs; usually a broken stripe along each side of back and sometimes reddish bands across back of each segment; no "pencil-like" tufts present. *Moths:* moderately robust, gray with irregu-

FIG. 9.14. Dagger tussock moth (*Acronicta* spp.). Most of the light-colored setae on the caterpillar are obscured by the light background leaving only the dark-colored "pencil" tufts distinct. Moth (X1¼), caterpillar (X1¾).

lar lighter and darker markings; antennae are thread-like on both species; many have a dagger-like mark near the inner posterior area of the forewings; hindwings usually darker and brownish.

LIFE CYCLE. Winter is passed in the pupal stage in a thin hairy cocoon spun in the ground litter. The moths appear during late spring from May to July, and the larvae are present during the remainder of the summer and fall until October.

CONTROL. Insecticidal control seldom if ever will be needed to control these insects under forest conditions, but sometimes it may be desirable to kill the caterpillars when on shade and ornamental trees. DDT sprays or mists probably would be effective. See page 101 for details of application.

HAIRY CATERPILLARS

Lepidoptera, Lymantriidae, Notodontidae, and Pieridae

In this group are included those hairy caterpillars that live exposed on the foliage and lack spines, tubercles, or other projections of the body wall. The setae never arise in dense tufts as they do on larvae of the tussock moths. Other hairy caterpillar species that are tent or web makers and those that are also characterized by having spines or tubercles are considered under these respective categories rather than here. See Table 9.8.

TABLE 9.8 COMMON SPECIES OF HAIRY CATERPILLARS*

1. Small, circular, convex spots or projections (glands) present on mid-back of the 6th and/or 7th abdominal segments 2
1. Glands as described in 1 above not present 4
2. Row of large whitish, circular to retangular spots along back; each spot constricted near middle (page 147) . . . . SATIN MOTH
2. Body not marked as in 2 above 3
3. Body dark gray with a pair of raised blue spots on back of each of the thoracic segments and the first 2 abdominal segments; paired red spots on back of each of the following 6 abdominal segments (page 141) . . . . GYPSY MOTH
3. Body and head brownish with a row of white tufts along each side (page 145) . . . . BROWN TAIL MOTH
4. Setae long and often of irregular lengths 5
4. Head and body covered with short, fine, closely set setae of uniform length which arise from small pupillae; body dark green with 2 white stripes along each side; occurs only in West (page 148) . . . . PINE BUTTERFLY
5. Pronotum (shield) differs in appearance from other thoracic segments; when disturbed larvae raise both the fore and posterior ends of bodies (page 150) . . . . DATANA spp.
5. Pronotum does not differ greatly from the other body segments 6
6. Body pear shaped; densly hairy with setae arising in groups from low warts; caterpillars somewhat resemble small tailless mice; 7 pairs prolegs, but only 5 pairs have crochets; fleshy projection present next to each spiracle . . . . PUSS CATERPILLARS
6. Body characteristics not as described in 6 above; typical cylindrical hairy caterpillars with setae not arising in groups from low warts; colored frequently with blue, red, or yellow; the forest tent caterpillar, has a row of keyhole or subtriangular-shaped light spots along back (pages 111 and 115) . . . . TENT CATERPILLARS

* If the setae arise in groups, they are not so dense that the body wall can not be seen through the hairs. See also Table 9.7.

GIPSY MOTH

Porthetria dispar **(L.) Lepidoptera, Lymantriidae**

In the United States probably more money has been spent for combating the gipsy moth than for the control of all other forest insects combined. These large expenditures for the control of a single insect species have been made because it was thought desirable to keep this introduced pest from spreading. Sometimes it appears that the insect population in the infested area has reached an unsteady equilibrium similar to that exhibited by some native pests, yet the trend may still be upward. During the last two of the four peak years since 1927 the total infested area defoliated has been greater than during the earlier two outbreaks.

INJURY. The caterpillars consume most of the leaf tissues. They feed during the early spring from the time the leaves start to unfold until they mature late in June, about 5 or 6 weeks later. Two thirds or more of the leaf destruction occurs during the last 2 weeks. The deciduous hardwoods, which are most commonly infested, are not readily killed by repeated annual defoliations, but wood production is greatly reduced.

HOSTS. General feeders. Deciduous broad-leaved species such as oaks, aspens, most willows, basswood, red, gray, and paper birches are favored. Older caterpillars also eat the foliage of various species of conifers such as pines, spruces, hemlock, and even red cedar. Unfavorable hosts are the black and yellow birches, hickories, butternut, walnut, maples, ashes, elms, locusts, sycamore, yellow poplar, and balsam fir.

RANGE. The New England States with the exception of the northern half of Maine. Several other infestations have been found in surrounding states with the most extensive occurring in northeastern Pennsylvania. Another outlying infestation has recently developed in southern Michigan. The gipsy moth is a European species that was introduced into the United States in 1869. The importer apparently was a French mathematician and astronomer who was trying to help the silk industry of his native France by hybridizing the tough gipsy moth with the silk worm moth so as to develop a disease-resistant silk producer.

INSECT APPEARANCE. *Caterpillars:* about 1½ to 2½ inches long when full grown; body dark gray marked on back of each of the 3 thoracic segments plus the first 2 abdominal segments with a pair of slightly raised blue spots; similarly shaped and arranged pairs of red spots on each of the following 6 abdominal segments; stiff hairs arise from these spots as well as from other numerous low tubercle-like spots making the insect rather hairy; small gland on mid-back of each of the 6th and 7th abdominal segment. *Pupae:* typical tan to dark brown lepidopterous type; never enclosed within a cocoon but are attached to objects by a few silken threads. *Moths:* males have slender brown bodies with brown wings irregularly crossed with darker lines, wingspread about 1½ inches; females have a heavy stout body; wingspread about 2½ inches; forewings whitish crossed by several dark irregular lines. *Eggs:* laid in flattened masses of 100 to 800 on various objects such as bark and stones; masses covered so as to conceal eggs with packed yellowish or tan hair from the female.

LIFE CYCLE. The moths appear from about mid-July through August. They mate immediately, and each female produces a mass of eggs. The fall and winter is passed in the egg stage. Hatching takes place the following spring at the time the leaves are unfolding. The caterpillars feed mostly during the night and rest during the day. Many of those that

FIG. 9.15. Gipsy moth (*Porthetria dispar*) (X1½). Upper left: white female moth with a mass of eggs; lower left: male moth; upper right: pupae; lower right: caterpillars showing characteristic paired spots along back.

are half grown or older move to sheltered places such as trunk cavities, under loose bark or the ground litter where they rest during the day. If such places are not available, however, they are forced to remain in the trees. They generally reach maturity late in June and pupation occurs during the following few weeks. There is only 1 generation per year.

NATURAL CONTROL. Egg mortality begins when the winter temperatures drop to about $-15°$ F, and temperatures lower than this cause increasingly heavier mortality. All the eggs are killed when $-25°$ F is reached. Sometimes late spring frosts destroy the new foliage and thereby cause heavy larval mortality due to starvation. At times a virus disease also has caused great caterpillar destruction, but this commonly occurs only after the insects are nearing maturity.

Since 1905 many species of parasites and predators have been introduced from Europe for the purpose of combating the gipsy moth and, of these, about 10 have been established and are considered of some importance. The most important of these are a predaceous ground beetle, two species of parasitic flies, and a small parasitic wasp.

SILVICULTURAL CONTROL. Encouraging the less favored hosts has been suggested as a method for preventing epidemics from developing; therefore, whenever practical, the favored tree species should be removed so that the quantity of foliage of these is less than 50 per cent.

It has also been found that a cover of loose, deep, ground litter may adversely affect the gipsy moth caterpillar population. This type of ground cover makes a good habitat for various predators that feed on the older caterpillars resting in the litter. When the ground litter is sparse or hot and dry, however, the caterpillars remain in the trees. Dense stands growing on the better moister sites should develop this desirable type of resistance, but it may be difficult to produce favorable litter conditions on those sites which are poor and dry.

INSECTICIDAL CONTROL. During many years of gipsy moth abundance, various methods of control were used such as creosoting the egg masses and mechanically destroying the caterpillars as they congregated under burlap bands. These expensive and laborious methods together with the use of lead arsenate sprays had only limited usefulness, even when only shade trees were protected. At the present time more practical methods are available. The best of these for treating forested areas is aerial application of DDT with the usual dosage of one pound of DDT per acre. When only shade trees are to be sprayed, ground hydraulic sprayers can be used; but mist blowers are more efficient. Details regarding the use of these methods are presented in the chapters dealing with insecticides and application methods.

One aspect of the general program conducted by the federal government and the cooperating states is to maintain a gipsy moth free barrier zone (25 to 30 miles wide) along the eastern border of New York State and the adjacent areas of the New England States. This border area is surveyed carefully every year in order to locate any infestations by using traps to catch the flying males. The attractant used consists of extracts from the abdominal tips of virgin females. All infested areas found in this way are treated the following spring with DDT. The objective of concentrating the main control efforts in the barrier zone is to prevent the insect from becoming established in the extensive areas of susceptible oak forests that occur to the west and south. The outlying infestations are also treated repeatedly with DDT aerial spraying for the purpose of eradicating this insect from these areas.

Behre, C. E., A. C. Cline, and W. L. Baker. 1936. *Mass. Forest and Park Assoc. Bull.* 157.
Bess, H. A., S. H. Spurr, and E. W. Littlefield. 1947. *Harvard Forest Bull.* 22.
Brown, R. C., and R. A. Sheals. 1944. *J. Forestry* 42:393–407.
Burgess, A. F., and W. L. Baker. 1938. *U.S.D.A. Circ.* 464.
Dowden, P. B., and H. L. Blaisdell. 1959. *U.S.D.A. Forest Pest Leaflet* 41.
Mosher, F. H. 1915. *U.S.D.A. Bull.* 250.
O'Dell, W. V. 1959. *J. Forestry* 57:271–273.
Perry, C. C. 1955. *U.S.D.A. Tech. Bull.* 1124.

BROWN-TAIL MOTH

***Nygmia phaeorrhoea* (Donov.), Lepidoptera, Lymantriidae**

The brown-tail moth is another tree pest that was introduced from Europe during the latter part of the nineteenth century, but the damage caused by this species has not been as serious as that caused by the gipsy moth. The brown-tail caterpillar are also troublesome because they have barbed, poisonous hairs, which cause an irritating skin rash on many people. Some are affected so severely that they must keep away from infested areas during times when the caterpillars are present.

INJURY. The more serious defoliation caused by these leaf-chewing caterpillars occurs during the spring, when they consume most of the leaf tissues. Late summer skeletonizing feeding also occurs, but this usually causes no appreciable damage. Sometimes, however, the summer injured leaves cause the trees to have a scorched appearance.

HOSTS. Various Rosaceae fruit trees such as apple, pear, plum, and cherry are most favored, but white oak and willow are also attacked.

RANGE. Throughout most of the New England States as far north as the southern third of Maine.

INSECT APPEARANCE. *Caterpillars:* about 1½ inches long when full grown; body color dark brown to almost black, with a broken white line formed by a tuft of short white hairs on sides of each abdomen segment; second and third thoracic segments each with a transverse row of 6 small irregular orange rings followed by a transverse row of irregular orange spots, thin broken double reddish orange lines along back; pair of short, dense, tan tufts on back of each of the abdominal segments, red glands on mid-back of each of the 6th and 7th abdominal segments. *Pupae:* typical; inside a loose cocoon; sometimes numbers of cocoons occur together to form a large mass; usually placed among the leaves at the tips of the wings. *Moths:* males with forewings tannish white and a wingspread about 1⅒ inches, small tuft of brown hairs at posterior end of abdomen; females with white wings; wingspread about 1½ inches; posterior of abdomen bulbous appearing, due to dense tuft of brown hairs that gives the insect its common name. *Eggs:* on undersides of leaves in elongate masses up to about ¾ inch long and containing as many as 300 eggs; these are covered with the brown hairs from the "tail" of the female.

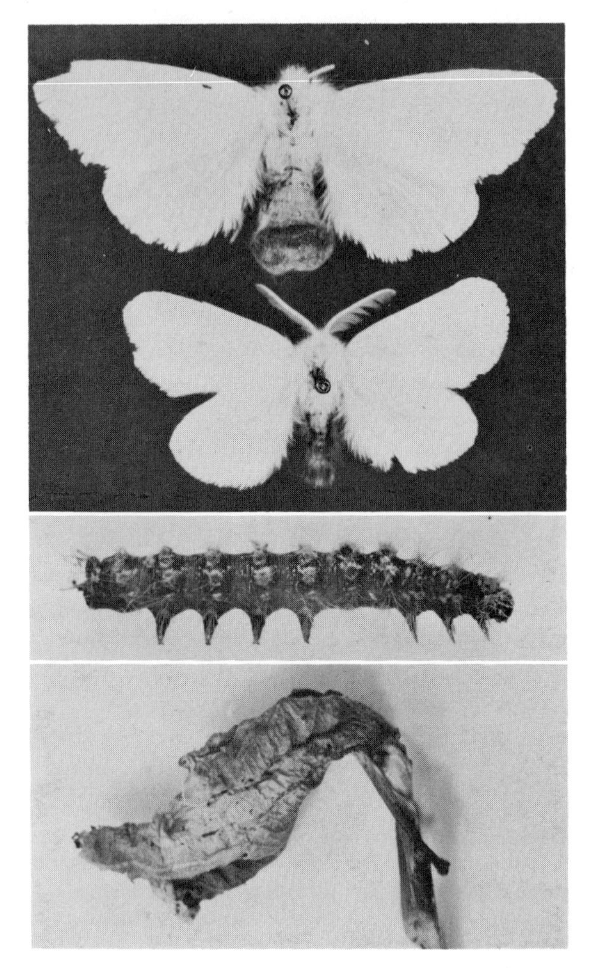

FIG. 9.16. Brown-tail moth (*Nygmia phaeorrhaea*). Moths and caterpillar (X2). Female moth above male. Below: winter web made by young caterpillars consists of enwebbed leaves (X¾).

LIFE CYCLE. The moths appear early in July. After mating, the females immediately lay their eggs, which hatch about a month later. The young caterpillars are leaf skeletonizers until cold weather stops their activity. The gregarious caterpillars then spin winter webs by fastening the leaves together at the branch tips. In each of these webs hundreds of small caterpillars hibernate together until the following spring. They emerge from these webs at the time the leaves are unfolding and immediately begin to feed. As would be expected, defoliation is greatest during June at the time the caterpillars are nearing maturity. Transformation to the moth stage takes place during the latter part of June. There is only 1 generation per year.

CONTROL. Winter temperatures below −25° F kill the hibernating larvae. Spraying with DDT at the usual dosage is the most effective artificial control. See page 101 for details of application.

Burgess, A. F., and W. L. Baker. 1938. *U.S.D.A. Circ.* 464.
Schaffner, J. V., Jr. 1950. *U.S.D.A. Misc. Publ.* 657. Pp. 412–414.

SATIN MOTH

Stilpnotia salicis (L.) Lepidoptera, Lymantriidae

This species is a shade tree pest which feeds during the spring and early summer. It is not a serious forest pest.

INJURY. The young caterpillars cause light skeletonizing leaf feeding during the late summer, but the severe damage occurs the following spring when the older larvae consume most of the leaf tissues.

HOSTS. Willows and poplars are favored.

RANGE. This European tree pest was introduced into North America on both coasts about 1920. It now occurs in the New England region and in the Pacific Northwest as far south as Oregon.

INSECT APPEARANCE. *Caterpillars:* 1¾ to 2 inches long when full grown; head blackish; body tan to blackish, mottled with lighter markings; along the back are 10 large whitish circular to somewhat rectangular blotches, each of which is constricted near the middle; a thin white broken line extends the sides of these blotches; transverse row of 6 to 8 low reddish-tan tubercles bearing yellowish hairs on each segment. *Pupae:* typical shape, color dark brown with orange or tan spots bearing long yellowish hairs; inside loose cocoons rolled in leaves on or among twigs. *Moths:* wingspread 1 to 2 inches; body dark brown with sparse to dense whitish, yellowish, or tan clothing hairs; wings white with satin-like luster; slightly tan near bases. *Eggs:* oval masses about ¾ to 1 inch long consisting of 100 to 400 eggs covered with a glistening silvery material.

LIFE CYCLE. The moths emerge during July. Mating occurs almost immediately and the females deposit their eggs on tree trunks, leaves, and branches. These hatch after about 2 weeks, and the resulting larvae immediately start skeletonizing the leaves. This feeding, however, lasts for only a couple of weeks and, therefore, causes little injury. The larvae then spin small individual web enclosures (hibernacula) in which they pass the winter. The following spring, when the leaves begin unfolding, they again become active, start feeding, and consume most of the leaf tissues except the petioles and larger veins. Pupation occurs inside

FIG. 9.17. Satin moth (*Stilpnotia salicis*) (X1⅓). Note the rectangular to hour-glass shaped white markings on the back of the caterpillar. An egg mass and the characteristic hairy pupae are shown below.

of cocoons formed in rolled leaves or among the twigs and lasts about 10 days. There is only 1 generation per year.

CONTROL. DDT is an effective poison for this insect. See page 101 for details of application.

Burgess, A. F., and S. S. Crossman. 1927. *U.S.D.A. Bull.* 1469.

PINE BUTTERFLY

Neophasia menapia (F. and F.), Lepidoptera, Pieridae

This defoliating insect is one of the most dangerous forest pests of ponderosa pine in the West. There have been several severe outbreaks, the worst of which occurred in the state of Washington during the 1890's, when almost a billion feet of timber was killed.

INJURY. Most of the needle tissues are consumed.

HOSTS. Ponderosa pine is the chief host, but western white and lodgepole pine are also infested when they are intermixed with the ponderosa pine.

RANGE. Throughout the West wherever the hosts grow.

INSECT APPEARANCE. *Caterpillars:* about 1 inch long when full grown; body pale to dark green with 2 white stripes along each side; appear velvety because they are covered with very short fine closely set hairs. *Pupae* (chrysalides): naked on low vegetation. *Butterflies:* wingspread about 1¾ inches; colored white with black markings, wings of females may be tinged with yellow, and some have orange spots along outer margin of each hindwing. *Eggs:* in rows on needles; always in tops of crowns.

LIFE CYCLE. Winter is passed in the egg stage. Hatching occurs early in the spring at the time the new shoots appear. The young caterpillars feed in groups and usually several simultaneously attack a single needle eating from the needle tip toward the base. Later they feed singly. By

FIG. 9.18. Pine butterfly (*Neophasia menapia*) (X1½). An enlarged view of the short body hairs of a caterpillar is shown below.

late June the caterpillars lower themselves to the ground by means of silken threads and pupate on the low vegetation. The butterflies are present during late summer from August to October.

CONTROL. Aerial spraying with DDT is effective. See page 101 for details of application.

Evenden, J. C. 1926. *J. Agri. Research* 33:339–344.

DATANA CATERPILLARS

Datana spp., Lepidoptera, Notodontidae

Of the various *Datana* species, the yellow-necked caterpillar and the walnut caterpillar are most commonly encountered.

INJURY. These defoliators sometimes become abundant in forests, but they cannot be considered very serious forest defoliators. They consume most of the leaf tissues with the exception of the larger veins.

HOSTS. Many species of deciduous broad-leaved trees.

RANGE. Throughout eastern United States, the adjacent parts of southern Canada, and in California, Idaho, and British Columbia.

INSECT APPEARANCE. *Caterpillars:* gregarious; when disturbed they raise their heads and posterior ends while holding to the twigs or leaves by their middle prolegs; about 2 inches long when full grown; bodies of full-grown larvae black with various other colors for the different species; bodies clothed with whitish or grayish hairs. These species are as follows: *Yellow necked caterpillar* [*D. ministra* (Drury)]—neck ringed with a thin yellow band; part of pronotal area (pronotal shield) yellow; body has 5 narrow, yellow longitudinal lines along each side with the lower one interrupted by the bases of the legs; another yellow line along the middle of the underside; legs yellow at base; *Walnut caterpillar* (*D. integerrima* G. and R.)—pronotal shield black; prolegs reddish on inner sides, partly grown larvae red to brown. *D. angusii* G. and R.—similar to *D. ministra,* except pronotal shield is black. *D. drexeli* Hy. Edw.—pronotal shield yellow; 10th abdominal segment yellow except for black tergal plate and anal prolegs; body with 11 thin longitudinal lines along body; bases of legs yellow; partly grown larvae brownish with yellow stripes with pronotal shield and legs black. *D. major* G. and R.—head, pronotal shield and legs dark red; 4 longitudinal rows of whitish or yellowish spots in each side and 3 similar rows on underside; partly grown larvae reddish to brownish black with white or yellow stripes and pronotal shield black or partly brown. *D. contracta* Wlkr.—pronotal shield yellow; body with 11 longitudinal yellowish stripes with the one below the spiracles wider than others; thoracic legs black with yellow bases; first 4 prolegs yellow with a black patch on outer side of each; anal pro-

FIG. 9.19. *Datana* spp. (X1¼). Yellow-necked caterpillar (*D. ministra*) moth and larvae above. *D. major* caterpillars below.

legs black. **Moths:** wingspread 1½ to 1¾ inches; wings light to dark tan; forewings crossed with 4 or 5 darker lines and with a darker edging along the outer margin; hindwings are yellowish and without cross lines; anterior part of thorax darker.

LIFE CYCLE. Winter is passed as pupae inside thin cocoons in the soil. The moths emerge during June and July and, after mating, the females deposit their eggs in masses on the underside of the leaves. The resulting caterpillars feed until late summer at which time they enter the soil to pupate. There is only 1 generation per year.

CONTROL. Insecticidal control will seldom, if ever, be needed for forest infestations. If only a few low-growing ornamental trees are infested, the gregarious caterpillars can be removed easily by hand. On taller-growing shade trees, DDT in sprays or mists should be effective. See page 101 for methods of application.

Schaffner, J. V., Jr. 1950. *U.S.D.A. Misc. Publ.* 657:397–400.

SPINY OR TUBERCLED CATERPILLARS

Lepidoptera, Various Families

The caterpillars of the various common families belonging to this group can be identified by referring to Table 9.9.

TABLE 9.9 COMMON FAMILIES OF LEAF-CHEWING LEPIDOPTEROUS CATERPILLARS THAT HAVE SPINES, ENLARGED PROJECTING TUBERCLES, OR LOBE-LIKE EXTENSIONS OF THE BODY WALL

1. Prolegs absent (page 164) . . . . SLUG CATERPILLARS
1. Prolegs present 2
2. Body with lobes of body wall extending out on sides; spines lacking (page 164) . . . . LAPPET CATERPILLARS
2. Body with spines or tubercles 3
3. Spines or tubercles present on head (page 162) . . . . . NYMPHALID CATERPILLARS
3. Spines or tubercles absent on head 4
4. Minute hooks (crochets) on ends of prolegs of two or more intermixed lengths; body usually armed with small to large spines 5
4. Crochets not of intermixed lengths; back of some abdominal segment with humps or tubercles or 2 tail-like projections (anal prolegs) extending back from posterior abdominal segment (page 163) . . . . PROMINENTS
5. Usually only spine present is a single horn-like projection or tubercle on mid-back of 8th abdominal segment; a few species with 2 pairs additional small spines on thorax; back of each abdominal segments crossed with 5 to 7 impressed lines (page 158) . . . . SPHINX CATERPILLARS
5. Abdominal segments not crossed with a number of impressed lines; body usually with numerous spines or tubercles on body 6
6. Spines or tubercles on thorax usually about as large as those on abdomen; if any thoracic outgrowths are larger, the last abdominal tergum (analplate) lacks spines or tubercles (page 152) . . . . GIANT SILKWORM CATERPILLARS
6. Anal plate with spines or tubercles; some spines or tubercles on mesothorax many times longer than those on abdomen (page 156) . . . ROYAL MOTH CATERPILLARS

GIANT SILKWORMS

Lepidoptera, Saturniidae

The caterpillar characteristics are presented in Table 9.10. Many of the large moths are beautifully colored and, therefore, sought by collectors. The pandora moths are the only serious forest pests belonging to this family. These are discussed separately. **HOSTS.** Species other than the pandora caterpillars are general feeders on broad-leaved deciduous trees, but some do favor certain hosts. **LIFE CYCLE.** There is usually only 1 generation per year. The moths appear during midsummer, and the females lay their eggs on the leaves. Late in the summer the larvae mature and spin large, papery cocoons fastened to the twigs and branches. The luna, Io, and buck moths differ in this respect,

FIG. 9.20. Giant silkworm moths and caterpillars. Upper left: luna moth (*Actias luna*) (X½); upper right: cecropia moth (*Hyalophora cecropia*) (X⅓); lower left: polyphemus moth and caterpillar (*Antheraea polyphemus*) (X½); lower right: io moths (female below) (X½) and caterpillar (X1) (*Automeris io*).

in that they form their cocoons beneath the ground litter. The pupae hibernate over the winter. The life cycles for the buck moth differ, in that the moths appear in September and the resulting eggs pass the winter as masses encircling the twigs. **CONTROL.** Insecticides are seldom, if ever, needed to control the species which infest broad-leaved trees. Should applied control ever be needed, however, DDT would be effective. See page 101 for details of application.

Holland, W. J. 1903. *The moth book.* Doubleday, Page and Co. Pp. 80–94.
Schaffner, J. V., Jr. 1950. *U.S.D.A. Misc. Publ.* 657:377–380.

TABLE 9.10 COMMON SPECIES OF GIANT SILK WORM CATERPILLARS*

Couplet	Characteristics	Result
1.	Feed on pine foliage; body with small spines; colored gray to tan with a single broad or double thin pale line along mid-back; occur west of the Great Plains (page 154)	PANDORA MOTH
1.	Feed on the foliage of deciduous broad-leaved trees	2
2.	Spines with long sharp branches	3
2.	Spines or tubercles not branched or with only short spines	4
3.	Head and body green; red and white line along each side (p. 153)	*Automeris io*† (F.)
3.	Head brown; body brown to black, covered with small pale spots; one species with a yellow stripe along each side	BUCK MOTHS (*Hemileuca* spp.)
4.	Four large, bright orange to red tubercles on thorax; body green	5
4.	Tubercles colored yellow, gold, pink, or silvery green but never bright orange to red; body green	6
5.	Large yellow to red tubercles on back of 8th abdominal segment	*Callosamia promethea*† (Drury)
5.	Yellow tubercles on back of first 8 abdominal segments; others are blue (page 153)	*Hyalophora cecropia*† (L.)
6.	Oblique yellow line on each side of seven abdominal segments; each segment with 6 yellow, gold or silvery tubercles (page 153)	*Antheraea polyphemus*† (Cram)
6.	Oblique yellow lines on abdomen absent	7
7.	Short, stout, golden tubercles on back; bluish tubercles on sides	CEANOTHUS SILK MOTH [*Hyalophora euryalus* (Bvd)]
7.	With pinkish or greenish tubercles	8
8.	Six pinkish or greenish tubercles on each segment (page 153)	*Actias luna*† (L.)
8.	With black spots between 2 rows of green tubercles; with black spines; occurs only in West	GLOVER'S SILK MOTH [*Hyalophora gloveri* (Strecher)]

* Characteristics are for the older, larger larvae.
† The specific names are also used as the common names.

PANDORA MOTHS

Coloradia pandora Blake and *C. doris* Barnes, Lepidoptera, Saturniidae

These insects are most serious defoliators of western pine forests, and in the past they have caused much damage.

INJURY. Most of the needle tissues are consumed.

HOSTS. Ponderosa, Jeffrey, and lodgepole pines.

RANGE. Throughout the West wherever the hosts grow.

INSECT APPEARANCE. *Caterpillars: C. pandora*—large, about 2½ inches long when full grown; head brown; body tan with a broken brown band

FIG. 9.21. Pandora moth (X1¼) (*Coloradia pandora*). Note the uniform size of the small spines on the caterpillars (X1).

along each side of back forming a light stripe along mid-back; darker irregular lines also along sides; each segment with a transverse row of 6 short (⅛ inch) hairy spines. *C. doris*—gray with reddish tinge; without longitudinal bands; thin double light line along mid-back; spines once compound with single hair from tip of each branch; 5 pairs prolegs. **Moths:** *C. pandora*—large and robust wingspread 3 inches or more; gray wings with inner portions of hindwings with reddish tinge; 2 thin dark irregular lines cross both forewings and hindwings; dark round spot near center of each wing, on forewings these are between the two cross lines. *C. doris*—wings less heavily scaled than species above; wing spots oval.

LIFE CYCLE. The moths appear during late June and July. After mating the females lay globular green eggs on the twigs, branches, and ground litter. These hatch during August, and the young caterpillars immediately start feeding on the needles. During the first winter, when the partially grown caterpillars are about 1 inch long, they hibernate at the base of the needles. The following spring they resume feeding and become full grown by late June. At this time they crawl down the trees and bore into the soil to a depth of 1 to several inches where they pupate. Here they transform to typical mahogany-colored pupae and remain until the following year. The moths then emerge to complete the cycle 2 years after the eggs were laid.

In past times the Indians utilized these insects for food. Those living in California built fires beneath the trees so that the resulting smoke irritated the caterpillars and caused them to drop. The Oregon Indians preferred the pupae.

CONTROL. Sometimes a wilt disease is a most important natural control agent. The numerous ground squirrels present in many of the western pine forests also feed extensively on these insects and, therefore, they probably are of importance in preventing epidemics from developing.

Insecticidal control has not been reported as having been tried against these pests, but probably aerial spraying with DDT at the usual dosages would be effective. See page 101 for details regarding formulations and application methods.

Patterson, J. E. 1929. *U.S.D.A. Tech. Bull.* 137.

ROYAL MOTH CATERPILLARS

Lepidoptera, Citheroniidae

The family characteristics of the caterpillars are presented in Table 9.9, and the specific characteristics of the common species are listed in Table 9.11. Some of the species are beautiful moths, and the more

FIG. 9.22. Giant royal moths (X½). Left: imperial moth and caterpillar; (*Eacles imperialis*); right: royal walnut moth and (hickory horned devil) caterpillar (*Citheronia regalis*). Note that two or more pairs of spines on thorax are larger than those on the abdomen.

common of these can be identified by referring to a moth book containing colored illustrations such as that of Holland's. None are serious forest pests, and only the oak worms are shade tree pests. Even these are not serious. The *Anisota* spp. restrict their feeding to certain hosts, but the other two species listed are general feeders. The other biological characteristics are similar to those for the oak worms, which are discussed here in detail.

TABLE 9.11 COMMON SPECIES OF ROYAL MOTH CATERPILLARS

1.	Only one pair of enlarged prominent spines on thorax; spine on mid-back of 9th abdominal segment small and inconspicuous (page 158)	*Anisota* spp. 2
1.	Two or more pairs of prominent spines on thorax; caterpillars 4 to 5 inches long when full grown	5
2.	Feeds only on foliage of maples; body yellowish green with 7 dark green or blackish longitudinal lines (page 160)	GREEN-STRIPED MAPLE WORM [*A. rubicunda* (F.)]
2.	Feeds on oaks	3
3.	Body without whitish granulated spots; color black striped with 8 narrow yellow or orange longitudinal lines	ORANGE-STRIPED OAK WORM [*A. senatoria* (A. and S.)]
3.	Body with whitish granulated spots	4
4.	Body tan to reddish; head red	SPINY OAK WORM [*A. stigma* (F.)]
4.	Body green to gray; head green to brown	BROWN OAK WORM [*A. virginiensis* (Drury)]

5. Two pairs of long stout, horn-like spines on each of the 2nd and 3rd thoracic segments; some of these almost as long as thorax is thick; body smooth; green to reddish (page 157)........HICKORY HORNED DEVIL [*Citheronia regalis* (F.)]
5. Length of longest spines on thorax only about one fourth the thickness of thorax; body greenish and thinly clothed with setae (page 157)..................... ....IMPERIAL CATERPILLAR [*Eacles imperialis* (Drury)]

ANISOTA OAK WORMS

Anisota spp. Lepidoptera, Citheroniidae

INJURY. Most of the leaf tissues are consumed.

HOSTS AND RANGE. Oaks throughout eastern United States and southeastern Canada.

INSECT APPEARANCE. *Caterpillars:* 1½ to 2 inches long when full grown; 5 pairs prolegs; row small spines across each segment; other characteristics as listed in Table 9.11. *Moths:** wingspread 1¾ to 2½ inches; reddish tan with purplish line crossing outer area of each wing; small white spot near center of each forewing.

LIFE CYCLE. Winter is passed as pupae in the soil. The moths emerge mostly during the months of May to July, and, after mating, the females deposit their eggs in groups on the undersides of the leaves of the host plants. These hatch after about a week and a half and the resulting caterpillars are present during the remainder of the summer and even early fall. Usually there is only one generation per year, but sometimes the green-striped maple worm has two in regions where the summer season is long.

CONTROL. Applied methods have never been needed to protect forests from the attacks of these insects. Sometimes, however, it is desirable to protect shade trees. This can be done by spraying with suitable DDT formulations. See page 101 for the details of application.

Holland, W. J. 1903. *The moth book.* Doubleday, Page and Co. Pp. 94–97.

SPHINX CATERPILLARS (HAWK MOTHS)

Lepidoptera, Sphingidae

The general characteristics of these large, naked, cylindrical caterpillars are presented in Table 9.9. When resting the caterpillars assume a characteristic position by arching the anterior part of their body upward so that they resemble a sphinx. The large moths have stout spindle-

* The moths of the green-striped maple worm differ in that the wings are yellowish with the basal and outer thirds of the forewings pinkish.

FIG. 9.23. Small royal moths (*Anisota* spp.) (X1⅓). Above: orange striped oak worm (*A. senatoria*) moth, caterpillars, pupa, and egg masses on willow oak leaves (X½); below: spiny oak worm caterpillars (*A. stigma*). Note that only one pair of spines on the thorax are much larger than the other body spines.

FIG. 9.24. Green striped maple worm (*Anisota rubicunda*) moth, caterpillars and pupa (X1½).

shaped bodies with strong, long, and rather narrow wings. The antennae are thickened near the middle or toward the tips with the latter recurved to form a hook. Their beaks are very long and are held coiled beneath the head. There are many species, but none of these is of much importance; therefore, only the catalpa species will be discussed here to illustrate the biology. Most species feed on deciduous hosts, but a couple of *Lapara* spp. feed on pines. Most have life cycles similar to that for the catalpa sphinx, which will now be described.

CATALPA SPHINX

Ceratomia catalpae (Bdv.), Lepidoptera, Sphingidae

INJURY. The caterpillars are leaf-chewing defoliators that consume most of the solid leaf parts. It is chiefly a shade tree pest for catalpa.

HOSTS AND RANGE. Species of *Catalpa* throughout eastern United States north to New Jersey and Illinois.

INSECT APPEARANCE. *Caterpillars:* large, about 3 inches long when full grown; head and posterior median spine black; body colored with various amounts of black and yellow; both dark and light forms occur; 5 pairs prolegs. *Moths:* large, 3 inch wingspread; body spindle shaped; forewings colored grayish brown with small white patches along outer edge make wings appear scalloped; several irregular darker cross lines also present; hindwings of rather uniform coloration with outer band dark and with small white spots along edge.

LIFE CYCLE. The moths appear in the early spring, and after mating, the females deposit eggs in large masses on leaves, stems, or branches. These eggs hatch in a few days, and the resulting larvae mature about a month later. They then enter the soil to pupate. In the northern part of the range there are 2 generations per year, whereas farther south there may be 3 or 4 generations. Winter is passed as pupae in the soil.

CONTROL. DDT sprays are effective. TDE (DDD) sprays probably also would be effective, for this insecticide is most satisfactory against other species of hornworms.

Holland, W. J. 1903. *The moth book.* Doubleday, Page and Co. Pp. 41–77.

FIG. 9.25. Catalpa sphinx (*Ceratomia catalpae*) (X¾). Note the small spine on the tergum at the posterior of the abdomen.

NYMPHALID DEFOLIATORS

Lepidoptera, Nymphalidae

The caterpillar characteristics are presented in Table 9.9. The adults are very colorful and attractive butterflies that can be identified most easily by referring to a manual such as that by Klotz (1951). Although none of the species is a serious pest of either forest or shade trees, both the moths and caterpillars often are encountered; therefore, some common species are listed in Table 9.12. Only one of these will be discussed here for the purpose of illustrating the biology of the group.

TABLE 9.12 COMMON SPECIES OF TREE INFESTING NYMPHALID CATERPILLARS

1. Head black or dark red 2
1. Head green or brown 4
2. Head black 3
2. Head dark red; body brown mottled with yellow; hosts are hackberries . . . . QUESTION MARK BUTTERFLY [*Polygonia interrogationis* (F.)]
3. Body mostly black with small whitish and reddish spots (page 162) . . . . SPINY ELM CATERPILLAR
3. Body greenish to reddish; blackish along back; hosts are birches, willows and poplars . . . . *Nymphalis j-album* (B. and L.)
4. Body striped with yellow on back or sides; hosts are hackberries . . *Asterocampa* spp.
4. Body brownish green with back of some thoracic and abdominal segments whitish or pinkish; main hosts are birches, willows, and poplars . . . . *Limenitis* spp.

SPINY ELM CATERPILLAR (MOURNING CLOAK BUTTERFLY)

Nymphalis antiopa (L.), Lepidoptera, Nymphalidae

HOSTS. Elms, willows, hackberry, and poplars are preferred. RANGE. Wherever the hosts grow. INSECT APPEARANCE. *Caterpillars:* 1½ to 2 inches long when full grown; head black, bilobed with many unbranched spines bearing setae; body black sprinkled with whitish dots; back of abdomen with 7 or 8 reddish spots; prolegs also reddish; transverse row of black branched spines across each segment; 5 pairs prolegs. *Butterfly:* wingspread 2 to 3 inches; wings brownish purple with yellowish band on outer edges; adjacent to each band is a row of blue spots. LIFE CYCLE. The butterflies hibernate over the winter. During early spring the females deposit eggs on the twigs, and the resulting caterpillars remain together defoliating one branch at a time. The pupae are naked chrysalids, which are attached to twigs and branches. There are often 2 generations per year. CONTROL. Insecticides are seldom, if ever, needed to hold these insects in check even on shade trees. The caterpillars are often gregarious; therefore, when they are found feeding on small shade trees, the simplest thing to do is remove the small branch containing the colony and destrov it together with the larvae. However, should sprays ever be needed, DDT is effective. See page 101 for details of application.

Klotz, A. B. 1951. *A field guide to the butterflies.* Houghton Mifflin Co., Boston, 349 pp.

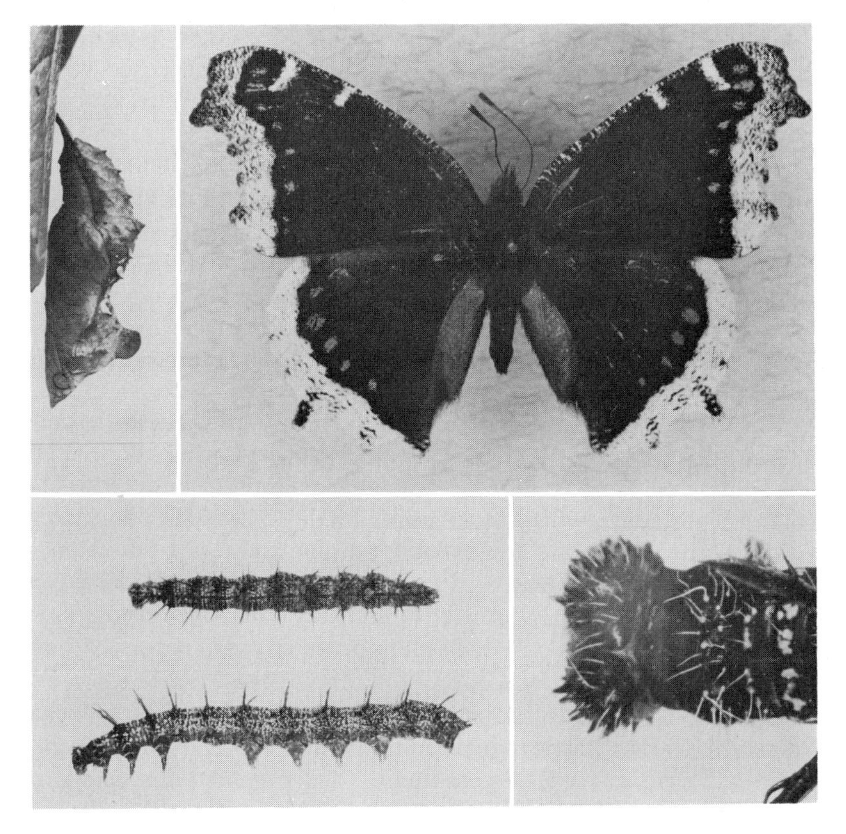

FIG. 9.26. Spiny elm caterpillar (X1) and mourning cloak butterfly (X1½) and pupa (X1½) (*Nymphalis antiopa*). Note the spines on the enlarged head.

SPINED OR TUBERCLED PROMINENTS

Schizura spp., Lepidoptera, Notodontidae

The red humped caterpillar [*Schizura concinna* (J. E. Smith)] and the unicorn caterpillar [*S. unicornis* (J. E. Smith)] are the most commonly encountered of this group, but other common notodontid species are discussed under "Naked Caterpillars." These species sometimes become abundant locally, but they are of only minor importance as forest pests.

INJURY. The older caterpillars are leaf-chewing defoliators that consume all the solid leaf parts except the mid-rib, whereas the young caterpillars are leaf skeletonizers.

HOSTS. Many species of deciduous broad-leaved trees, but cherry, elm, poplar, and willow are most commonly infested.

RANGE. Throughout the United States and southern Canada.

INSECT APPEARANCE. *S. concinna—Caterpillars:* about 1 inch long when full grown; head red; body with longitudinal black and yellow lines; back of first abdominal segment enlarged to form a conspicuous red hump; back with double row of short, stout, black spines and with smaller spines along sides of body; 5 pairs prolegs. *Moths:* wingspread 1 to 1⅜ inches; forewings grayish each crossed with a darker line and with hind margins rusty; hindwings light gray and unmarked except for irregular darker spots along outer edge. *S. unicornis—Caterpillars:* about 1⅓ inches long when full grown; prominent pointed projection on mid-back of 1st abdominal segment; color varigated brown, orange, and green. *Moths:* somewhat similar to *S. concinna,* but forewings rather uniform brownish gray.

LIFE CYCLE. There may be either 1 or 2 generations per year depending on the length of the summer. The moths appear during May or June and again in July and August if a second generation is produced. The females deposit their white eggs in masses on the under side of the leaves. Newly hatched larvae are gregarious for a time and feed by skeletonizing the underside of the leaves; later the older larvae consume all except the mid-rib. Usually they completely defoliate one branch before moving to another. Winter is passed as full-grown larvae in papery cocoons among the ground litter.

CONTROL. A DDT spray at the usual dosage should be effective against this species whenever needed on shade trees. Forests probably never will need an applied treatment. See page 101 for details of application.

SLUG CATERPILLARS

Lepidoptera, Limacodidae

The caterpillars of these moths are of no importance as shade or forest pests, but they are commonly noticed because some have venomous spines that cause severe burning and skin rash. *Caterpillars:* all lack prolegs and crochets, but sucker-like discs may be present. Most species have lobe-like or tubercle-like outgrowths from the body. LIFE CYCLE. The pupae hibernate over the winter in papery cocoons attached to twigs. Each cocoon has a lid closing the opening through which the moths subsequently emerge.

LAPPET MOTHS

Epicnaptera americana (Harr.), Lepidoptera, Lasiocampidae

In addition to the species above, there are several others, but none has caused any trouble either to forests or shade trees. HOSTS. General feeders on deciduous broad-leaved trees. RANGE. Throughout eastern North America. INSECT APPEARANCE. *Caterpillars:* about 2½ inches long when full grown; body flattened with sides extended to form lobed projections; colored mottled bluish gray with red cross bands on back of 2nd and 3rd thoracic segments; each band with 3 black dots; 5 pairs prolegs. *Moths:* reddish tan with 2 irregular broad bands forming a "V" on each wing; edges scalloped. **LIFE CYCLE.** The pupae hibernate over the winter in tough cocoons placed on the bark. The moths emerge in early summer and the larvae are present for a couple of months thereafter. There is usually only 1 generation per year, but there may be more farther south. **CONTROL.** This insect has never caused enough trouble to need applied control measures.

NAKED CATERPILLARS

Lepidoptera, Various families

This group includes various species in which the caterpillars are relatively free of setae, spines, tubercles, or other distinct outgrowths of the body wall. Larvae that have parts of segments swollen, however, are included here. The more common species are presented in Table 9.13 and in the text that follows.

TABLE 9.13 COMMON NAKED TREE FOLIAGE FEEDING LEPIDOPTEROUS CATERPILLARS

1. Small caterpillars, 1/4 inch long when full grown; leaf skeletonizers; cocoons with longitudinal ridges (page 169) . . . . LEAF SKELETONIZERS
1. Older larvae 1 to many inches long when full grown 2
2. Back of 8th abdominal segment enlarged and swollen 3
2. Not as in 2 above 4
3. Head and back of 8th abdominal segment red; body black and yellow striped; eastern North America (page 165) . . . . RED-HUMPED OAK WORMS
3. Body dark green with black and yellow stripes; covered with minute roughness; California (page 168) . . . . CALIFORNIA OAK WORM
4. Minute hooks on ends of prolegs (crochets) consist of 2 intermixed lengths (page 152) . . . . GIANT SILK WORMS (Table 9.10)
4. Crochets not as in 4 above but they may become shorter at ends of the groups . . . 5
5. Bodies gray to brown, mottled or striped (page 169) . . . . CUTWORMS
5. Bodies greenish with distinct reddish, brownish, or purplish markings 6
6. Reddish to purplish saddle-shaped patch on back (page 167) . . . . SADDLED PROMINENT
6. Body color variable but often there is a light line along mid-back bordered on each side by a band of variable width (page 167) . . . . VARIABLE OAK WORM

RED-HUMPED OAK WORMS

Symmerista albicosta (Hbn.) and *S. albifrons* (A. and S.), Lepidoptera, Notodontidae

INJURY. Sometimes these species are common to abundant and may cause local forest defoliation for a year or so, but they are not of great importance as forest pests. The caterpillars devour most of the leaf tissue.

HOSTS. Oaks, especially the white oaks. Sometimes they also feed on basswood, beech, elm, and maple.

RANGE. Eastern United States and adjacent parts of southeastern Canada.

INSECT APPEARANCE. *Caterpillars:* 1½ to 1¾ inches long when full grown; head orange red, wider than thorax; body smooth and naked with back of 8th abdominal segment enlarged to form an orange red hump; body yellowish with 5 dark narrow lines extending along the back on *S. albicosta,* but *S. albifrons* has only 3 of these darker lines; 3 similar lines along each side and a band of dark irregular spots present just above legs; legs, orange to red; 5 pairs prolegs. *Moths:* wing expanse 1½ to 2 inches; forewings gray with white band along outer ⅔ of front margin, area immediately behind white band darker than remainder of wings; other parts with irregular darker markings; hindwings lighter gray and unmarked.

LIFE CYCLE. The moths are present from May to July during which time they mate and the females lay pale green eggs in small groups on the underside of the leaves. The resulting caterpillars are present during the remainder of the summer and fall. Winter is passed as pupae

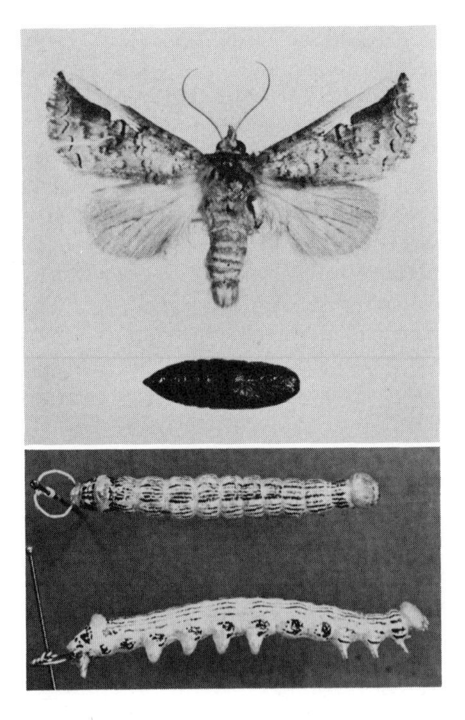

FIG. 9.27. Red-humped oak worm (*Symmerista albifrons*) (X1½). Note the swollen tergum near the posterior of the abdomen.

enclosed in thin, white, oval cocoons spun in the ground litter. In the North there is only 1 generation per year, but in the South there may be 2. The first generation then pupates on the leaves in the trees.

CONTROL. Applied control seldom, if ever, will be needed in forests. DDT as a spray or mist should be effective if this insect ever becomes abundant on shade trees. For details of application methods see page 101.

SADDLED PROMINENT

Heterocampa guttivitta (Wlk.), Lepidoptera, Notodontidae

INJURY. Small to moderate-sized outbreaks of this insect occur frequently, but these usually are of short duration. Loss of woody growth is the main damage caused. The young larvae are leaf skeletonizers, but most of the damage is done by the older caterpillars that consume most of the leaf tissues.

HOSTS. The foliage of beech is favored, but sugar maple and other deciduous broad-leaved trees are also eaten.

RANGE. Eastern United States, southern Canada, and west to Colorado.

INSECT APPEARANCE. *Caterpillars:* about 1½ inches long when full grown; body somewhat cylindrical but is thicker near middle; color variable yellowish to bluish green with a reddish brown or purplish saddle-shaped patch on back; head large with broad reddish band on each side; 5 pairs prolegs. *Moths:* wingspread 1½ to 2 inches; color brownish gray crossed with indistinct variable markings.

LIFE CYCLE. The moths appear during the spring and the females lay the eggs singly on the leaves. Each female may lay as many as 500 eggs. These hatch in the usual time of a week to 10 days. The caterpillars mature about 5 or 6 weeks after hatching and then descend to the ground where they pupate in the leafmold. Here they spend the winter. There is only 1 generation per year.

CONTROL. Aerial spraying with DDT should be effective when outbreaks continue for several years and when the values involved warrant treatment. See page 101 for details regarding application methods.

VARIABLE OAK WORM

Heterocampa manteo (Dbldy.) and several related species, Lepidoptera, Notodontidae

Nearly everything presented for *H. guttivitta* is also true for the variable oak worm. The differences are as follows: HOSTS. General feeder on deciduous broad-leaved trees. INSECT APPEARANCE. *Caterpillars:* about 1½ inches long when full grown; color yellowish green with pale longitudinal line along the mid-

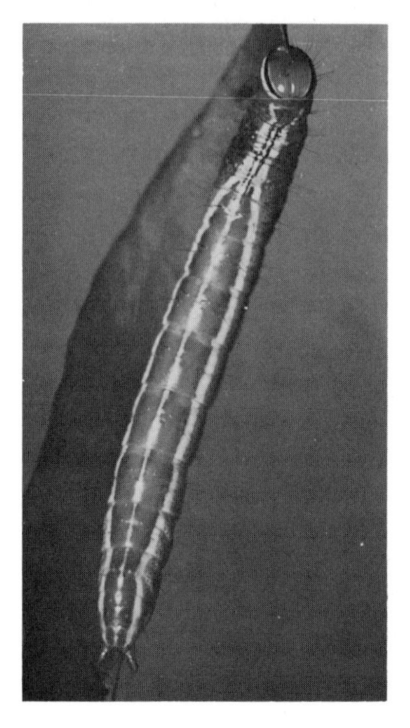

FIG. 9.28. Variable oak worm caterpillar (*Heterocampa manteo*) (X2).

back, which is bordered on each side with variable reddish-brown bands; head with a broad brown or black band and another whitish on each side of head. *Moths:* wingspread about 1⅔ inches; color ashy gray with indistinct variable dark markings.

CALIFORNIA OAKWORM

Phryganidia californica Pack., Lepidoptera, Dioptidae

INJURY. This species is primarily a pest of shade and ornamental trees. The smaller larvae skeletonize the leaves by feeding mostly on the upper surfaces, whereas the larger caterpillars consume most of the leaf parts except the larger veins.

HOSTS AND RANGE. Species of oaks and occasionally other trees such as Eucalyptus throughout the oak-growing regions of California.

INSECT APPEARANCE. *Caterpillars:* about 1 inch long when full grown; body naked, dark olive green with conspicuous black and yellow stripes along the back and sides; back of the 8th abdominal segment slightly humped; 5 pairs prolegs. *Moths:* wingspread about 1¼ inches; light brown with rather prominent dark wing veins.

LIFE CYCLE. Winter is passed either as young larvae on the foliage or as eggs. The spring caterpillar generation develops mostly from March to May. Pupation occurs on the tree trunks, fences, or bushes. They do not enclose themselves in cocoons. The moths for the spring brood appear during June, and the females lay their eggs in groups of up to 40 on the undersides of the leaves. These produce the summer generation that matures, pupates, and develops into moths during November. There are 2 generations per year.

CONTROL. DDT emulsion sprays or mists applied when the larvae are small should be effective for protecting shade trees. See page 101 for the details regarding application methods.

Essig, E. O. 1958. *Insects and mites of western North America.* Macmillan Co. Pp. 688–691.

BUCCULATRIX LEAF SKELETONIZERS

***Bucculatrix* spp. Lepidoptera, Lyonetiidae**

INJURY. These are not serious forest pests because the defoliation occurs so late in the summer that most of the growth has already been produced. The young larvae are leaf miners, whereas those older produce the more significant skeletonizing damage. HOSTS. Various species infest birches, oaks, apple, and hawthorn. RANGE. Eastern North America wherever the hosts grow. INSECT APPEARANCE. *Caterpillars:* small, ¼ inch long when full grown; body stout and greenish; 5 pairs prolegs. *Cocoon:* whitish with longitudinal ridges. *Moths:* small, wingspread 5⁄16 inch; wings narrow; hindwings with a broad fringe of hairs; forewings brown or black with light markings; hindwings gray. LIFE CYCLE. The pupae hibernate over the winter in the characteristic ribbed cocoons in the ground litter. The moths emerge in the summer, and the resulting larvae are leaf miners for about a month. From then on they are leaf skeletonizers. CONTROL. Applied methods have never been used against these insects because the tree damage is never severe.

CUTWORMS

Various species, Lepidoptera, Phalaenidae (Noctuidae)

INJURY. Cutworms sometimes are troublesome in forest nurseries, where they can cause much damage. Some species simply cut off the plants, whereas others climb and eat the foliage. Many species feed only at night. HOSTS. All species of plants. RANGE. Wherever the hosts grow. INSECT APPEARANCE. *Caterpillars:* drably colored gray to brown; some species striped; 5 pairs prolegs. *Moths:* drably colored gray often mottled with lighter markings. LIFE CYCLE. The caterpillars usually hibernate over the winter under ground litter. They mature and pupate in the early summer. The moths emerge in a couple of weeks and the females lay the eggs for the next generation. There is usually only 1 generation per year. CONTROL. DDT applied as a dust, wettable powder or emulsion concentrate at the rate of 2 pounds per acre is an effective poison.

CASE BEARERS AND BAG WORMS

Lepidoptera, Coleophoridae and Psychidae

Some caterpillars have the distinctive habit of constructing small portable cases consisting of silk and leaf parts. One larva occupies each case, and only when feeding or moving do they partially emerge and expose the anterior parts of their bodies. Various species infest many conifers and deciduous broad-leaved hosts, but only two that infest conifers are of sufficient importance to be discussed here. These are the larch case bearer and the evergreen bagworm. The general moth appearance for the two families differs greatly as indicated in the discussions which follow.

LARCH CASE BEARER

Coleophora laricella (Hbn.) Lepidoptera, Coleophoridae

The larch case bearer is the only species of this group that is of any importance as a forest pest. Some infestations have been severe and persistant over several successive years, thereby causing reduced woody growth and even severe tree mortality.

INJURY. The caterpillars, which are needle miners during the early caterpillar life, later become the more conspicuous case formers. Damage caused by the late summer needle mining is seldom conspicuous or serious, but the defoliation during the spring by the case inhabiting caterpillars causes most of the tree injury. During this spring period the larvae always live individually within the cases they construct from part of a mined leaf, which is lined with fine silk.

HOSTS. Species of larch.

RANGE. This case bearer is a European species that was introduced into Massachusetts during the last century. Since then it has spread over eastern North America wherever larch grows.

INSECT APPEARANCE. *Caterpillars:* about 3/16 inch long when full grown; body color brown with head, pronotal shield and anal plate black. *Case:* somewhat cigar-shaped; made from only part of a mined needle, lined with silk; gray with a yellowish or brownish patch on outer side, 1/4 to 3/8 inch long. *Pupae:* typical, located inside the cases. *Moths:* small, 3/8 inch wingspread; silvery to brown color; wings narrow and fringed with long hairs. *Eggs:* minute, brown, and resemble an inverted jelly mold when observed with the aid of a lens.

LIFE CYCLE. During the autumn, the winter hibernation period, and the spring feeding period the larvae live individually inside the cases, and whenever they move the cases are carried along. Pupation occurs in the early summer so that the moths emerge mostly during June. After mating, the females deposit the eggs rather promiscuously over the foliage. These soon hatch and the resulting larvae mine within the needles from this time until autumn. In September the larvae construct the individual cases in which they live the remainder of their larval life. The larvae hibernate over the winter inside these cases fastened to the twigs. There is only 1 generation per year.

CONTROL. Insecticidal control has not been tried under forest conditions, but for shade trees dormant lime sulphur spray has been recommended. It should be applied just before spring needle growth starts and mixed at the rate of 1 part lime sulphur solution to 9 parts water. Very likely some of the newer synthetic insecticides such as DDT would also be effective against this insect.

Herrick, G. W. 1912. *Cornell Univ. Agri. Exp. Sta. Bull.* 322.
Schaffner, J. V., Jr. 1937. *Mass. Forest and Park Asso. Tree Pest Leaflet* 12.

BAGWORM

Thyridopteryx ephemeraeformis (Haw.) Lepidoptera, Psychidae

INJURY. This insect is chiefly a shade tree pest. The caterpillars construct cases or bags from small pieces of the foliage by fastening the pieces together with silk. Only one individual lives in each bag.

HOSTS. The foliage of many species of conifers and broad-leaved trees are eaten but the cedars and arborvitaes are preferred. Sometimes it even feeds on herbaceous plants.

RANGE. Throughout eastern United States except the northern tier of states and in the region adjacent to the Gulf of Mexico.

INSECT APPEARANCE. *Caterpillars:* stout, ¾ to 1 inch long when full grown; tan to brown with head and thorax spotted with black; setae sparse so larvae appear naked; always live individually within the bags. *Bags:* maximum size during the autumn when the larvae are maturing 1 to 2 inches long and about ¾ inch diameter, both ends pointed with only one end open. *Pupae:* inside bags. *Moths:* males are hairy, black, slender, and wasp-like; wingspread of about 1 inch; wings almost free of scales; females wingless and worm-like, remain inside the bags during their short adult life following pupation.

LIFE CYCLE. The eggs are deposited inside the old female bags where they remain over the winter. Late in the spring the eggs hatch and each larva immediately makes a small case in which it lives throughout its larval life. From time to time as the larvae grow, they enlarge the size of these cases to accommodate their increasing size. In order to feed or move about the larvae must partially emerge from the case so that heads and thoracic legs are exposed. Pupation occurs within the bags during late summer and the moths emerge about three weeks later. The

FIG. 9.29. Bagworm (*Thyridopteryx ephemeraeformis*). Left: bag, which caterpillar makes and inhabits throughout larval life, with pupal case extending through opening (X1½); upper right: male moth showing wings nearly devoid of scales (X2); and lower right: caterpillar (X2).

males then fly to the female cases and mate. Each female deposits 800 to 1,000 eggs inside the bag, and then her shriveled remains drop to the ground and she dies. There is only 1 generation per year.

CONTROL. Insecticidal controls have never been needed to protect forests. For killing the larvae on ornamental trees, however, malathion is an effective poison. It should be used at the rate of one pound of the actual insecticide, either as an emulsion or as a wettable powder, per 100 gallons of spray. Other organic phosphate insecticides are also effective but more dangerous for the spray operator. Lead arsenate at the rate of 4 pounds per 100 gallons of spray is also effective when used before the larvae are half grown. Double this dosage is needed to kill the older larvae. DDT and the other chlorinated hydrocarbon insecticides are not sufficiently toxic to this insect to warrant their use.

Haseman, L. 1912. *Univ. Missouri Agri. Exp. Sta. Bull.* 104.
Howard, L. O., and F. H. Chittenden. 1916. *U.S.D.A. Farmers' Bull.* 701.

LEPIDOPTEROUS LEAF MINING CATERPILLARS

Lepidoptera, Gelechiidae, Gracilariidae and others

Many species of small caterpillars are leaf miners, but most of these do not cause serious injury; therefore, it is usually sufficient for us to simply recognize them as foliage miners. Some leaf rollers, needle tiers, and case makers are also leaf miners during their early larval life. Other needle miners also enweb the shoots. Thus these sections should be checked when identifying a species that causes foliage mining damage. The only serious leaf mining forest pest is the lodgepole needle miner. This species is discussed in detail. Other commonly encountered species of *Recurvaria* spp. infest cypress, hemlock spruces, arbor vitae, and cedar. An eastern hard pine needle miner is *Exoteleia pinifoliella* (Chamb.), and *Epinotia nanana* (Tr.) is a spruce needle miner that causes a type of damage resembling the needle tiers. All of these have only 1 generation per year. The partially grown larvae hibernate over the winter in the mines. Pupation occurs in the spring and the moths emerge during early summer.

Needham, J. G., S. N. Frost, and B. H. Tothill. 1928. *Leaf mining insects.* Williams and Wilkin Co., Baltimore, Md. 351 pp.

LODGEPOLE NEEDLE MINER

***Recurvaria milleri* Busck, Lepidoptera, Gelechiidae**

The lodgepole needle miner is the only important foliage mining forest insect pest. It has caused extensive defoliations and some timber losses in the northern Rockies and in California.

INJURY. The tiny caterpillars feed by mining the insides of the needles so that only the outer shells remain. These soon turn brown, which makes the crowns of infested trees appear as though fire scorched. After several years of successive defoliation the tops start to die and then general tree decadence follows. Bark beetles usually cause the final death of the weakened trees.

HOSTS. Lodgepole pine is the main host, but other species of white and hard pine may also be infested during epidemics.

RANGE. Canadian Rockies south into the northern Rockies of the United States and in California.

INSECT APPEARANCE. *Caterpillars:* small, 5/16 inch when full grown; body naked; yellow to orange with red along mid-back line; head and pronotal shield dark brown, 5 pairs of prolegs. *Moths:* wingspread about 1/2 inch, wings narrow and with long hairs on hind margins; color silvery gray with forewings irregularly marked with black; antennae ringed with black and white; a large dark spot on back of mid-abdominal region.

LIFE CYCLE. This insect requires 2 years to complete 1 generation. The moth appears mostly during July. After mating, the females lay their eggs in groups of 3 to 16, either in mined needles or in or around the needle sheaths. The larvae bore singly in the needles and usually mine from the tip toward the petiole. Each mines about 3 needles during its larval life. Hibernation during both winters is passed in the larval stage inside the hollowed needles with one larva per needle. Pupation also occurs inside the needles during a 3 to 4 week period from June to August of the second year.

CONTROL. Insecticides have not been used against this needle miner under forest conditions. Keen, however, suggests that aerial spraying with benzene hexachloride may be effective. The rate suggested is 1/10 pound of the gamma isomer per acre. Parathion or malathion applied at the rate of ½ pound per 100 gallons of spray probably would be more effective.

Keen, F. P. 1952. *U.S.D.A. Misc. Publ.* 273:109–111.
Patterson, J. E. 1921. *J. Agri. Research* 21:127–142.
Stark, R. W. 1954. *Can. Entomol.* 86:1–12.

HYMENOPTEROUS DEFOLIATORS

There are only two groups of insect defoliators that belong to the order Hymenoptera. These are the sawflies and the ants. They can be identified by noting the characteristics as presented in Table 9.1 (page 97).

SAWFLIES

Hymenoptera, Tenthredinoidea

Sawflies are so named because the female wasps each has a saw-toothed ovipositor. This structure is used for cutting slits in plant tissues in which the eggs are placed. There are many species of sawflies which, in the larval stage, feed on the foliage of trees. The more important of these infest conifers. Characteristics of some of the common species are presented in Tables 9.14, 9.15, and 9.16.

TABLE 9.14 COMMON SPECIES OF SAWFLY CATERPILLAR THAT INFEST CONIFERS

1. Infest pines	Table 9.15
1. Infest conifers other than pines	2
2. Infest species of Cupressaceae	3
2. Infest species of Pinaceae	4

3. Infest arbor vitae and juniper; occur in northeastern North America; head light brown body greenish with 3 longitudinal strips; 8 pairs prolegs . . . . ARBOR VITAE SAWFLY [*Monoctenus melliceps* (Cress)].
3. Infest Monterey cypress in California . . . CYPRESS SAWFLY (*Susana cypressi* Roh.)
4. Infest spruces, firs, or hemlocks; caterpillars have 7 or 8 pairs of prolegs 7
4. Infest larch; caterpillars have 7 pairs of prolegs 5
5. Head black; body not striped (page 183) . . . . LARCH SAWFLY
5. Head brown; body with longitudinal strips; occur only in the West 6
6. Two narrow dark green strips along each side (page 184) . . . . TWO-LINED LARCH SAWFLY
6. One dark line along back (page 184) . . . . WESTERN LARCH SAWFLY
7. Infest spruces 8
7. Infest species of fir or hemlock 10
8. Caterpillars with 7 pairs prolegs; body dark yellowish green with many gray-green stripes along back and sides 9
8. Caterpillars with 8 pairs prolegs; head brown, body light green; all except last larval instar with 5 narrow white lines along body (page 185) . . . . EUROPEAN SPRUCE SAWFLY
9. Head yellowish (page 185) . . . . YELLOW-HEADED SPRUCE SAWFLY
9. Head green (page 186) . . . . GREEN-HEADED SPRUCE SAWFLY
10. Infest hemlocks (page 186) . . . . HEMLOCK SAWFLY
10. Infest firs (page 186) . . . . FIR SAWFLY

INJURY. Only the larvae feed on the foliage. A few species are leaf miners, some are skeletonizers, and a few construct webs, but most feed openly on the foliage and consume the great part of the leaf tissues. Some are gregarious and remain together in groups as they defoliate one twig or branch at a time, whereas others are solitary, especially as they grow older. During the first few larval instars the insects may eat only the outer parts of the leaves or needles, but later they consume all of the foliage tissues. Three fourths or more of the feeding damage occurs during the last larval instar. Those that feed on conifers feed from the tips toward the bases of the needles and thereby waste little foliage.

INSECT APPEARANCE. *Caterpillars:* most are about ¾ to 1 inch long when full grown; body naked, cylindrical, and worm-like commonly lined or spotted; heads globular and usually colored darker than the bodies; with 3 pairs of short jointed thoracic legs and usually 8 pairs (occasionally 6 or 7 pairs) of fleshy, unjointed abdominal legs (prolegs), which are never armed with groups of minute hooks (crochets). A few species (slugs) secrete a slimy material over their bodies. *Pupae:* within tan to brown, oval-shaped parchment-like silk cocoons that are about ⅛ to ⅜ inch long; usually constructed in the litter or soil beneath the host trees. *Wasps:* stout, short, and thick waisted, somewhat resemble

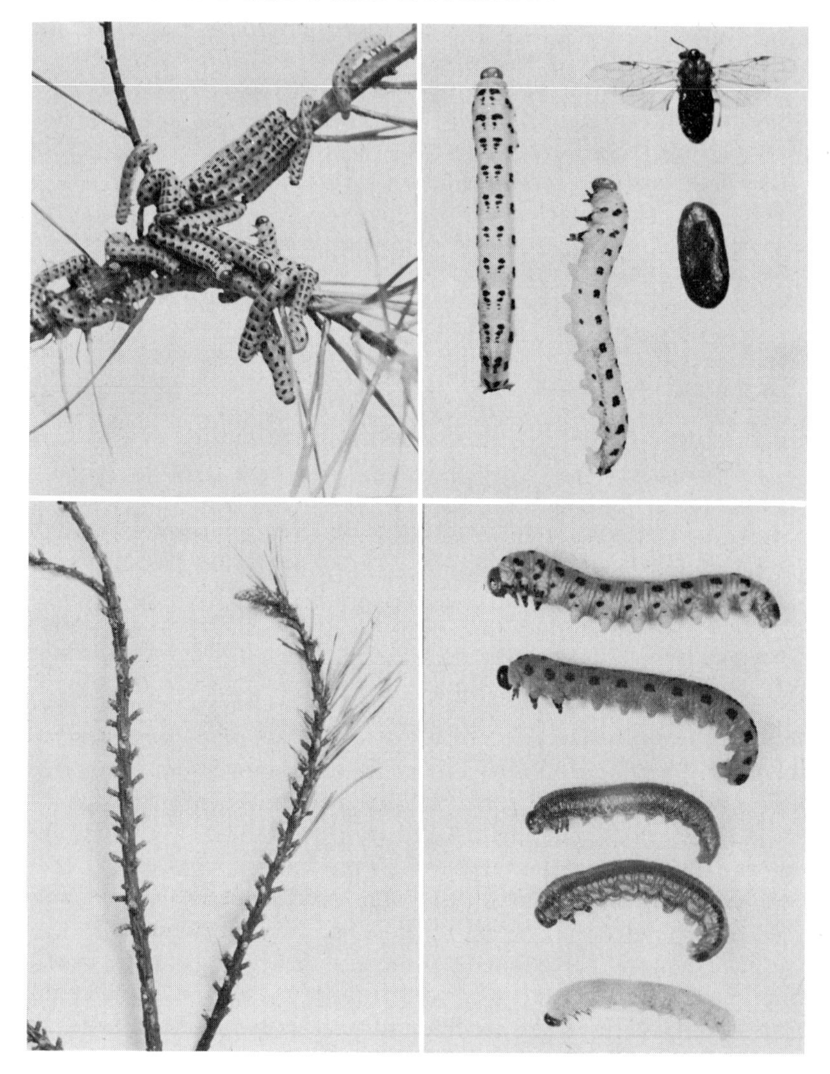

FIG. 9.30. Sawflies. Above left: group of redheaded pine sawfly (*Neodiprion lecontei*) caterpillars feeding (X¾); upper right: redheaded pine sawfly wasp, caterpillars, and cocoon (X2); below: sawfly damage to pine (X⅓) and five species of sawfly caterpillars (X2).

flies; with four membranous wings but generally they are poor fliers; forewings with a conspicuous, dark, thickened spot near the middle of the front edge; females with ovipositer at tip of abdomen shaped, toothed, and used like a saw to cut small slits in plant tissue usually along the edges of the leaves or needles; commonly colored black, but some have yellow, orange, or red on various parts of the body; antennae rather slender on females, whereas on the males they are feathery. The adults, seldom seen in the field, are very difficult to identify. Therefore, their identifying characteristics are not presented here. The larvae, on the other hand, can be tentatively identified in the field on the basis of host and larval color patterns.

LIFE CYCLE. There are two common types of seasonal developmental patterns. In one type the eggs overwinter and hatch in the spring. The resulting larvae develop during the early 3 to 8 weeks of the season and when mature they spin individual cocoons in the litter or soil where they remain until fall. The adults then emerge, mate, and the females deposit the eggs for the next brood. For these there is only 1 generation per year. For the second type of development the prepupae (mature larvae within cocoons) hibernate over the winter. Pupation and adult emergence occur in the spring or summer. These species commonly have 1 to several generations per year depending on the length of summer. For both types there are usually 4 or 5 larval instars with the females commonly having one more than the males. The eggs usually are laid individually in slits cut by the females.

The prepupae commonly enter a resting state (diapause) where they may remain for several months to several years before development is resumed. Usually, however, only a small percentage of the population exhibits the extremely long diapause.

Reproduction is usually bisexual, but the females of many species can reproduce parthenogenetically. For some, the unfertilized eggs always develop into males, whereas for other species only females result. Commonly 100 to 150 eggs are laid per female. Usually only 1 egg is placed in each slit.

CONTROL. When conditions warrant the use of applied control DDT is the most practical insecticide to use. Aerial applications require the usual dosage of 1 pound per acre with the DDT dissolved in oil solvents. For treating small trees such as spot infestations in plantations wettable powder or emulsion formulations containing 1 pound of actual DDT per 100 gallons of spray applied with hand sprayers is effective. Sometimes mist blowers also may be used to advantage. See pages 66, and 72.

Inasmuch as sawflies commonly form their cocoon in the litter they are very vulnerable to predation by the mice and shrews. Especially during the fall the ground inhabiting rodent populations may be very high.

GENERAL REFERENCES

Atwood, C. E., and O. Peck. 1943. *Can. J. Research* 21 (D) 109–144.
Hetrick, L. A. 1956. *Forest Sci.* 2:181–185.
Keen, F. P. 1952. *U.S.D.A. Misc. Publ.* 273 (revised).

Ross, H. H. 1955. *Forest Sci.* 1:196–209.
Schaffner, J. V., Jr. 1950. *U.S.D.A. Misc. Publ.* 657.
Yuasa, H. 1920. *Univ. Illinois Biological Monographs* 7:325–490.

PINE SAWFLIES

Hymenoptera, Diprionidae

Many species of sawfly caterpillars feed on the foliage of pines. Some of these are serious forest pests and, thereby, cause much concern to foresters.

The taxonomy of this insect group has been so confused that, in the past, few species could be identified readily, but recently this has been rectified somewhat by the work of Ross (1955). Larval characteristics of the commoner species are presented in Table 9.15. These can be classified roughly into two groups on the basis of their habits, for example, the spring feeders and the summer feeders. All species of *Neodiprion* and *Diprion* have 8 pairs of prolegs, whereas the pine webbing sawflies have none.

TABLE 9.15 COMMON SPECIES OF SAWFLY LARVAE THAT INFEST PINES*

1. Caterpillars build and live in loose silk webs that become filled with much excrement and needle fragments; fleshy abdominal legs (prolegs) absent (page 183) . . . . . PINE-WEBBING SAWFLIES
1. Caterpillars do not construct and live in webs; 6 to 8 pairs prolegs present . . . 2
2. Occur east of the Great Plains 7
2. Occur west of the Great Plains 3
3. Occur in the northern Rockies 4
3. Occur in parts of the West other than the northern Rockies 5
4. Caterpillar head dark brown; host lodgepole pine (page 182) . . . . . LODGEPOLE PINE SAWFLY
4. Caterpillar head black; hosts lodgepole and ponderosa pines (page 180) . . . . . JACK PINE SAWFLY
5. Infests pinyon pines; occurs in southern Rockies and in California . . . . . PINYON SAWFLY [*Neodiprion rohweri* (Midd.)]
5. Infests pines other than pinyon 6
6. Infests mostly sugar and western white pines . . . . . SUGAR PINE SAWFLY [*Neodiprion edwardsii* (Nort.)]
6. Infest ponderosa pines . . . . PONDEROSA PINE SAWFLIES [*Neodiprion fulviceps, N. edwardsii* (Nort.)] and *N. gilletei* (Roh.)
7. Four to 6 rows of spots along body 8
7. Back of body not spotted but sides may be 9
8. Head orange to brown with 6 rows of black spots along body (page 182) . . . . . REDHEADED PINE SAWFLY

8. Head black; body yellowish white with rows of black spots along body (page 182) . . . . . . WHITE PINE SAWFLY
9. Head black 10
9. Head orange to dark brown 19
10. Sides of body with dark spots or dark mottling 11
10. Sides not spotted or mottled but may have dark longitudinal lines 15
11. Body yellow to greenish mottled with irregular dark brown depressed areas and similar shaped light raised areas sometimes darkened areas along back appear as 2 lines (page 182) . . . . . . INTRODUCED PINE SAWFLY
11. Larvae with definite black spots on sides of body 12
12. Occur in northeastern United States and southeastern Canada (see 14 below) 13
12. Occur in southeastern United States (page 180) . . . . . . *Neodiprion hetricki* and *N. pratti pratti*
13. Two light lines border a dark line along each side, the dark line tends to break into spots; light line along mid-back, tergum of last abdominal segment black (page 180) . . . . . . EUROPEAN PINE SAWFLY
13. Two dark lines along back; spots on sides sometimes faded or absent on middle of body 14
14. Occurs in southeastern Canada and bordering parts of the United States, (page 182) . . . . . . SWAINE'S SAWFLY
14. Occurs in northeastern United States west to Minnesota and south to North Carolina (page 180) . . . . . . *Neodiprion pratti paradoxicus*
15. Occurs in northeastern North America 16
15. Occurs in southeastern United States 17
16. Body yellowish green with 4 rows of dark stripes along body; host chiefly jack pine (page 180) . . . . . . JACK PINE SAWFLY
16. Back gray-green to black with a light green mid-back stripe; fainter lines on each side; blackish stripe at base of legs; host chiefly red pine (page 182) . . . . . . RED PINE SAWFLY
17. Front of head marked with a light spot (page 182) . . . . . . ABBOTT'S SAWFLY
17. Front of head not marked with a light spot 18
18. Black spot on back of last abdominal segment; body greenish, with 4 dark longitudinal stripes (page 182) . . . . . . *Neodiprion exitans*
18. Last abdominal tergite without black spot; body otherwise colored as in 18 above (page 180) . . . . . . *Neodiprion hetricki* and *N. pratti pratti*
19. Body with dark spots along sides 20
19. Sides not spotted 22
20. Occurs in southeastern Canada and bordering parts of the United States; 2 dark lines along back and a line of spots along each side, spots may be absent on middle of body (page 182) . . . . . . SWAINE'S SAWFLY
20. Occur in southeastern United States 21
21. Two dark lines along back; each side with a broken stripe and a double row or spots; known host is pitch pine (page 182) . . . . . . PITCH PINE SAWFLY
21. Two dark stripes along back, and 1 row spots along each side; hosts are short leaf and loblolly pines (page 180) . . . . . . SPOTTED LOBLOLLY PINE SAWFLY
22. Front of head marked with either a prominent light or dark spot 23

22. Front of head not marked as in 18 above (page 180). . . . . ARKANSAS SAWFLY
23. Front of head with a light spot (page 182). . . . . ABBOTT'S SAWFLY
23. Eye spots and blotch on front of head dark; light V-mark near center of blotch occurs in northeastern United States (page 182). . . . . *Diprion frutetorum*

* Larval and biological characteristics are not available for most of the western pine-infesting sawflies.

SPRING-FEEDING PINE SAWFLIES

Neodiprion spp. Hymenoptera, Diprionidae

The species, range, and chief hosts of the sawflies that belong to this spring feeding group are as follows:

1. ARKANSAS SAWFLY (*N. taedae linearis* Ross); south central United States; loblolly and short-leaf pines.
2. SPOTTED LOBLOLLY PINE SAWFLY (*N. t. taedae* Ross); southeastern United States (Virginia); loblolly, short-leaf and virginia pines.
3. JACK PINE SAWFLY (*N. pratti banksianae* Roh); eastern Canada and the Lake States region; jack pine occasionally on other pines.
4. *N. p. pratti* (Dyar); southeastern United States; loblolly, short-leaf and scrub pine.
5. *N. p. paradoxicus* Ross; northeastern United States from Maine to North Carolina; jack, Virginia, pitch, and shortleaf pines.
6. RED PINE SAWFLY (*N. nanulus nanulus* Schedl.); southeast Canada and northeast United States and Lake States; red and jack pines.
7. EUROPEAN PINE SAWFLY [*N. sertifer* (Geoff.)]; northeast North America west to Illinois; red, pitch, short-leaf, and occasionally other pines.
8. HETRICK'S SAWFLY (*N. hetricki* Ross); southeast United States; loblolly and possibly other pines.

HOSTS AND RANGE. This information was presented briefly previously. It is very likely, however, that these sawflies will be found over a somewhat wider range than indicated, and also the larvae may be found feeding on pine species other than those listed.

INSECT APPEARANCE. *Caterpillars:* typical; see page 175. Specific characteristics for the various species are presented in Table 9.15. *Adults:* see Ross (1955). *Cocoons:* typical; see page 175. *Eggs:* inserted in slits in rows along the edges of the needles.

LIFE CYCLE. Winter is passed as eggs in the needles of the hosts. These hatch in the early spring and the resulting larvae feed on the year-old needles during the following 4 to 8 weeks. When mature, the larvae descend to the ground and spin their cocoons in the soil. The insects live inside the cocoons throughout the summer as prepupae.

FIG. 9.31. Crown of short-leaf pine defoliated by sawflies.

Pupation and adult emergence occurs in the late fall. After mating, the females lay their eggs by inserting them in individual slits cut in the needles. There is only 1 generation per year.

CONTROL. Applied control has never been applied to forests, but very likely aerial applications of DDT at the usual dosage of 1 pound per acre would be effective. See page 72 for details of application.

Coyne, J. F. 1959. *U.S.D.A. Forest Pest Leaflet* 34.
Ewan, H. G. 1957. *U.S.D.A. Forest Pest Leaflet* 17.
Hetrick, L. A. 1956. *Forest Sci* 2:181–185.
Ross, H. H. 1955. *Forest Sci.* 1:196–209.

SUMMER FEEDING PINE SAWFLIES

Neodiprion spp. and *Diprion* spp., Hymenoptera, Diprionidae

Species, ranges, and chief hosts of the summer-feeding pine sawflies are listed here. The caterpillars of these sawflies also may be found feeding during the spring of the year, but usually they are more abundant during the summer and fall seasons. They also differ from the spring-feeding group in that they pass the winter in the soil as prepupae inside cocoons. See Table 9.15 for the larval characteristics of the commoner species.

1. RED-HEADED PINE SAWFLY [*N. lecontei* (Fitch)]; eastern half of North America; all hard (2- or 3-needle) pines, sometimes the foliage of other conifers is eaten by older larvae. Usually infest smaller-sized trees up to about 15 feet tall.

2. WHITE PINE SAWFLY [*N. pinetum* (Nort.)]; northeast and north central North America; white pine.

3. PITCH PINE SAWFLY [*N. pini rigidae* (Nort.)]; northeast United States; pitch pine.

4. LODGEPOLE PINE SAWFLY (*N. burkei* Midd.); western North America; lodgepole pine.

5. ABBOTT'S SAWFLY [*N. abbottii* (Leach)]; eastern North America; all hard pines.

6. SWAINE'S SAWFLY (*N. swainei* Midd.); southeast Canada and the Lake States; jack pine.

7. *N. excitans* Roh.; southeast United States; most hard pines.

8. INTRODUCED PINE SAWFLY [*D. similis* (Htg.)]; northeastern North America west to the Lake States; chiefly the soft 5-needled pines.

9. *D. frutetorum* (F.); northeast North America; red and scotch pines.

INSECT APPEARANCE. *Caterpillars:* typical; see page 175; species characteristics are presented in Table 9.15. *Wasps:* see Ross (1955) for the adult characteristics of *Neodiprion*. *Eggs:* inserted in slits in the needles. *Pupae:* typical; see page 175.

LIFE CYCLE. Winter is passed as mature larvae (prepupae) within cocoons in the ground litter or soil. Pupation occurs at various times from spring through the summer. This instar lasts about 2 weeks. Each female deposits her eggs in individual slits, which she cuts with her saw-like ovipositor along the edges of the needles. Six to twelve eggs are placed in each needle, and all the eggs on the needles of one twig are usually all laid by one female. There is only 1 generation per year in the northern part of the range, but farther south there may be 2 or more. Frequently the prepupae become dormant (diapause) and remain inactive for one to several years before pupating and emerging as adults.

CONTROL. DDT sprays have been used successfully against the red-headed pine species. Spraying should be used when 10 to 20 per cent of the trees are infested in plantations 6 feet tall. Hand spraying with compressed air sprayers can often be used efficiently, because the gregarious caterpillars congregate on certain trees, and these infested trees are easily seen. Sometimes, however, aerial spraying may be more efficient especially for those species that infest taller trees. See page 72 for the details of application.

Benjamin, D. M. 1955. *U.S.D.A. Tech. Bull.* 1118.
Burke, H. E. 1932. *U.S.D.A. Circ.* 224.
MacAloney, H. J. 1957. *U.S.D.A. Forest Pest Leaflet* 14.
Middleton, W. 1938. *U.S.D.A. Farmer's Bull.* 1259.
Ross, H. H. 1955. *Forest Sci.* 1:196–209.

PINE-WEBBING SAWFLIES

Acantholyda spp., Hymenoptera, Pamphiliidae

INJURY. Caterpillars form loose silk webs which they fill with much excrement and needle fragments. HOSTS. Pines. RANGE. Wherever the hosts grow. *Caterpillars:* ¾–1 inch long; no prolegs; *A. erythrocephala* (L.) head yellow, top marked with many small dark brown spots; *A. zappei* (Roh.) head brown, body green with a dark stripe along back; *A. sp.* on Monterey pine, body dark green or brownish, head black. LIFE CYCLE. The mature larvae (prepupae) hibernate over winter in soil. They pupate in the spring, and the adults soon emerge. The females immediately lay eggs exposed on the tissues so that the larvae are present during late spring. They then enter the soil where they remain until the next spring. There is only 1 generation per year.

LARCH SAWFLY

Pristiphora erichsonii (Htg.), Hymenoptera, Tenthredinidae

INJURY. This sawfly is a serious forest pest. Many repeated annual defoliations weaken the trees so that they finally are attacked and killed by the inner-bark boring eastern larch beetle.

HOSTS. All species of larch.

RANGE. Throughout the northern parts of North America as well as in Europe and Asia. Some entomologists think that this insect has been introduced into North America, but if this is so, the transplantation occurred a long time ago.

INSECT APPEARANCE. *Caterpillars:* typical; see page 175; about ¾ inch long when full grown; head and thoracic legs black; body dull grayish green above and paler underneath; 7 pairs prolegs. *Cocoons:* typical, dark brown. *Wasps:* black with basal half of abdomen orange; legs yellowish to orange.

LIFE CYCLE. Winter is passed as mature larvae (prepupae) inside individual cocoons scattered in the moss or litter beneath the trees. A small percentage of the population may remain in a diapause in this stage for 2 or more years before pupating and emerging as adults. Pupation occurs in the spring, and the adults emerge over an extended period of time during May, June, and early July. Parthenogensis is the rule, and less than 1 per cent of the insects are males. The eggs are laid in slits cut in the new shoots and not in the needles as is the common oviposition site for most conifer infesting sawflies. Eggs hatch about a week after being laid, and the gregarious larvae feed for about 3 weeks before maturing and spinning cocoons. Generally there is only 1 generation per year.

CONTROL. No satisfactory insecticidal control has been developed for treating large forested areas. A single aerial application of DDT has not been very effective (60 to 70 per cent mortality), because the larvae vary so much in age and their activity extends over a long season. This occurs because the wasps emerge over an extended period of time. Probably two applications would increase the larval mortality sufficiently to make the treatment effective. It has been suggested that ornamentals can be protected by spraying with benzene hexachloride wettable powder or emulsion at the rate of ¼ pound of the gamma insomer per 100 gallons of spray.

Larch sawfly populations commonly suffer severe decimations when excessively wet weather occurs, either during the summer when the larvae are cocooning or during the spring period of adult emergence. At these times flooding of the swampy sites commonly occupied by tararack drowns the insects.

Butcher, J. W., and C. B. Eaton. 1952. *U.S.D.A. Bur. Entomol. Plant Quarantine* E-841.
Coppel, H. C., et al. 1955. *Can. Entomologist* 87:103–140.
Drooz, A. T. 1956. *U.S.D.A. Forest Pest Leaflet* 8. 1960. *U.S.D.A. Tech. Bull.* 1212.
Hewitt, C. G. 1912. *Can. Dept. Agri. Bull.* 10 (2nd series) or *Entomol. Bull.* 5.
Lejeune, R. R., W. H. Fell, and D. P. Burbridge. 1955. *Ecology* 36:63–70.
Muldrew, J. A. 1956. *Forestry Chronicle* 32:20–29.
Reeks, W. A. 1954. *Can. Entomologist* 86:471–480.

TWO-LINED LARCH SAWFLY

Anoplonyx occidens Ross, Hymenoptera, Tenthredinidae

HOSTS. Western species of larch. **RANGE.** Northwestern United States. *Caterpillars:* about ⅜ inch long when full grown; head dark brown; eyes black; body brownish green with 2 narrow dark green stripes along sides; 7 pairs prolegs; coxae of thoracic legs partly colored brown. **LIFE CYCLE.** Similar to that for the larch sawfly.

Keen, F. P. 1952. *U.S.D.A. Misc. Publ.* 273 (revised). Pp. 120–121.

WESTERN LARCH SAWFLY

Anoplonyx laricivorus Roh. and Midd., Hymenoptera, Tenthredinidae

This species is similar in most respects to *A. occidens,* except that the larvae have only a single dark green line along middle of back.

EUROPEAN SPRUCE SAWFLY

***Diprion hercyniae* (Htg.), Hymenoptera, Diprionidae**

INJURY. During the 1930's this introduced defoliating insect was seriously injuring trees and causing heavy tree mortality in eastern Canada and in the New England States. A polyhedral virus disease then suddenly appeared and decimated the sawfly caterpillar population throughout its North American range. Since then this insect has caused little damage.

HOSTS AND RANGE. Spruces in eastern Canada and northeastern United States. This insect was introduced into North America sometime during the 1920's.

INSECT APPEARANCE. *Caterpillars:* about ¾ inch long when full grown; head brown; body light green with 5 narrow white lines along body; these disappear in the last feeding larval instar; 8 pairs prolegs. *Wasps:* female typical dark-colored sawflies with yellow markings.

LIFE CYCLE. Somewhat similar to that for the redheaded pine sawfly. The mature larvae overwinter inside cocoons (prepupae) in the ground litter. Pupation and adult emergence occur during the spring. Males are very rare and parthenogenic reproduction is the rule. The eggs are laid singly in slits cut in the old needles. These soon hatch, and the resulting needle-feeding larvae mature in 3 or 4 weeks. They then descend to the ground where they spin cocoons. There is only 1 generation per year in Canada, but in the United States there are 2. A prepupal dispause lasting 2 or more years is common, and sometimes a large proportion of the population might exhibit this phenomenon.

CONTROL. Insecticidal control has not been used against this defoliator in forests, because, since the time when the new more effective insecticides were developed, there has been no need for applied control methods. The polyhedral virus disease aided by the ground-feeding rodents that eat the cocooned larvae apparently have kept the insect population low.

Reeks, W. A., and G. W. Barter. 1951. *Forest Chronicle* 27:1–16.
Schaffner, J. V., Jr. 1950. *U.S.D.A. Misc. Publ.* 657:559–562.

YELLOW-HEADED SPRUCE SAWFLY

***Pikonema alaskensis* (Roh.), Hymenoptera, Tenthredinidae**

HOSTS AND RANGE. Spruces in Canada and northern United States south to Colorado. *Caterpillars:* about ¾ inch long when full grown; head yellowish; body dark yellowish green above and lighter underneath; 2 gray-green stripes along back and 2 or 3 similar stripes along each side; 7 pairs prolegs; linear spot near base of legs on all except last 3 segments. LIFE CYCLE. Somewhat similar to that for the European spruce sawfly, except that the eggs are laid between the new needles and are not inserted in them. Cocooning occurs in the mineral soil. There is only 1 generation per year.

GREEN-HEADED SPRUCE SAWFLY

Pikonema dimmockii (Cress.), Hymenoptera, Tenthredinidae

This insect is similar in most respects to the yellow-headed spruce sawfly, except that both the head and body are green.

HEMLOCK SAWFLY

Neodiprion tsugae Midd., Hymenoptera, Diprionidae

INJURY. The hemlock sawfly is a serious defoliator of many western forests.

HOSTS. Western hemlock is the main host. Occasionally mountain hemlock and silver fir also are infested.

RANGE. The coastal areas of the Pacific Northwest and the northern Rockies.

INSECT APPEARANCE. *Caterpillars:* head black; body gray to yellow green with dark longitudinal stripes; 8 pairs prolegs. *Wasps:* yellow brown to dark brown, thick-waisted wasps.

LIFE CYCLE. Similar to that for the spring-feeding pine sawflies. See page 180.

CONTROL. Applied controls have not been used against this insect, but should conditions ever warrant treatment, aerial application of DDT at the usual rate of one pound per acre would be effective.

Downing, G. L. 1959. *U.S.D.A. Forest Pest Leaflet* 31.

FIR SAWFLY

Neodiprion abietis (Harr.), Hymenoptera, Diprionidae

INJURY. This insect has caused some serious defoliation both in the East and the West. HOSTS. Species of fir. RANGE. Throughout North America wherever the hosts grow. *Caterpillars:* about ¾ inch long when full grown; head and outer sides of thoracic legs black; body dull green marked with broad stripes along body; 2 of these along back dark green with the side stripes brownish or blackish; 8 pairs prolegs. LIFE CYCLE. Similar to that for the spring feeding pine sawflies.

Brown, C. E. 1953. *Can. Forest Biol. Proc. Publ.* 2.
Struble, G. R. 1957. *Forest Sci.* 3:306–313.

SAWFLIES THAT INFEST BROAD-LEAVED TREES

Hymenoptera, Tenthredinoidea

Characteristics for some of the common sawfly caterpillars that feed on the foliage of angiosperm trees are presented in Table 9.16. In addition, the biological characteristics for a few of these are presented here.

Should these sawfly caterpillars of minor importance ever become troublesome, all except the leaf miners can be killed readily with DDT. The DDT should be applied at the usual dosages as directed on page 177. The leaf miners can be controlled by using benzene hexachloride as indicated below.

TABLE 9.16 COMMON SPECIES OF SAWFLY CATERPILLARS THAT FEED ON THE FOLIAGE OF DECIDUOUS ANGIOSPERMES

1. Caterpillars form webs in the foliage of plum and cherry; larvae without prolegs; antennae each with 7 segments . . . . PLUM WEB-SPINNING SAWFLY [*Neurotoma inconspicua* (Nort.)]
1. Webs are not formed; larvae with prolegs; each antenna has less than 7 segments 2
2. Leaf miners; larvae with vestigial prolegs (page 188) . . LEAF-MINING SAWFLIES
2. Caterpillars either skeletonize or consume most of the leaf tissues 3
3. Caterpillars with 8 pairs of prolegs 4
3. Caterpillars with 6 or 7 pairs prolegs 8
4. Each antenna with only one segment; infest elm, willow, poplar, and other species (page 188) . . . . ELM AND WILLOW SAWFLIES
4. Each antenna usually has 5 segments 5
5. Body armed with some spines 7
5. Body without spines; infest ash; body yellowish or greenish white 6
6. Head brown and much smaller than body . . . . BROWN-HEADED ASH SAWFLY [*Tomostethus multicinctus* (Beauv.)]
6. Head black . . . . BLACK-HEADED ASH SAWFLY [*Tethida cordigera* (Beauv.)]
7. Body covered with tufts of white hairs that rub off easily; head white; body green; infest butternut, walnuts and hickories . . . . WOOLLY BUTTERNUT SAWFLY [*Blennocampa caryae* (Nort.)]
7. First thoracic segment usually fringed with spines; infest mostly oaks . . . . *Periclista* spp.
8. Posterior of abdominal segment without projections; underside of many abdominal segments each with an eversible gland 9
8. Projections present on posterior abdominal segment; glands as described in 8 above absent 10
9. Larvae hairy; setae of two lengths arise from prominent tubercles; body with 4 rows of spots; infest mostly poplars and willows . . . . *Trichiocampus* spp.
9. If setae arise from tubercles they all are of about the same length; antennae conical, segment 4 as wide at base as long; infest mostly willow, mountain ash, cherry, and oak . . . . *Pristiphora* spp.
10. Posterior of abdomen with one pair of projection (cerci) 11
10. Posterior of abdomen with more than two projections . . . . ALDER SAWFLY [*Hemichroa crocea* (Fourcroy)]
11. Leaf rollers or folders; infest mostly willow and poplar . . . . *Pontania* spp.
11. Larvae do not roll or fold leaves 12

12. Cerci about as wide at base as long; infest mostly species of birch . . . . DUSKY BIRCH SAWFLY (*Croesus latitarsus* Nort.)
12. Cerci longer than width at base; infest mostly poplars and willows but also feed on other species much as cherry, alder, blue beech . . . . *Pteronidea* spp.

LEAF-MINING SAWFLIES

Fenusa spp. *Phyllotoma* spp. and *Profenusa* spp., Hymenoptera, Tenthredinidae

INJURY. Mine inside the leaves. HOSTS AND RANGE. Various species infest birch, elm, cherry, poplars, and other tree species wherever these trees grow. *Larvae:* about ¼ to ⅓ inch long when full grown; body somewhat flattened and whitish colored; prolegs vestigial. LIFE CYCLE. Winter is passed as prepupae within cocoons either inside the leaves or in the soil. The adults emerge during the late spring, mate and the females lay their eggs in the leaves. The larvae mature late in June. There are 1 to 3 generations per year. CONTROL. Spray with benzene hexachloride emulsion, 1 pint 25 per cent gamma isomer per 100 gallons of water.

Dowden, P. B. 1941. *U.S.D.A. Tech. Bull.* 757.
Schread, J. C. 1954. *Conn. Agri. Exp. Sta. Circular* 185.

SLUG SAWFLIES

Caliroa spp., Hymenoptera, Tenthredinidae

INJURY. Skeletonize the foliage usually by feeding only on the upper epidermis. HOSTS AND RANGE. Oaks and Rosaceae species wherever the hosts grow. *Larvae:* about ⅓ to ¼ inch long when full grown; body is yellow, but usually appears green due to a slimy covering; body shape is broad anteriorly and tapering toward the rear; 7 pairs prolegs with none present on last abdominal segment. LIFE CYCLE. Similar to that for *N. lecontei,* except that they cocoon or form pupation cells in the soil rather than in the litter. Usually there are 2 generations per year.

ELM AND WILLOW SAWFLIES

Climbex americana Leach and others, Hymenoptera, Cimbicidae

HOSTS. Elms and willows and sometimes various other species. RANGE. Northern North America south to Colorado, Nevada, and Calif. *Caterpillars:* about 1¾ inches long when full grown; yellowish or greenish white with black spiracles; black stripe along mid-back; skin with a pebbled texture and with many transverse wrinkles; antennae with only 1 segment; 8 pairs prolegs. When resting the larvae usually coil the posterior of abdomen around a leaf or stem in a characteristic manner. *Wasps:* ¾ to 1 inch long; antennae knobbed; head and thorax black; abdomen bluish, females with 3 or 4 yellowish spots on sides; antennae and tarsi have an orange tinge; wings smoky brown. LIFE CYCLE. Similar to that for *N. lecontei,* except that the period of adult emergence usually is extended until mid-August, and there is only 1 generation per year.

ANTS

Hymenoptera, Formicidae

In addition to the town ants discussed here, there are many other species of ants that may be very injurious to young tree seedlings. Also see page 395. Most of these species probably can be controlled by dusting the soil with 3 to 5 pounds per acre of either dieldrin or heptachlor. This treatment might be practical to use in nurseries or in other areas where the values are high, but probably would be to costly for use on general forest lands.

TEXAS LEAF-CUTTING ANT (TOWN ANTS)

***Atta texana* (Buckl.) Hymenoptera, Formicidae**

INJURY. This ant pest is a serious defoliator of young planted pines, wherever the ants occur. Most damage occurs in the fall and early spring when other green vegetation is scarce or absent. During these periods they may attack and defoliate pines within a radius of 400 feet of their nests. The needles are removed in pieces and carried to the subterranean nests where the foliage is concentrated in loose masses to form the substrate on which the ants cultivate a fungus.

Leaf-cutting ants, like other ants, are social insects that live in colonies and make their homes in the ground. The surface of the soil over these burrows is mounded several inches high, and one mound may cover several square feet. Large numbers of these mounds commonly occur together in one area so that the colonies may cover a thousand or more square feet. The underground burrows extend for many feet below the surface and consist of a maze of tunnels of varying sizes. Some of these cavities may be over a foot in diameter, and it is here that the agricultural activities of the ants are conducted.

HOSTS AND RANGE. Most species of plants in Texas and Louisiana, except those plants with milky sap.

INSECT APPEARANCE. *Reproductive caste:* red-brown ants with heads narrower than either thorax or abdomen; ¾ to ⅞ inch long; unmated individuals with 2 pairs of membranous wings; these removed shortly after mating. *Worker caste:* size varies greatly from 1/16 to ½ inch long; body reddish brown; wingless; head bilobed, very large, about 2 times wider than thorax; a pair of prominent spines on back of each of the thoracic segments, 1 pair spines on the top of the head, and a smaller pair on the front of the head; 2 rounded nodes (small segments) with erect scales located between the thorax and the abdomen; all workers are sterile females. *Larvae:* present only within the nests; legless, cylindri-

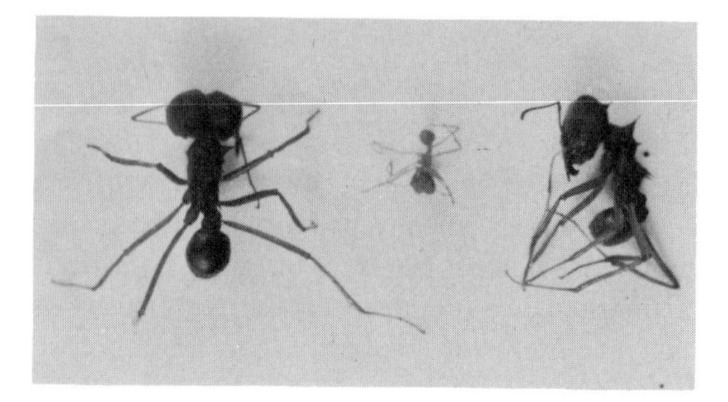

FIG. 9.32. Leaf cutting or town ants (*Atta texana*) (X2). Note the enlarged heads and the variation in size of these adult workers.

cal grubs with globular heads and weakly developed mouthparts. **Pupae:** also present only within the nests; inside papery cocoons.

LIFE CYCLE. This leaf-cutting ant, like all ants, is a social insect that lives in a colony often consisting of hundreds of thousands of individuals. The nests in which these colonies live have been described previously. The two castes in each colony are the reproductives and the workers. The winged reproductives leave the parent colony during the spring or early summer for the purpose of mating and starting new colonies. Mating occurs away from the nests. Each female mates only once, and from this mating she receives sufficient sperm to last throughout her life even though she may produce thousands of offsprings. The males die shortly after mating. Each female immediately finds or makes a small cavity in the soil, seals herself in, and sheds her wings. She then proceeds to lay a number of eggs and care for the eggs and the resulting larvae and pupae until they transform to adult workers. These workers then take over all the work of the colony. Thereafter the female is relegated to function only as an egg producer. Most of the offspring are workers, but, if a colony prospers for a few years, some reproductive forms are also produced. Although young can be found in the colonies throughout the year, they are more abundant during the warmer months. The length of time it takes for an individual to develop is unknown, but probably it is from 1 to 2 months.

These ants are also known as agricultural ants because they use the macerated leaf masses as a substrate on which they grow a pure fungus culture. Apparently the fungus, and not the leaves, forms the food for both the adults and larvae. It is remarkable that these ants can grow a pure fungus culture in this way, for whenever one of these gardens is separated from the ants it immediately becomes contaminated with many other fungi.

CONTROL. Before areas infested with this defoliating insect can be regenerated to pine, the ant colonies must be destroyed. For this purpose it has been found that the fumigant, methyl bromide, is the most effective poison. The gas is introduced into the colonies in several places 2 or more feet below the surface by means of special applicators. Treatments should be made on cool wet or cloudy days during late fall or early winter. Colonies of average size (cover about 600 square feet of surface soil) usually require about 1 pound of the fumigant. When these are located in porous soils, however, more toxicant may be needed. Two weeks after treatment any colonies still active should be retreated.

Bennett, W. H. 1955. *Texas Forest Serv. Circ.* 44.
Holt, W. R. 1957. *U.S.D.A. Forest Serv. Southern Forest Pest Reporter* 19.
Johnston, H. R. 1944. *J. Forestry* 42:130–132.
Walter, E. V., L. Seaton, and A. A. Mathewson. 1938. *U.S.D.A. Circ.* 494.

BEETLE DEFOLIATORS

Coleoptera

Adult leaf-eating beetles can be readily recognized by the presence of wing covers (elytra) that usually cover the abdomen rather completely. The larvae are fleshy and grub-like. Some species have three pairs of jointed thoracic legs whereas others are legless. Prolegs (fleshy abdominal legs) are absent, but sometimes projections on the last abdominal segment may resemble prolegs. The antennae are small and inconspicuous. Characteristics of the various families of beetles that contain leaf chewers are presented in Table 9.17.

TABLE 9.17 THE MAIN GROUPS OF LEAF-FEEDING BEETLES AND THEIR LARVAE

1. Wing covers (elytra) present; antennae conspicuous (beetles) 2
1. Elytra absent; antennae inconspicuous (larvae) 5
2. Beetles with front of head prolonged into a beak (page 195) . . . . WEEVILS
2. Front of head not prolonged into a beak 3
3. Antennae clubbed with the enlarged portion flattened and projected on only one side of the antennal axis (lamellate) (page 196) . . . . CHAFER BEETLES
3. Beetles or larvae without clubbed antennae as in 3 above 4
4. Elongate beetles with soft elytra; posterior tarsi with 4 segments; others with 5 segments; tarsal claws toothed or cleft; head constricted behind to form a neck (page 195) . . . . BLISTER BEETLES
4. Hemispherical to somewhat elongate beetles; all tarsi with 5 segments; 3rd segment bilobed with the 4th segment often small and inconspicuously located in the notch formed by 3rd segment (page 192) . . . . LEAF BEETLES
5. With 3 pairs thoracic legs (page 192) . . . . LEAF BEETLE LARVAE
5. Legless (page 195) . . . . WEEVIL LARVAE

LEAF BEETLES

Coleoptera, Chrysomelidae

Leaf beetles are the most important of the beetle defoliators, but even these are only of minor importance as tree pests. Both the adults and the larvae are foliage chewers and frequently pupation also occurs on the leaves and needles.

INJURY. The larvae are commonly leaf skeletonizers, whereas the adults consume much of the leaf tissues. They often feed by cutting holes in the leaves. The larvae of a few species are leaf miners.

INSECT APPEARANCE. *Beetles:* body hemispherical to somewhat elongate; elyta covers abdomen; antennae thread-like; tarsi 5 segmented with the 3rd segment bilobed and the 4th segment inserted in the notch of the lobe; 4th segment often so small it can not be seen readily; 5th segment elongate. *Larvae:* mouthparts well developed; 3 pairs of well-developed thoracic legs; fleshy abdominal legs (prolegs) generally absent; sometimes a pair of fleshy protuberances at the tip of abdomen resembles a pair of prolegs.

LIFE CYCLE. The adults usually hibernate over the winter in protected places such as the ground litter, in buildings, or other protected places. They emerge in the spring and the females lay eggs to start the next generation. There are commonly only 3 larval instars. During the warm season the life cycle is completed in 5 to 7 weeks; therefore, usually there are 2 to 3 generations per year.

CONTROL. Insecticidal control seldom, if ever, has been needed to control leaf beetle infestations in forests. For protecting ornamentals or nursery trees DDT sprays or mists applied at a dosage of ½ pound of actual insecticide per 100 gallons of spray will provide good control.

MacAloney, H. J. 1950. *U.S.D.A. Misc. Publ.* 657:271–279.
Keen, F. P. 1952. *U.S.D.A. Misc. Publ.* 273:122–125.

COMMON SPECIES OF LEAF BEETLES

Coleoptera, Chrysomelidae

For the species now to be discussed, except as noted otherwise, the type of leaf injury, the life cycle, and control are typical for leaf beetles as outlined previously. Insecticides have never been needed to control these insects in forests.

1. ELM LEAF BEETLE [*Galerucella xanthomelaena* (Schr.)]. **HOSTS.** Species of elms. **RANGE.** Introduced from Europe and now occur throughout United States and Canada. *Larvae:* about ½ inch long when full grown; head, legs, and tubercles black; body dull yellow with 2 stripes along back. *Pupae:* about ¼ inch long; bright orange-yellow with a few black bristles; formed in bark crevices or at base of trees. *Beetles:* about ¼ inch long; yellowish to olive green; black stripe

FIG. 9.33. Elm leaf beetle (*Galerucella xanthomelaena*) (X3). Larvae left, pupae center, and adults right.

sometimes present along outer margin of each wing cover; eyes black; antennae and legs yellowish. *Eggs:* yellow to orange, spindle shaped; attached endwise in groups of 5 to 25 to underside of leaves. Each female may lay 400–800 eggs.

Forest Insect Division. 1952. *U.S.D.A. Leaflet* 184.

2. LOCUST LEAF MINER (*Chalepus dorsalis* Thunb.). **INJURY.** The larval stage produces blister-like mines. **HOST.** Black locust. **RANGE.** Throughout eastern half of North America. *Beetles:* about ¼ inch long; body black with outer margin of elytra and thorax reddish orange; thorax and elytra densely punctured; elytra ridged. *Larvae:* white to yellowish; always inside leaves. *Pupae:* within old leaf mines. The eggs are laid in groups of 3 to 6 on the underside of the leaves. **CONTROL.** Insecticidal control has never been used to control this insect in forests. For protecting more valuable trees DDT should be effective when applied at the usual dosages in the spring after most of the beetles have emerged from hibernation. Killing the leaf-mining larvae would be more difficult and would probably require a fumigating type of insecticide such as benzene hexachloride.

3. BASSWOOD LEAF MINER [*Baliosus ruber* (Web.)]. **INJURY.** The adults skeletonize the upper leaf surface, whereas the larvae are leaf miners. **HOSTS.** Basswood is preferred but oaks and apple may also be infested. **RANGE.** Throughout eastern North America. *Beetles:* about ¼ inch long; broad reddish, and wedge-shaped with indistinct darker markings on sides and posterior half of elytra. *Larvae:* typical; light colored. *Pupae:* within the leaf mines. **LIFE CYCLE.** This species has only 1 generation per year.

Hodson, A. C. 1942. *J. Econ. Entomol.* 35:570–573.

4. PINE COLASPIS (*Colaspis pini* Barber). **INJURY.** The beetles eat only the edges of the year-old needles on smaller trees. The adjacent green needle parts then turn brown causing the infested trees to appear fire scorched. Seldom are the trees permanently injured. Feeding is heaviest during May and June. **HOSTS.** Pines. **RANGE.** Throughout the Gulf States. ***Beetles:*** about 3/16 inch long; rusty yellow or brown with faint greenish iridescence. **LIFE CYCLE.** Differs from the usual type in that the eggs are laid in the soil, and the larvae feed on the roots of various grasses and herbaceous plants. Pupation occurs in the spring and shortly thereafter the needle-feeding adults appear.

Kowal, R. J. 1951. *Forest Farmer* 10 (5).

5. COTTONWOOD LEAF-MINING BEETLE (*Zeugophora scutellaris* Suffr.). **INJURY.** The larvae are solitary leaf-miners of poplars, whereas the adults skeletonize the lower surfaces of the young leaves. **HOSTS.** Poplars and willows. **RANGE.** Most of United States. ***Beetles:*** about 1/8 inch long; elytra and abdomen black; remainder of body yellow. **LIFE CYCLE.** Typical except that the eggs are laid singly in punctures in the lower leaf surfaces, and pupation occurs in the soil.

6. IMPORTED WILLOW LEAF BEETLE [*Plagiodera versicolora* (Laich.)]. **HOSTS.** Willows and poplars. **RANGE.** Northeastern United States. ***Beetles:*** about 1/8 inch long; metallic blue or green sometimes tinged with red or bronze. ***Eggs:***

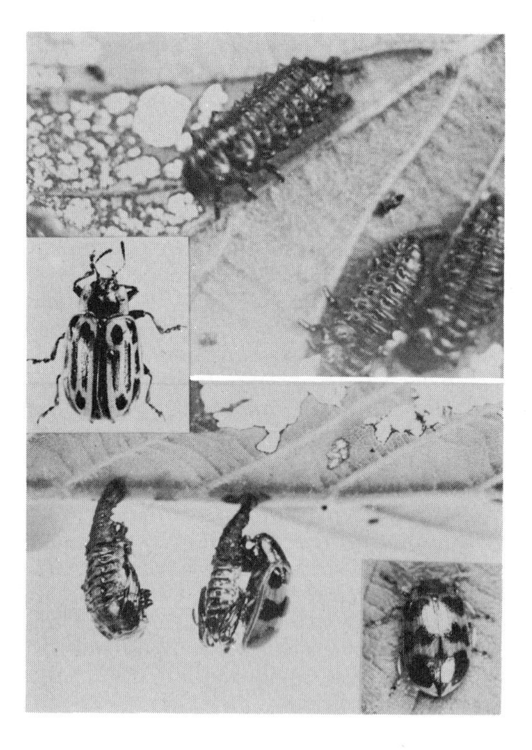

FIG. 9.34. Cottonwood and willow leaf beetles, larvae, and pupae. *Chrysomela scripta* above and *C. interrupta* below. (X3)

yellow, spindle shaped; laid in irregular masses of up to 30 in number; on leaves. *Larvae:* 3/16 inch long when full grown; head and legs black; body cream colored with black and brown markings; 2 rows of dark brown tubercles on each side of body; a yellowish sucker-like disc at the posterior end. *Pupae:* attached to leaves; yellowish turning almost black.

Dowden, P. B. 1939. *Mass. Forest and Park Assoc. Tree Pest Leaflet* 36.
Hood, C. E. 1940. *U.S.D.A. Circ.* 572.

7. COTTONWOOD AND WILLOW LEAF BEETLES (*Chrysomela* spp.). HOSTS. Poplars and willows. RANGE. Throughout United States and Canada, except *C. lapponica* L., which occurs only in the Pacific Northwest. (*Chrysomela scripta* F.)—*Beetles:* about 1/4 inch long; head and thorax black, with thorax margins yellow or red; elytra gold to black but usually yellowish with black interrupted stripes. (*C. interrupta* F.)—*Beetles:* like *C. scripta* above, except elytra deep yellow or red variously spotted with black spots. [*C. tremula* (F.)]—*Beetles:* about 1/4 inch long; head and thorax green with rest reddish brown with no spots on elytra. (*C. lapponica*)—*Beetles:* about 1/4 inch long, reddish spotted with black. *Larvae:* all three species are colored black to dirty yellow.

8. ALDER FLEA BEETLES [*Altica ambiens* (Lec.)]. HOSTS. *Alnus* spp. as well as poplar and willow. RANGE. Throughout the United States and southern Canada. *Beetles:* about 1/4 inch long, dark shiny blue to greenish blue; thorax broader than long; with elytra wider than thorax. *Larvae:* dark brown to almost black and dark yellow beneath.

BLISTER BEETLES

Coleoptera, Meloidae

INJURY. Blister beetles rarely cause tree damage, and if they do, it is to ornamentals or only to trees growing in nurseries. HOSTS. General foliage feeders of herbaceous species, but broad-leaved trees are also attacked. RANGE. These insects are widely distributed, but tree damage has occurred chiefly in the Midwest. *Beetles:* gray to black; elongate; elytra soft; back of head constricted to form a neck; posterior tarsi each with 4 segments, other tarsi have 5 segments each. *Larvae:* seldom seen for they are soil inhabiting predators on grasshopper eggs. LIFE CYCLE. Blister beetles winter as full grown larvae in the soil. Pupation occurs in the spring and the adults soon emerge. The beetles feed ravenously to cause the plant damage. After mating the females lay the eggs in the soil. Usually there is only 1 generation per year. CONTROL. DDT used at the usual dosages is effective. See page 177.

WEEVIL DEFOLIATORS

Coleoptera, Curculionidae

INJURY. Only a few species of weevils feed on the foliage of trees, and none of these is of much importance. The larvae of some are leaf miners, others are leaf rollers and some feed by cutting holes in the foliage. HOSTS. Both broad-leaved trees and conifers. RANGE. Wherever the hosts grow. *Weevils:* small beetles with front of head prolonged into beak; antennae attached to the beak, elbowed with basal segment very long. *Larvae:* body soft, cylindrical to flattened (miners), whitish; head globular with mouthparts directed downward toward neck; legless. *Pupae:* resemble adults but are whitish and lack fully developed elytra. LIFE CYCLE. See page 192. CONTROL. Insecticidal methods have never been needed to hold these insects in check even on shade trees; however, insecticides such as dieldrin or heptachlor should be effective if control ever is needed.

CHAFER BEETLES

***Phyllophaga* spp., *Macrodactylus* spp., and *Pachystethus* spp., Coleoptera, Scarabaeidae**

INJURY. Only the adult beetles are foliage feeders, but the larvae sometimes may be important root destroyers. Generally they cause only local defoliation. HOSTS. Both conifers and deciduous broad-leaved trees. RANGE. Wherever the hosts grow. *Beetles:* clubbed (lamellate) antennae; club flattened, and consists of 3 or more plates that extend out from only one side of the antennal axis. *Phyllophaga* spp.—large; robust; ⅝ to ⅞ inch long; reddish to dark brown, nocturnal feeders commonly known as May or June beetles. *Macrodactylus* spp.—long legged, tan, chafer beetles about ⅜ inch long. *Pachystethus*—about ¼ inch long; light tan with greenish bronze head and prothorax and dark tan wing covers. LIFE CYCLE. *Phyllophaga* spp. require 2 to 5 years to complete 1 generation. See page 397. Member of the other genera have 1 generation per year. The larvae overwinter in the soil, and pupation occurs in the spring; therefore, the adult beetles appear during the spring and early summer. The females lay their eggs in the soil where the larvae live and feed on the roots of various plants. CONTROL. Applied control never has been needed in forests but it has been necessary in forest nurseries and on ornamentals. DDT sprays or mists at the usual dosages are satisfactory foliage applications for killing the beetles.

GRASSHOPPERS AND WALKING STICKS

Orthoptera

Of the many species in this order only the walking sticks and grasshoppers are of much importance as foliage feeders of trees. Sometimes, however, crickets, particularly the mole crickets, destroys many tree seedlings.

WALKING STICKS

***Diapheromera femorata* (Say), Orthoptera, Phasmatidae**

INJURY. Valuable stands of timber are seldom defoliated by these insects for usually only the scrub oak forests are heavily infested. These weed trees occupy space and interfere with the growth of the more valuable conifers; therefore, any defoliation damage done to these scrub-oak stands usually is considered beneficial. Usually however, the tree damage is not enough to do much good. When walking sticks occur in

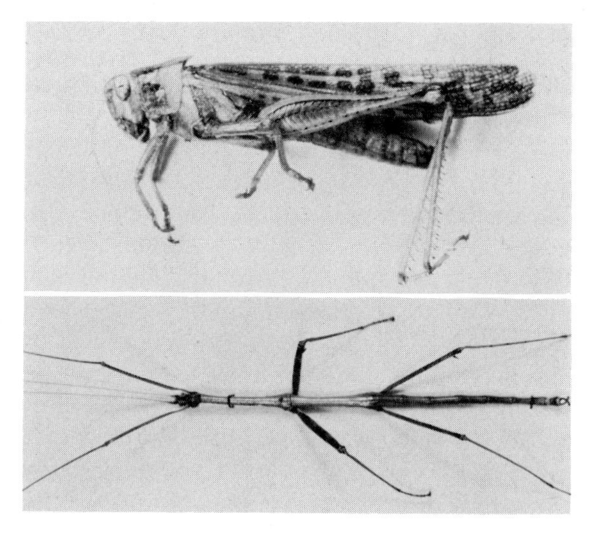

FIG. 9.35. Grasshopper (*Schistocerca* spp.) (X1) and walking stick (*Diapheromera femorata*) (X½).

large numbers they become most annoying pests by simply tickling people and other animals as they crawl over them.

HOSTS. The foliage of black oaks and wild cherry is preferred, but they also feed on many other species of hardwoods.

RANGE. Everywhere east of the Rocky Mountains, but heavy infestations have occurred chiefly in the North on scrub oaks.

INSECT APPEARANCE. *Nymphs:* body green when young becoming dark green to brown when older; elongate slender stick-like insects that resemble twigs; mesothorax and metathorax about 5 times longer than the short prothorax; legs also slender and stick-like, all of about equal size. *Adults:* similar to nymphs but larger; wingless.

LIFE CYCLE. The females drop their eggs indiscriminantly on the ground during the late summer and fall, where they become dispersed among the ground litter. Here the eggs remain throughout the winter, the following summer, and the second winter. They then hatch the second spring. The nymphs develop during the summer and mature in August. Consequently, only 1 generation develops every 2 years.

CONTROL. Generally insecticidal control is not used in forested areas for the purpose of protecting the trees but may be used in resort areas to prevent the annoyance caused to the people by the crawling insects. Aerial spraying with DDT at the usual dosage of 1 pound per acre has been found effective.

GRASSHOPPERS

Melanoplus spp., *Camnula* spp. and others, Orthoptera, Acrididae (Locustidae)

INJURY. Grasses and other herbaceous plants form the usual food of grasshoppers, but after these hosts have been consumed the hoppers frequently eat the leaves and even the tender bark on both deciduous and evergreen trees.

RANGE. Grasshoppers are widely distributed, but tree damage usually occurs only in the Great Plains region and the adjacent forested areas.

INSECT APPEARANCE. These insects are well known by everyone. They have chewing mouthparts, hind legs enlarged and specialized for jumping, antennae thread-like and shorter than head and thorax, and 3 segmented tarsi. They undergo gradual metamorphosis; therefore, the appearance of the nymphs is similar to that of the adults, except that the wings are not fully developed.

LIFE CYCLE. The eggs are deposited in the surface of the soil where they remain over winter. These eggs hatch in the spring, and the nymphs feed and develop until late summer when they mature and become adults. Usually there is only 1 generation per year.

CONTROL. Insecticidal control has never been used to protect forest plantings, but should conditions ever warrant the use of applied methods, a number of insecticides are effective. These include various cyclodienes such as chlordane, 1 pound; aldrin, 2 ounces; dieldrin, 1 ounce; heptachlor, 4 ounces; or toxaphene, 1½ pounds. Any of these is effective if the amounts indicated here are applied uniformly on a per acre basis. Usually emulsion formulations are very satisfactory.

DIPTEROUS LEAF MINERS

Phytomyza spp. and *Agromyza* spp. Diptera, Agromyzidae

INJURY. There are many species of maggots that mine in the foliage of woody plants, but they cause significant damage only to a few ornamental trees. Of these, the holly leaf miner [*P. ilicis* (Curt.)] probably is the most injurious. The leaf damage consists of serpentine or blotch-like mines in the leaves. HOSTS. Various trees are attacked by the different fly species. Elm, catalpa, holly, *Prunus,* and oak are common hosts. RANGE. Wherever the hosts grow. INSECT APPEARANCE. *Maggots:* small (less than 3⁄16 inch long), whitish, cylindrical, and soft bodied; legless with indistinct head at pointed end of body; mouthparts are 1 or 2 parallel toothed hooks. *Flies:* small, dark or yellowish colored. LIFE CYCLE. The insects hibernate over the winter inside puparia within the leaf mines. The flies emerge in the spring, and, after mating, the females lay eggs on the foliage. There may be 1 or 2 generations per year. CONTROL. For killing the miners on ornamentals two applications of a spray containing benzene hexachloride (1 per cent gamma isomer) applied about mid-June and early July is effective. A DDT spray applied just before the flies emerge is also effective. It should be used at the usual dosage of about 1 pound actual DDT per 100 gallons of spray. One or two subsequent applications at two-week intervals insure better control. Dieldrin (⅓ pound per 100 gallons of spray) is also most effective for killing the flies.

Needham, J. G., S. W. Frost, and B. H. Tothill. 1928. *Leaf mining insects.* Williams and Wilkins Co., Baltimore, Md. 351 pp.

CHAPTER 10

Inner-Bark Boring Insects

The succulent inner bark is one of the more nutritious tissues found in trees; therefore, it is not surprising that many insects infest this region. In order to reach and feed on these tissues, the insects must be borers. Their activities destroy the carbohydrate-rich tissues, and when the galleries encircle the stems they girdle the trees. This girdling blocks the downward translocation of foods and causes the roots to die. Subsequently the crowns deteriorate and die because dead roots can no longer absorb and supply water to the other tree parts. Inner-bark borers such as bark beetles also can cause the rapid death of trees because they make numerous holes through the bark and because they introduce blue-staining fungi. The holes allow gaseous air to reach the water-conducting wood cells and there the gaseous air breaks the "threads" of water extending from the roots to the leaves. When this disruption is sufficiently extensive, the water supply to the tree's crown is effectively blocked. The possible actions of the blue-staining fungi will be discussed later.

Inner-bark boring damage is caused chiefly by beetles, but there are also a few caterpillars and a few maggots that have this habit. The various types of inner-bark borers can be determined by referring to Table 10.1. Many of the species that are very injurious as wood borers also bore in the inner bark for a period of time before they enter the wood. The reader should seek additional information on this topic in Chapter 11. The common families of beetle larvae found beneath the bark of trees and logs are included in Table 10.2.

TABLE 10.1 THE MAIN GROUPS OF INNER BARK BORERS

1. Galleries of 2 types; egg galleries bored by the adult beetles are of rather uniform width; the more numerous individual larval galleries arise from and increase in size as they extend away from the egg galleries; larvae of a few species live together and bore a single common gallery; beetles cylindrical; antennae with head-like club (capitate); larvae cylindrical, legless, somewhat curved in long axis, heads globular with mouthparts directed downward (page 203) . . . . BARK BEETLES
1. Only larval galleries present; these always increase in width as the larvae grow . . 2
2. Galleries narrow, never wider than 1/16 inch; narrow streaks formed in the wood are more often seen than the tunnels in the inner bark; larvae, whitish, legless, and with indistinct pointed heads; mouthparts (parallel hooks) at tip or inside but appear through the translucent body wall (page 264) . . . . MAGGOTS
2. Galleries wider than above; larvae always with distinct heads 3
3. Borings granular; globular excrement pellets and silk webbing often present in the galleries; caterpillars with 3 pairs of short jointed thoracic legs plus 5 pairs of fleshy abdominal legs (prolegs) that are armed at ends with groups of minute hooks, or crochets (page 260) . . . . CATERPILLARS
3. Borings granular or fibrous; globular excrement pellets and silk webbing absent; larvae never have prolegs or crochets 4
4. Borings fine and tightly packed in galleries; frequently curved, arc-like lines appear in borings; larvae always legless, whitish, and elongate with enlarged prothorax flattened and widened; always wider than abdominal segments; plate-like areas on back and under side of prothorax distinct and are of about equal size; back plate always with a distinct median longitudinal line, V- or Y-shaped marks or grooves; head flattened with mouthparts directed forward; spiracles cresent shaped (page 244) . . . . FLAT-HEADED BORERS
4. Borings granular and/or fibrous; generally not extremely tightly packed in galleries; larvae various, but if prothorax widened and flattened then larvae have legs and/or the plate-like areas underneath are less well developed than those of the back; spiracles oval or round 5
5. Borings always granular; galleries usually less than 1/4 inch; larvae whitish, cylindrical, and somewhat curved in long axis; head globular with mouthparts directed downward (page 258) . . . . WEEVILS
5. Borings granular and/or fibrous; galleries of older larvae wider than 1/4 inch; larvae whitish and elongate, cylindrical to flattened; with 3 pairs of small thoracic legs, or legless; head flattened with mouthparts directed forward; prothorax with indistinct plate-like areas on back and underneath; those below less well developed than those on back (page 255) . . . . ROUND-HEADED BORERS

TABLE 10.2 LARVAE OF COMMON FAMILIES OF BEETLES (COLEOPTERA) FOUND BENEATH THE BARK OF RECENTLY KILLED TREES OR FRESH LOGS

1. Legs absent 2
1. With 3 pairs of jointed, thoracic legs 5

2. Head rather globular; mouthparts located close to body; body cylindrical, often somewhat curved in the long axis 3
2. Head usually flattened, often somewhat retracted into prothorax; mouthparts always directed forward; body flattened or cylindrical 4
3. Parent beetles also commonly present in galleries; beetles small; body cylindrical; antennae each with head-like club (capitate) . . . . . . (bark beetles) SCOLYTIDAE
3. Adult beetles never present in galleries but occasionally may be found in pupal chambers . . . . (weevils) CURCULIONIDAE
4. Hardened prothoracic area on back (pronotal plate) and underneath (prosternal plate) well developed and of about equal size; body always somewhat flattened; prothorax enlarged, flattened, and much wider than the other body segments; spiracles cresent-shaped . . . . (flat-headed borers) BUPRESTIDAE
4. Pronotal plate larger than prosternal plate; body cylindrical to rather flat; spiracles oval-shaped . . . . (round-headed borers) CERAMBYCIDAE
5. Head and body very hairy and usually dark colored; head shape rather globular; scavengers . . . . (skin beetles) DERMESTIDAE
5. Head and body not densely covered with setae 6
6. Each spiracle located at tip of a short projection . . . . (sap beetles) NITIDULIDAE
6. Spiracles not projected 7
7. Legs well developed; always longer than half the thickness of body 10
7. Legs poorly developed; always less than half the thickness of body 8
8. Head poorly developed; soft and light colored; small recurved hooks (cerci) present on back of last abdominal segment; parasitic . . . . CATOGENIDAE
8. Head hard and colored darker than body; cerci absent if head light colored . . 9
9. Posterior mouthparts on under side of head attached considerably behind place where mandibles are attached (retracted); head rather globular with mouthparts directed forward . . . . MELANDRYIDAE
9. Mouthparts not retracted; head flattened with mouthparts directed forward . . . . (round-headed wood borers) CERAMBYCIDAE
10. Mandibles sickle-shaped, several times longer than width at base; legs with 5 segments plus one or 2 claws; paired projections (cerci) occur on the posterior abdominal segment; predaceous . . . . (ground beetles) CARABIDAE
10. Mandibles not sickle-shaped; legs with 4 or fewer segments plus a claw; with or without cerci 11
11. A pair of jointed projections (cerci) on the posterior of abdomen 12
11. Cerci absent or, if present, never jointed 14
12. Each spiracle has 2 slit-like openings; predators . . . . HISTERIDAE
12. Each spiracle has only one opening 13
13. Each mandible with a grinding surface (mola) . . . . . . (carrion beetles) SILPHIDAE
13. Mandibles without molar surfaces . . . . (rove beetles) STAPHYLINIDAE
14. Body wall tough, hard, and colored tan to brown; spiracles each with 2 slit-like openings; sutures on head form lyre-shape pattern; upper lip (labrum) absent . . . . (wire worm) ELATERIDAE
14. Characteristics not as for wire worms 15
15. Posterior abdominal segment colored brown and nearly semicircular in shape; without paired projecting structures (cerci) . . . . . (*Corticous* spp.) TENEBRIONIDAE

15. Cerci present 16
16. Body very flattened (depressed); 3 or 4 times wider than thick; paired rigid projections (cerci) present on back of posterior abdominal segment . . . . (flat bark borers) CUCUJIDAE
16. Body not extremely depressed; cylindrical to somewhat flattened 17
17. Backs of 2 thoracic and 5 or 6 abdominal segments roughened with small points; cerci form recurved spines . . . . SYNCHROIDAE
17. Backs of thoracic and abdominal segments not roughened 18
18. Posterior edge of mouth parts on underside of head attached considerably behind where mandibles are attached (retracted) 19
18. Mouthparts not retracted . . . . (checkered beetles) CLERIDAE
19. Sac-like depression between pair of recurved spines (cerci) on back of last abdominal segment . . . . (cylindrical bark beetles) COLYDIIDAE
19. Cerci present, but no sac-like depression between the two . . . . (ostomids) OSTOMIDAE

BARK BEETLES

Coleoptera, Scolytidae

It has been estimated that in the United States 90 per cent of the insect-caused tree mortality and more than 60 per cent of the total loss of wood growth is due to bark beetles. Ninety per cent of this loss has occurred in the West, where overmature timber is still abundant. It may be that after most of this old timber has been removed, the younger stands will not suffer as severely; nevertheless, bark beetles will always be very injurious forest pests.

INJURY. The crowns of trees heavily infested with bark beetles turn from the characteristic green to yellow and then brown. Depending on temperature and moisture, this color change occurs after a period of about one month to as much as a year following attack. Close inspection of infested tree boles will show that they have one or more of the following characteristics; small irregular globules of resin (pitch tubes); piles of reddish, granular boring dust in the bark crevices and/or small (1/32 to 3/32 inch diameter) circular holes through the bark. Bark removed from infested trees will have two types of galleries on the inner surface. These are egg galleries and larval galleries. The egg galleries are bored by the adult beetles; therefore, they are rather uniform in width, whereas the larval galleries increase in size as the grubs grow. The small holes through the bark are made by the beetles when entering or leaving or for ventilating the egg galleries. The gallery characteristics of the more important genera of bark beetles are presented in Table 10.3.

FIG. 10.1. A stand of pine killed by the southern pine beetle (*Dendroctonus frontalis*).

FIG. 10.2. A pitch tube on bark. These commonly form on living resinous trees when attacked by bark beetles. Also note the small exit holes in the bark made by the emerging beetles (X1).

TABLE 10.3 GALLERY AND BEETLE CHARACTERISTICS OF THE MORE IMPORTANT GENERA OF BARK BEETLES THAT INFEST THE MAIN BOLES, ROOTS, AND LARGER BRANCHES

1. Infest conifers 2
1. Infest hardwoods (angiosperms) 14
2. Infest cedars, cypress, junipers, and redwoods; beetles with heads visible from above, antennal club longer than wide; part of antenna between basal segment and club (funicle) with 5 segments (page 242) . . . . . *Phloeosinus*
2. Infest pines, spruces, firs, larch, hemlock, and Douglas-fir 3
3. Small egg galleries that originate from egg galleries made by larger beetles; less than 1/16 inch wide; beetles with heads visible from above; 3rd segment of fore tarsi neither notched nor bilobed; antennal funicle with 2 or 3 segments . . . . . *Dolurgus* and *Crypturgus*
3. Egg galleries that originate from separate entrance holes 4
4. Only 1 or 2 galleries originate from each entrance hole; the tunnels may criss-cross, but enlarged chambers are not formed at the crossing points; sometimes they may fork near ends; beetles with heads visible from above; front of pronotum not roughened and posterior of elytra not concave 5
4. Usually 3 or more egg galleries start from a central nuptial chamber 9
5. Only one egg gallery originates from each entrance; galleries longer than 6 inches or, if shorter, they are very broad, 1/4 to 1 inch wide; infest Pinaceae species except the true firs; antennal club is as wide or wider than long, funicle 5 segmented; with pair of small projections above mouth (page 214) . . . . . *Dendroctonus*
5. Single egg galleries usually shorter than 6 inches; less than 1/8 inch wide; 1 or 2 galleries may originate from an entrance (nuptial) chamber; beetles with 7 segmented antennal funicle 6
6. Usually attack the bases and roots of dying trees or bottom side of logs in contact with soil; bases of elytra not regularly toothed; hosts are chiefly pines, but other Pinaceae species sometimes infested 7
6. Usually attack above the bases of the trees; hosts are chiefly true firs and Douglas-firs but other Pinaceae species sometimes infested 8
7. Eggs laid in individual niches; beetles with 3rd tarsal segments little widened and shallowly notched . . . . . *Hylastes*
7. Eggs laid together in grooves at sides of galleries; larvae feed together gregariously; beetles with 3rd tarsal segments much widened and bilobed . . . . . *Hylurgops*
8. Beetles with abdominal sterna ascending abruptly behind thorax or may even be concave; front tibia with a prominent spine on outer edge of tip (page 235) . . . . . *Scolytus*
8. Beetles with abdominal sterna not ascending but almost horizontal; head visible from above; body covered with scales; base of elytra with fine teeth (page 240) . . . . . *Pseudohylesinus*
9. Egg galleries wider than 1/16 inch 10
9. Egg galleries less than 1/16 inch wide 13

10. Beetles each with 2 pairs of eyes; head visible from above; hosts chiefly spruces (page 241) . . . . . *Polygraphus*
10. Eyes not divided so as to form 2 pairs; head hidden beneath the prothorax, not visible from above 11
11. Posterior of elytra convex or flattened; never concave; without teeth but sometimes granulated; hosts spruces and firs (page 241) . . . . . *Dryocoetes*
11. Posterior of elytra concave and/or with small teeth-like spines around edges of concavity 12
12. Posterior elytral concavity deep; posterior edge of elytra forms a small shelf-like ridge; 4 to 6 teeth on each elytron bordering the concavity; hosts chiefly pines and spruces (page 229) . . . . . *Ips*
12. Posterior elytral concavity shallow or absent; posterior edge of elytra with an acute ridge but not shelf-like; preferred hosts are firs and Douglas-fir (page 241) . . . . . *Pityokteines*
13. Egg galleries short; 2 to 6 eggs placed in each niche; hosts are pines, larch, spruces, and true firs . . . . . *Orthotomicus*
13. Egg galleries long with only one egg placed in each egg niche; hosts are chiefly pines; female beetles with front of head deeply concave . . . . . *Pityogenes*
14. Egg galleries extend across or obliquely around tree boles 15
14. Egg galleries longitudinal along tree boles or several radiate from a common nuptial chamber 18
15. Infest only elms (page 243) . . . . . *Hylurgopinus*
15. Infest other hardwood species 16
16. Infest only ash (page 244) . . . . . *Leperisinus*
16. Infest other hardwood species 17
17. One arm of egg gallery longer than the other . . . . . *Phthorophloeus*
17. Arms of egg galleries of nearby equal length (page 243) . . . . . *Pseudopityophthorus*
18. Infest only alder (page 244) . . . . . *Alniphagus*
18. Infest other hardwood species 19
19. Egg galleries longer than 6 inches . . . . . *Dryocoetes*
19. Egg galleries shorter than 6 inches (page 235) . . . . . *Scolytus*

In addition to girdling trees and allowing air to reach the water-conducting vessels, bark beetles also act as vectors of blue-staining fungi (*Ceratostomella* spp.). These fungi have been implicated in causing the characteristic rapid crown dying following bark beetle attack. Blue-staining fungi are not virulent pathogens, but when they are introduced over much of a tree bole they apparently play some role in causing rapid foliage death. Suggested theories regarding their pathogenicity include the following: toxin production, mycelium plugging the tracheids, releasing gas bubbles in the tracheids and/or by producing particles that block the pit openings by causing tori aspiration. The first two theories do not appear sound. When only one side of a tree bole is infested, the

FIG. 10.3. Blue stain invading wood after having been introduced by bark beetles.

foliage is not partially killed. In addition, microscopic examinations of the "blued" xylem shows that few hyphae occur in the vertical tracheids. Most occur in the rays. Therefore, at the present time it appears that the last two theories listed are most tenable.

INSECT APPEARANCE. *Beetles:* small, 1/32 to 3/8 inch long; hard bodied; cylindrical; color tan (callow) to almost black; each antenna has an elongate basal segment (scape) connected to a series of short segments (funicle); the other end of the funicle is attached to a flattened head-like (capitate) club; underside of head with only one (gular) suture extending from lower lip (labium) to posterior edge; wing covers (elytra) cover the back of abdomen. *Larvae:* segmented, soft bodied, whitish, elongate, cylindrical, legless grubs; long axis of body somewhat curved; head color tan to brown, rather globular, with mouthparts directed downward; thoracic segments enlarged. *Pupae:* somewhat similar to adults except they are soft bodied, whitish, and lack fully developed wings and wing covers (elytra); appendages not tightly appressed to body; pupae rather immobile, but they can wiggle abdomens. *Eggs:* fragile, globular to ovoid in shape, white to translucent; can be recognized by location where found in niches or grooves along sides of egg galleries.

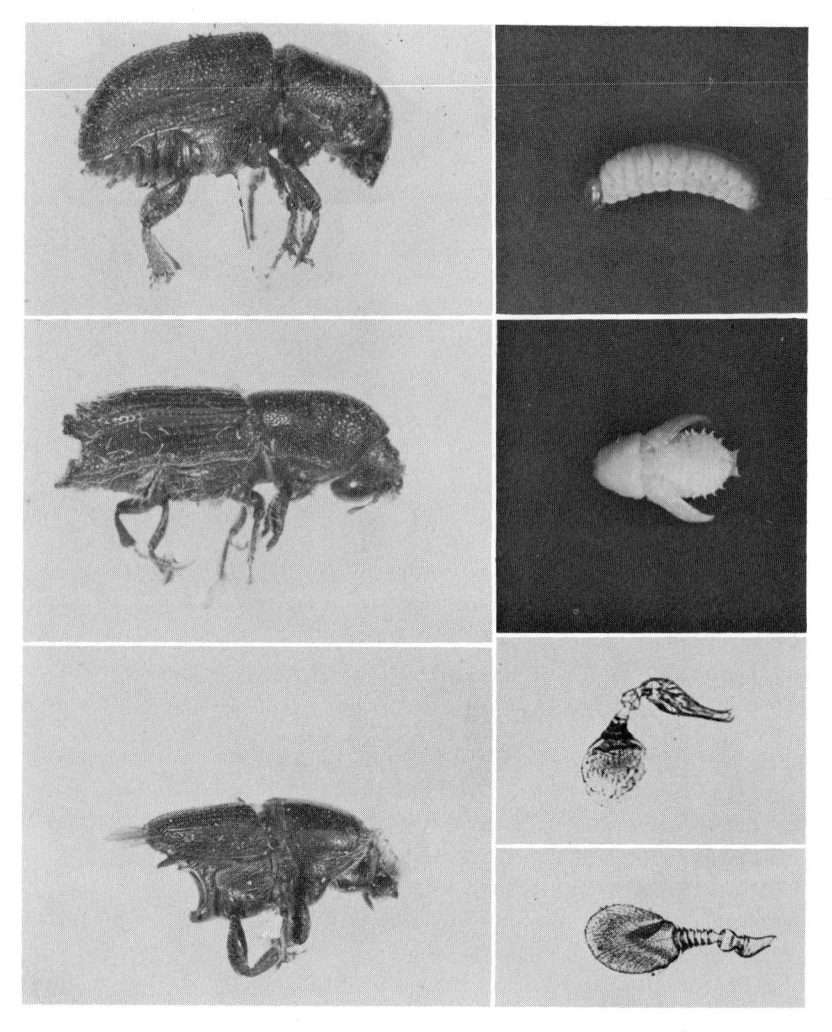

FIG. 10.4. Three most important genera of bark beetles. Top left: *Dendroctonus* (X7); middle left: *Ips* (X9); lower left: *Scolytus* (X7); top right: larva (X4); middle right: pupa (X4); lower right: typical scolytid beetle antennae.

LIFE CYCLE. Bark beetles commonly emerge and attack new hosts in the evenings, on cloudy days, or on sunny days during cool weather. Sometimes they actually appear in swarms. The attacking beetles bore their egg galleries in the inner bark where the females deposit eggs along the sides in small, individual niches or in grooves. Larval galleries start as minute tunnels from the sides of egg galleries and increase in width as they increase in length. Larvae of a few species feed together in a common burrow. At the ends of these larval galleries small pupal chambers are formed where the larvae transform to adults. Before emerging, the young, yellowish to tan (callow) beetles remain for a time in the inner bark where they bore and destroy much of the larval gallery pattern. Species occurring in the South may have 5 to 6 generations per year, whereas those occurring in the far North or at high elevations may have only 1, or even less. The developmental sequence for a species which complete its life cycle in about 2 months is about as follows: The first eggs are laid about 7 days after the attack starts. These hatch a week later and the resulting larvae mature after 3 more weeks. Pupation requires about 10 days and the callow young adults feed in the host tree for another week before emerging.

TREE SELECTION. The mechanism of tree selection by bark beetles is imperfectly understood. It appears, however, that pioneer beetles determine host suitability by boring trial entrances. As soon as a beetle finds and attacks a suitable host tree, it immediately attracts (probably by producing an odor) other individuals. These also attack and produce more attractants; consequently, the intensity of attraction increases as the attack develops. This attracting mechanism explains why bark beetles, especially the more important *Dendroctonus* and *Ips* species, tend to concentrate their attacks in a few trees rather than lightly infesting a great many. When many beetles are attracted to one of these attraction centers, and there is insufficient room for all in the original

FIG. 10.5. Dorsal view of the three most important bark beetle genera. Left: *Dendroctonus* (X7); middle: *Ips* (X9); right: *Scolytus* (X7). Drawings by Claire P. King.

attracting tree, the excess beetles move to and attack the adjacent trees. The large numbers of beetles attacking these trees make it possible for them to overcome tree resistance quickly. In this way bark beetles attack and kill groups of trees. These "group kills" are very characteristic for many species of bark beetles.

TREE CONDITION. It is well known that vigorous growing trees are more resistant to the attacks of many boring insects than are slow growing trees. Nevertheless, after large bark beetle populations have developed in suitable host material they can successfully attack even the most vigorous trees. The number of offspring produced in vigorous growing hosts, however, usually is small; therefore, such epidemics often decline rapidly. Any condition that weakens trees appears to increase their vulnerability to bark beetle attacks. The more common of these are droughts, flooding, windstorms, and fires. Trees that have been partially uprooted or broken are especially vulnerable. The actual physiological mechanism involved in this relationship of tree vigor and suitability for bark beetles development is not known, but various suggestions can be made. These suggested mechanisms include the quality and quantity of resin and/or sap flow, inner bark water content, and the chemical composition of the inner bark. In resinous trees the quantity of resin produced is an obvious type of protective device. At times the attacking beetles are trapped and killed—"pitched out" by the sticky material. Some species, however, appear to be able to tolerate heavy flows of resin. High inner-bark water content alone may not always be significant because some bark beetles readily attack trees with excessive inner-bark moisture. Moisture or sap moving in the trees, however, may have a more pronounced effect. One factor that frequently is associated with suitable host conditions is the great reduction or temporary cessation of radial growth during the growing seasons.

SURVEY METHODS. Aerial surveys are used extensively for detecting and appraising bark beetle activity. Ground surveys also must be used to obtain certain types of information to supplement that obtained from the air. For example, recently infested trees with green crowns can not be seen from a distance; therefore, in order to obtain good estimates by using aerial counts it is often necessary to determine the ratio of trees with discolored crowns to those infested but still green. Determination of brood success is also essential. Additional information on survey methods is presented in Chapter 8.

NATURAL CONTROLS. Various factors act adversely on bark beetle populations. These include tree vigor as discussed previously, parasites and predators, and competition for food and space. The more important predators are checkered beetles (Cleridae), ostomid beetles (Osto-

FIG. 10.6. Bark beetle predators (X3); Upper left: checkered beetle; lower left: Ostomid beetle; right: larvae. The latter are rather similar appearing for the two families.

midae), ants, and woodpeckers. Woodpeckers sometimes leave rather spectacular evidence (debarked trees) of their activity; therefore, at times, they appear to be rather effective. Competition among various species of inner-bark borers and sometimes among individuals of the same species is difficult to evaluate but, nevertheless, may also be of importance.

APPLIED CONTROL. There are various direct and indirect methods that can be used to reduce the size of populations or prevent large bark beetle populations from developing. When breeding material is susceptible because of natural senescence or unfavorable growing conditions, management methods sometimes are practical. This procedure entails removing the slow growing, vulnerable trees so that the remaining trees will have less competition and, therefore, become more vigorous and more insect resistant.

The best direct control method is to salvage the infested trees before the broods emerge. The fading, dying trees are located and mapped from the air. These are removed immediately by special salvage crews so that the logs either can be milled or stored in water. If the logs are sawed, the insects in the bark of the slabs and edgings must be destroyed. This destruction is usually accomplished by burning the debris. Water storage results in drowning most of the brood except that located on the top side of large logs.

Less aggressive bark beetles readily infest felled trees; therefore, they can be concentrated in trap trees or logs and the resulting broods destroyed. The number of trap trees used must always be large enough so that almost all of the beetles in the area will be absorbed. Sometimes two or more successive cuts may have to be made several weeks apart in order to accommodate the insects that mature and emerge at different times. These trap logs or bolts must then be removed and processed before the beetle broods mature and emerge. In pulpwood stands the cuttings can be made in the form of strips so as to provide easier truck access for trap bolt removal.

Whenever any of the previously described procedures are not feasible, the use of fire or insecticides may be used. Up until about 1940, burning was the most common method. Since then the use of insecticides has largely replaced the burning method, especially for treating thinner-barked trees.

In the West, orthodichlorobenzene was used in large quantities for killing the Engelmann spruce bark beetle but now it has been replaced by another fumigant, ethylene dibromide. One gallon (8 pounds) of the ortho was diluted with 5 gallons of diesel fuel as the carrier. The ethylene dibromide is formulated either as an oil solution or as an emulsion. For the solution, 1½ to 2 pounds of the insecticide is dissolved in 5 gallons of fuel oil, whereas the emulsion concentrate consists of 3 pounds of the fumigant dissolved in 1 gallon of fuel oil plus the emulsifier. Water is added to the latter so that the final quantity of emulsion is also 5 gallons. Benzene hexachloride has been the preferred insecticide for bark beetle control in the South. It is used as an oil solution consisting of ¼ to 1 per cent by weight of the gamma isomer. Any of the above insecticides must be applied heavily at a rate of about 1 gallon per 100 square feet of bark area. This cannot be accomplished by using aerial application; therefore, either ground hydraulic or compressed air sprayers must be used.

Felling and peeling or burning are used less now than formerly. These methods are more expensive and burning frequently also is hazardous; therefore, they are used now for treating nonsalvagable, thick-bark western pines and Douglas-fir or when only a few trees are to be treated under special conditions. Removing the bark is sufficient for those species that develop and mature in the inner bark. This treatment exposes the immature insects so that they are destroyed either by predators or by drying. These immature bark beetles are incapable of entering other trees or logs. The larvae of a few species, however, bore into the outer bark while still young; therefore, these are not destroyed by removing the bark.

Techniques of burning consist of piling and burning the smaller material, whereas the larger trees must be treated individually. For the latter the branches are piled along the bole and then fired. Sometimes fuel oil also is used to insure good ignition and subsequent burning. Even then, however, the bark on the top sides of the logs may not become hot enough to kill the brood. In such cases the bark must be peeled from this area.

Solar radiation has also been used in a limited way to kill beetle broods in thin-barked trees. The method consists of exposing the felled, infested boles to the summer sun. After one side has been sunned for a few days, the logs are partially turned so as to expose another bark face. The method has not proven very practical because the trees must be located in unshaded locations and the boles have to be turned several times.

It must be understood that in order for any type of control to be effective a large proportion of the insect population must be destroyed by the combined action of the applied method and by natural controls. Partial controls may be completely ineffective. See Chapter 5.

GENERAL BARK BEETLE REFERENCES

Chrystal, R. N. 1949. The barkbeetle problem in Europe and North America. *Forestry Abstracts* 11:3–12.

INSECT-HOST RELATIONS

Anderson, R. F. 1948. Host selection by the pine engraver. *J. Entomol.* 41:596–602.

Caird, R. W. 1935. Physiology of pines infested with bark beetles. *Botan. Gaz.* 96: 709–733.

Craighead, F. C. 1928. The interrelation of tree killing bark beetles and bluestain. *J. Forestry* 26:886–887.

Forest Insect Division, Bureau of Entomology. 1927. The relation of insects to slash disposal. *U.S.D.A. Dept. Circ.* 411.

Graham, S. A. 1924. Temperature as a limiting factor in the life of subcortical insects. *J. Econ. Entomol.* 17:377–383.

Leach, J. G., L. W. Orr, and C. Christensen. 1934 and 1937. The interrelationships of bark beetles and blue staining fungi in felled Norway pine timber. *J. Agri. Research* 49:315–342 and 55:129–140.

Nelson, R. M. 1934. Effect of bluestain fungi on southern pines attacked by bark beetles. *Phytopathologische Zeitschrift* 7:327–426.

Nelson, R. M., and J. A. Beal. 1929. Experiments with bluestain fungi in southern pines. *Phytopath.* 19:1101–1106.

TAXONOMY

Beal, J. A., and C. L. Massey. 1945. *Duke Univ. School Forestry Bull.* 10.
Blackman, M. W. 1922. *Miss. Agri. Exp. Sta. Tech. Bull.* 11.
Chamberlin, W. J. 1939. *The bark and timber beetles of North America.* Oregon State Coll. Coop. Asso., Corvallis, Oregon.
Dodge, H. R. 1938. *Univ. Minn. Tech. Bull.* 132.
Hopkins, A. D. 1909. *U.S.D.A. Bur. Entomol. Bull.* 83 (part 1).
———. 1915. *U.S.D.A. Bur. Entomol. Tech. Series* 17 (part 2).
Swaine, J. M. 1918. *Can. Dept. Agri. Entomol. Branch Bull.* 14. (Tech. Bull.).

DENDROCTONUS BARK BEETLES

Dendroctonus spp., Coleoptera, Scolytidae

In the United States, the *Dendroctonus* bark beetles have been the most injurious forest insect pests. Most of the damage has occurred in the West where the losses have been several times more than that caused by fire and even more than that caused by fungi.

HOSTS. Pines, spruces, larches, and Douglas-fir.

GALLERY PATTERN. *Egg galleries:* usually longer than 6 inches with only one gallery originating from each entrance; when shorter than 6 inches they are very wide (¼ inch or more); always packed with borings except for a short section occupied by the adults. Other gallery characteristics vary for the different species as indicated in Table 10.4.

TABLE 10.4 THE MORE IMPORTANT SPECIES OF DENDROCTONUS

1. Infest pines* 2
1. Infest conifers other than pines; egg galleries broad, ¼ inch wide, straight or slightly winding and extend longitudinally with the grain of the wood; eggs always laid together in grooves 14
2. Eggs laid together in grooves along the sides of the galleries; larvae feed together to form common brood chambers in inner bark; egg galleries broad, ¼ inch or wider; usually infest only lower 3 to 6 feet of tree boles 3
2. Larvae construct individual galleries; egg galleries less than 3⁄16 inch wide; infest most of tree boles 5
3. Egg galleries ½ to 1 inch wide; larvae with spines on back of the 8th and 9th abdominal segments; beetles large 3⁄16 to ⅜ inch long (page 223) . . . . TURPENTINE BEETLES
3. Egg galleries about ¼ to ⅜ inch wide; larvae with plate-like areas on back of 8th and 9th abdominal segments; beetles about 3⁄16 to ¼ inch long 4
4. Infest pines in the Lake States region; rare (page 225) . . . . RED-WINGED BEETLE
4. Infest bases and stumps of lodgepole pine in Rocky Mountain region; rare (page 225) . . . . LODGEPOLE PINE BEETLE
5. Egg galleries very winding; frequently forming S-shaped patterns with much crossing of galleries 6

5. Egg galleries rather straight to somewhat winding, extend longitudinally along tree boles 10
6. Occurs in southeastern United States north to southern Pennsylvania and Missouri (page 216) . . . . SOUTHERN PINE BEETLE
6. Occurs west of the Great Plains 7
7. Occurs in the Pacific Coastal States and the Northern Rockies (page 218) . . . . WESTERN PINE BEETLE
7. Occurs in the Southern Rockies north to Utah and the Black Hills 8
8. All egg galleries very winding 9
8. Some egg galleries long and only slightly winding 12
9. Beetles with long hairs on posterior of wing covers; larva with a rounded elevation on front of head (page 221) . . . . ARIZONA PINE BEETLE
9. Beetles without long hairs on posterior of elytra; larva lacks a rounded elevation on front of head (page 221) . . . . SOUTHWESTERN PINE BEETLE
10. Occurs in the southern Rockies north to Colorado, Utah, and the Black Hills . . 12
10. Occurs in the Pacific Coastal States and the northern Rockies south to Wyoming 11
11. Infests and successfully breeds only in Jeffrey pine; beetle pronotum with shallow, fine punctures (page 221) . . . . JEFFREY PINE BEETLE
11. Infest chiefly lodgepole, sugar, and ponderosa pines; beetle pronotum with moderately coarse and deep punctures (page 221) . . . MOUNTAIN PINE BEETLE
12. Larval galleries close to outer bark; and never exposed on inner surface of inner bark; beetles with a depressed groove along front of head (page 221) . . . . COLORADO PINE BEETLE
12. Larval galleries adjacent to inner surface of inner bark; beetles with front of head without a groove along front of head 13
13. Posterior of wing covers (elytra) with long hairs; pupation occurs mostly in outer bark (page 221) . . . . ROUND-HEADED PINE BEETLE
13. Posterior of elytra without long hairs (page 221) . . . . BLACK HILLS BEETLE
14. Egg galleries long, often extend 2 or more feet; hosts are chiefly Douglas-fir and larch 15
14. Egg galleries usually less than 9 inches long; hosts are chiefly spruces 16
15. Douglas-fir is chief host (page 228) . . . . DOUGLAS-FIR BEETLE
15. Eastern larch is the only host (page 228) . . . . EASTERN LARCH BEETLE
16. Occurs in northeastern North America; hosts are spruce (page 227) . . . . EASTERN SPRUCE BEETLE
16. Western species 17
17. Occurs throughout the Rocky Mountain region; host chiefly Engelmann spruce (page 225) . . . . ENGELMANN SPRUCE BEETLE
17. Occurs in the Pacific northwest coastal region or in Alaska and northern Canada 18
18. Infests sikta spruce in the Pacific Northwest (page 227) . . SITKA SPRUCE BEETLE
18. Occurs in Alaska and northwest Canada (page 227) . . ALASKA SPRUCE BEETLE

* Occasionally may infest other Pinaceae hosts especially when these are intermixed with the preferred hosts.

INSECT APPEARANCE. See the general description for bark beetles and Figures 10.4 and 10.5. *Beetles:* body cylindrical; rather stout; 1/16 to 3/8 inch long; tan to almost black; head broad, rounded, and visible from above, with a well-developed pair of small projections above mouth (epistoma); antennae with 5-jointed funicle and a short, compact, segmented club; pronotum long; broad and punctured; elytra posterior (declivity) convex and descends abruptly; elytra with longitudinal lines (striae) small to moderately deep; 3rd tarsal segment bilobed; fore coxae close together. *Larvae:* typical; some species have hardened plates or spines on back of the 8th and 9th abdominal segments; others have a rough or rounded elevation on front of head. *Pupae:* similar to adults except they are whitish and have wing pads instead of functional wings and wing covers (elytra); abdomen with 2 spines on posterior segment; some species also have spines on back and sides of other abdominal segments.

LIFE CYCLE. Winter generally is passed as adults or larvae within the host trees. All species are monogamous. The individuals of each pair remain together as they extend their egg gallery through the inner bark. Otherwise the life cycle is similar to that already described for bark beetles.

Hepting, G. H., and G. M. Jemison. 1958. *U.S.D.A. Forest Resource Rept.* 14:209–214.
Hopkins, A. D. 1909. *U.S.D.A. Bur. Entomol. Bull.* 83 (part 1).
Wygant, N. D. 1959. *J. Forestry* 57:274–277.

SOUTHERN PINE BEETLE

Dendroctonus frontalis Zimm., Coleoptera, Scolytidae

The southern pine beetle is the most injurious forest insect pest in the Southeast. It is greatly feared because epidemics develop with great rapidity, and much timber is killed. Weakening of the trees by flooding, windstorms, and especially droughts commonly precedes these outbreaks. Trees of all sizes are attacked, but usually those larger than 6 inches are infested first. Logs are seldom infested. The trees attacked commonly occur in groups. During epidemics, each of these "group kills" generally are about one acre in size, but sometimes they may be much larger. Usually only the mid and upper parts of the boles are infested. Even though the trees die quickly following summer attacks, the broods usually have left by the time the foliage has turned red. Accurate diagnosis can be made only by removing the bark and examining the boring pattern and the beetles responsible.

HOSTS AND RANGE. Pines and occasionally spruces in Maryland and West Virginia south and west to Florida and eastern Texas.

GALLERY PATTERN. *Egg galleries:* winding; often form S-shaped patterns; 1/16 to 3/32 inch wide; in inner bark; frequently cross each other. *Eggs:* in individual niches. *Larval galleries:* short, 1/4 to 1/2 inch long; first in inner bark and then they usually depart into the outer bark where pupal chambers are formed.

INSECT APPEARANCE. *Beetles:* 3/32 to 3/16 inch long; brown to black; head with a vertical median groove; elytra as wide as pronotum with long hairs on posterior area. *Larvae:* typical; head with rounded convex elevation on front. *Pupae:* similar to adult but whitish, and with wing cases instead of elytra. The reader should also refer to the general bark beetle and *Dendroctonus* characteristics presented previously.

LIFE CYCLE. Individuals in all stages of development may hibernate over winter but generally most are larvae. Emergence and new attacks, which begins in the early spring at about the time plant growth starts, continue until midsummer. During hot summer weather the life cycle can be completed in about a month or 5 weeks, but longer developmental periods occur when the weather is cooler. There commonly are 3 to 5 generations per year.

FIG. 10.7. Winding egg galleries in the inner bark made by the southern pine beetle (*Dendroctonus frontalis*) (X1/2 and X1). Also note the larvae galleries.

CONTROL. The most effective natural control appears to be the resistance developed in rapidly growing host trees. Droughts, which commonly precede outbreaks, apparently cause temporary to prolonged cessation of radial tree growth. This makes the trees more suitable for beetle development; therefore, any practical silvicultural treatment that will keep the trees growing longer during periods of stress should be used.

Special efforts should also be made to repress the beetle population development especially during times when conditions are favorable for the insect. This can be accomplished best by detecting the dying trees from the air and then using rapid salvage methods as described previously. Whenever rapid salvage methods are not practical, the infested trees are small, or the areas involved are inaccessible, insecticidal sprays can be used. The recommended insecticide, benzene hexachloride (formulated as an oil solution containing ¼ per cent by weight of the gamma isomer dissolved in fuel oil No. 2) should be applied heavily to the bark of the felled, infested trees. Careful supervision of the work is required so that only the boles containing living broods are treated.

Anon. October 1955. *The lumberman.* P. 67.
Beal, J. A. 1933. *J. Forestry* 31:329–336.
Craighead, F. C. 1925. *J. Econ. Entomol.* 18:577–586.
Craighead, F. C., and R. A. St. George. 1940. *Phytopath.* 30:976–979.
Heller, R. C., J. F. Coyne, and J. L. Bean. 1955. *J. Forestry* 53:483–487.
Hetrick, L. A. 1949. *J. Econ. Entomol.* 42:466–469.
Lee, R. E. 1954. *J. Forestry* 52:767.
Osgood, E. A. 1957. A bibliography. *Southeastern Forest Exp. Sta. Paper* 80.
St. George, R. A., and J. A. Beal. 1929. *U.S.D.A. Farmers' Bull.* 1586.

WESTERN PINE BEETLE

Dendroctonus brevicomis Lec., Coleoptera, Scolytidae

The western pine beetle has been the most destructive pest of mature and overmature trees throughout the ponderosa pine region of the Pacific Coastal States. In this area the average timber loss during a 25-year period from 1921 to 1945 was about 1 billion board feet per year. This is almost 1½ times the growth and more than 10 times greater than the loss caused by fire. The old, mature, and overmature, senescent trees are most vulnerable.

HOSTS AND RANGE. Ponderosa and Coulter pines in the Pacific Coastal States and the northern Rockies.

GALLERY PATTERN. Similar to that for the southern pine beetle. Pupation always occurs in the dry outer bark; egg gallery width ⅛ inch.

INSECT APPEARANCE. See the general descriptions for bark beetles and for *Dendroctonus.* Beetles: cylindrical; body stout; ⅛ to ⅕ inch

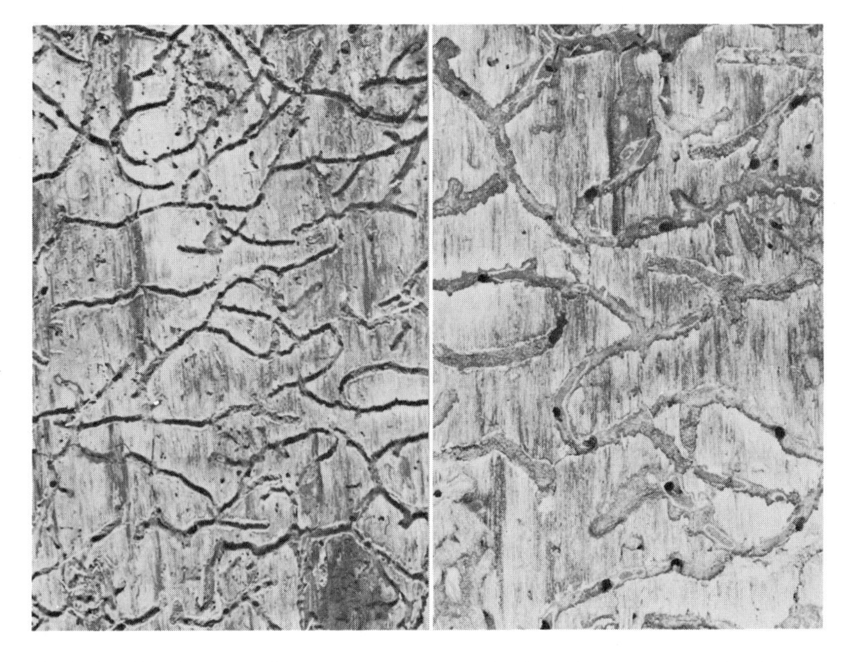

FIG. 10.8. Winding egg galleries made in the inner bark by the western pine beetle (*Dendroctonus brevicomic*) (X⅓ and X¾).

long; brown to black; posterior of elytra without long hairs. **Larvae** and **Pupae:** typical; have no specific identifying characteristics.

LIFE CYCLE. Typical for *Dendroctonus* with 2 or 3 generations per year being most common.

CONTROL. Woodpeckers and predaceous beetles are the most commonly encountered natural enemies of this beetle. At times low (−20° F and lower) winter temperatures have caused heavy brood mortalities but the actual killing temperatures vary because the insect races that occur in the colder climates are more resistant. Vigorous tree growth is the best natural control.

The best and most used type of applied control is based on using better forest management practices. One method consists of selectively logging the stands to remove the beetle susceptible, slow growing, poor vigor trees. These can be recognized readily because they have short, sparse, yellowish foliage. The amount of timber cut is adjusted to make the operation profitable yet kept small enough so that large areas can be covered rapidly. Usually 10 to 25 per cent (2,000 to 4,000 board feet per acre) of the volume is removed.

FIG. 10.9. The poor, thin-crowned, ponderosa (high risk) pine on the right is very vunerable to attacks by the western pine beetle (*Dendroctonus brevicomis*), whereas the vigorous appearing tree on the left is resistant.

Another similar method developed by Keen (1943) prescribes the use of a heavier cut removing 40 to 60 per cent of the timber volume. All the mature and overmature trees plus others having poor crowns are logged. This method probably is even more effective than the sanitation-salvage method, but proportionately less area can be covered in a given period of time.

Direct control methods used during past epidemics relied on the fell, peel, and burn techniques. Recently the use of ethylene dibromide sprays have showed some promise even on the thick barked ponderosa pines.

Johnson, P. C. 1949. *J. Forestry* 47:277–284.
Keen, F. P., and K. A. Salmon. 1942. *J. Forestry* 40:854–858.
Keen, F. P. 1943. *J. Forestry* 41:249–253.
———. 1950. *J. Forestry* 48:186–188.

————. 1955. *U.S.D.A. Forest Pest Leaflet* 1.
Miller, J. M. and F. P. Keen. 1960. *U.S.D.A. Misc. Pub.* 800.
Salmon, K. A., and J. W. Bongberg. 1942. *J. Forestry* 40:533–539.
Sowder, J. E. 1951. *Pacif. Northwest Forest and Range Exp. Sta. Research Rept.* 2.
Whiteside, J. M. 1951. *U.S.D.A. Circ.* 864.
Yuill, J. S. 1941. *J. Econ. Entomol.* 34:702–709.

SOUTHWESTERN PINE BEETLE (*Dendroctonus barberi* Hopk.)
ARIZONA PINE BEETLE (*D. arizonicus* Hopk.)
ROUND-HEADED PINE BEETLE (*D. convexifrons* Hopk.)
COLORADO PINE BEETLE (*D. approximatus* Dietz.)
Coleoptera, Scolytidae

The species listed here have characteristics similar to but have not been as injurious as the western pine beetles. They occur only in the central and southern Rockies. The distinguishing characteristics of these are presented in Table 10.4.

MOUNTAIN PINE BEETLE (*Dendroctonus monticolae* Hopk.)
BLACK HILLS BEETLE (*D. ponderosae* Hopk.)
JEFFREY PINE BEETLE (*D. jeffreyi* Hopk.)
Coleoptera, Scolytidae

The above species are considered together because they are very similar in both appearance and habits. They are all forest pests of major importance and are even more aggressive than the western pine beetle. They commonly infest young trees; therefore, they probably will continue to be troublesome even after all the virgin timber has been cut. All trees of more than about 3 inches in diameter are attacked. The trees attacked usually are in groups of 3 to 10, but during chronic outbreaks these "group kills" increase in size so that serious stand depletion occurs. Three fourths of the epidemics has occurred during drought periods.

HOSTS AND RANGES. *Mountain pine beetle:* sugar, western white, and lodgepole pines are preferred, but other pines also are attacked. It occurs in the Pacific Coastal States and northern Rockies from southern Canada south to western Wyoming. *Black Hills beetle:* Ponderosa pine is preferred, but limber, white bark, and lodgepole pines also rank high as hosts. It occurs in the southern Rockies from Colorado and Utah southward and in the Black Hills of South Dakota. *Jeffrey pine beetle:* only Jeffrey pine it attacked. The insect occurs only in California south into northern Mexico.

GALLERY PATTERN. *Egg galleries:* long, 2½ feet or more; placed longitudinally along bole; straight to slightly sinuous; ⅛ to 3⁄16 inch wide; frequently with a hook-like bend at the lower end adjacent to the entrance. *Eggs:* in individual niches. *Larval galleries:* individual; always completed in the inner bark; 1 to 5 inches long. *Pupal cells:* in inner bark.

FIG. 10.10. Long longitudinal egg galleries, short larval galleries, and pupal chambers made in the inner bark by the Black Hills beetle (*Dendroctonus ponderosae*) (X¾).

INSECT APPEARANCE. Refer to the general descriptions of bark beetles and of *Dendroctonus*. **Beetles:** the 3 species are very similar; body stout and cylindrical; ⅛ to ¼ inch long; brown to black; front of head without a depressed groove; pronotum decidedly narrowed toward the front. **Larvae:** typical; with front of head having a rounded convex elevation. **Pupae:** elytra cases somewhat roughened with sparse granules; head deeply grooved; long prominent spines on sides of abdomen.

LIFE CYCLE. Winter is passed beneath the bark mostly in the adult and larval stages. The beetles emerge from June to August. Over most of the range there is only 1 generation per year but in warmer climates (below 7,000 feet elevation and south of 40° latitude) there may be 2 per year. The beetles often attack two host trees during their life.

CONTROL. The best natural control is rapid continuous tree growth throughout the growing season.

Salvage and rapid utilization of infested and wind-thrown timber is the most desirable direct control method. Direct insecticidal control consists of heavily spraying the thinner-barked trees with ethylene dibromide. The formulation preferred is an oil solution containing 1½ pounds of the dibromide per 5 gallons of diesel oil. Usually the taller infested trees must be felled in order to spray the tops of the boles. An emulsion formulation (3 pounds ethylene dibromide, 8 ounces emulsifier, 1 gallon diesel oil, and the remainder water) has also been recommended for treating thin-barked lodgepole pine. The advantage of using this formulation is that the insecticide hauling job may be greatly reduced provided water is available in the areas to be treated. This insecticidal treatment is not effective on the old, thick-barked sugar and white pines; therefore, such logs must be burned or peeled in order to kill the broods.

Beal, J. A. 1939. *U.S.D.A. Farmers' Bull.* 1824.
————. 1943. *J. Forestry* 41:359–366.
Blackman, M. W. 1931. *N.Y. State College of Forestry Bull.* IV (4), *Tech. Publ.* 36.
Deleon, D., W. D. Bedard, and T. T. Terrell. 1934. *J. Forestry* 32:430–436.
Eaton, C. B. 1956. *U.S.D.A. Forest Pest Leaflet* 11.
————. 1941. *J. Forestry* 39:710–713.
Hopping, J. R., and W. G. Mathis. 1945. *Forestry Chronicle* 21:98–108.
Kinghorn, J. M. 1955. *J. Econ. Entomol.* 48:501–504.
Massey, C. L., R. D. Chisholm, and N. D. Wygant. 1952. *J. Econ. Entomol.* 45:861–862.
Struble, G. R., and P. C. Johnson. 1955. *U.S.D.A. Forest Pest Leaflet* 2.
Stevens, R. E. 1957. *U.S.D.A. Cal. Forest and Range Exp. Sta. Forest Research Notes* 122.

TURPENTINE BEETLES

Dendroctonus terebrans Oliv. and *D. valens* Lec., Coleoptera, Scolytidae

Up until the late 1940's the turpentine beetles were considered troublesome but not aggressive tree killers. Since then the black turpentine beetle (*D. terebrans*) has caused serious damage throughout much of the southeastern coastal plain. Prolonged drought conditions preceded the general outbreak, but stand disturbances such as turpentining and logging appear to increase tree susceptibility. In some areas, as much as 10 to 25 per cent of the trees have been killed in one year. The infested trees die slowly, so that fading of the crowns starts as much as 4 to 8 months after the initial attack.

The red turpentine beetle (*D. valens*) has not caused as much damage as its close relative, but it is not entirely innocuous. Both species usually limit their attacks to the basal 3 to 6 feet of the bole and the adjacent larger roots, although sometimes the attacks may be as high as 10 feet. A large pitch tube consisting of congealed resin, and reddish boring dust usually marks each entrance.

HOSTS. All species of pines. Fresh stumps as well as living trees are attacked.

RANGES. *D. valens*—most of the pine-growing regions of North America, except southeastern United States, but even here it occurs in the Appalachian region. *D. terebrans*—from southern New England south and west to the Gulf and eastern Texas.

GALLERY PATTERN. *Egg galleries:* very broad, ½ to 1 inch wide; extend downward from the entrance holes, often are in the roots; a few inches to several feet long; sinuous and always packed with borings and resin. *Eggs:* placed together in rows in grooves along sides of egg galleries. *Larval galleries:* broad, cave-type, fan shaped in which the larvae feed together. *Pupal cells:* formed either in the larval galleries or in the outer dry corky bark.

INSECT APPEARANCE. *Beetles:* large, ⅕ to ⅜ inch long; robust; reddish brown to black; pronotum almost as broad as wing covers (elytra); head large; *D. valens* beetles have long hairs on basal parts of elytra and mature beetles have a reddish color; *D. terebrans* beetles are blackish but may be reddish when young. *Larvae:* differ from many other *Dendroctonus* species by having prominent spines on back of both the 8th and 9th abdominal segments.

LIFE CYCLE. In the South all stages are present throughout the winter, but in colder climates the hibernating stages are mostly larvae and adults. There may be 1 or 2 generations per year depending on the length of the summer season.

FIG. 10.11. Portion of a broad, winding, turpentine beetle (*Dendroctonus terebrans*) egg gallery (X1).

CONTROL. Inasmuch as these beetles attack the tree bases they are very accessible for insecticidal treatment. Benzene hexachloride (1 per cent by weight) dissolved in No. 2 fuel oil (diesel oil) is the recommended formulation. The loose rough bark should be removed first and then the infested area thoroughly sprayed. Whenever the insect is abundant, stump spraying in logged areas is also recommended. Infested turpentined trees should be sprayed immediately and streaking stopped.

A rule commonly used by entomologists working on black turpentine beetle control is that trees having more than one attack per inch of diameter usually die. Therefore, such trees should be salvaged and the stumps sprayed.

Smith, R. H., and R. E. Lee. 1957. *U.S.D.A. Forest Pest Leaflet* 12.
Smith, R. H. 1958. *J. Forestry* 56:190–194.

LODGEPOLE PINE BEETLE (*Dendroctonus murrayanae* Hopk.), RED-WINGED BEETLE (*D. rufipennis* Kirby) — Coleoptera, Scolytidae

These two bark beetle species seldom are encountered. Their appearance, gallery patterns, and habits apparently are somewhat similar to those for turpentine beetles. The differences are presented in Table 10.4.

ENGELMANN SPRUCE BEETLE

Dendroctonus engelmanni Hopk., Coleoptera, Scolytidae

During the destructive Engelmann spruce outbreak in Colorado between 1940 and 1950, more than 4 billion board feet of timber was killed. This epidemic, as well as most of the others, started in windthrown timber. Large beetle populations developed in the damaged trees and then moved to the adjacent uninjured stands. From here the epidemic spread over a large area during the following ten years.

Large, mature trees are preferred as hosts, but as the epidemic progresses, successively smaller trees are attacked so that eventually all size trees are killed. Only the lower parts of the tree boles up to 30 to 50 feet are infested.

HOSTS. Engelmann spruce is the principal host. Other spruces and lodgepole pine are also infested when these are intermixed or adjacent to the chief host. Broods are unsuccessful in pine.

RANGE. Throughout the Rocky Mountain region.

GALLERY PATTERN. *Egg galleries:* short, 2½ to 9 inches long; broad, about ¼ inch wide; straight to slightly curved; frequently with a hook-like bend at the lower end adjacent to the entrance; bored longitudinally along bole; sometimes with a short Y-shaped fork at upper end; commonly 6 to 9 egg galleries per square foot of bark. *Eggs:* placed together

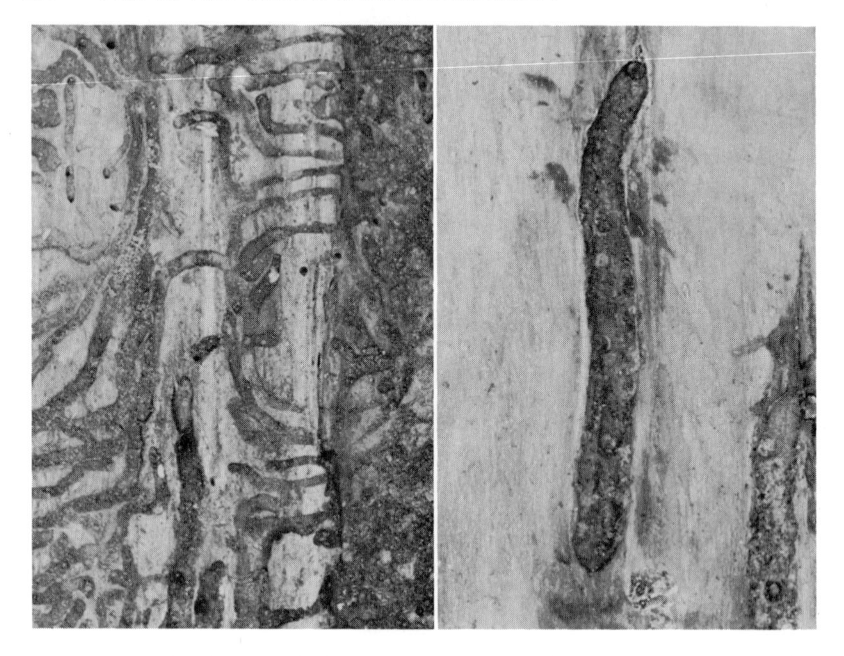

FIG. 10.12. Right: short longitudinal egg gallery of Engelmann spruce beetle (*Dendroctous engelmanni*) (X1); left: larval galleries (X1).

in grooves that are about 1½ inches long; usually 3 or 4 grooves per gallery. *Larval galleries:* common brood galleries are formed until larvae are about ⅓ grown; thereafter the larvae bore individual galleries that wind through the inner bark. *Pupal cells:* in inner bark.

INSECT APPEARANCE. *Beetles:* 3/16 to ¼ inch long; reddish brown to black; fine punctures on posterior of elytra. *Larvae:* typical; back of either or both 8th and 9th abdominal segments with hardened plates. *Pupae:* elytral pads smooth; abdomen has prominent spines along sides.

LIFE CYCLE. Either 1 or 2 years are required for this insect to complete 1 generation. Winter is passed as large larvae, pupae, or young adults. Beetles which attack new trees during late June or early July produce broods that mature and pupate by October. These pupae hibernate, overwinter, and emerge as beetles the following summer; therefore, they have about a 1-year life cycle. Other beetles that attack after mid-July produce broods that are about ¾ grown by the end of summer. These overwinter and mature the following summer. They emerge during the fall and bore small feeding wintering tunnels into the bark at the bases of the same trees from which they emerged. The second winter is passed in this location. They then emerge early the following summer to produce the early brood.

CONTROL. Rapid salvage of wind-thrown or other damaged timber is the best preventive method. This salvage often is difficult to carry out because of the inaccessibility of these high altitude, western spruce forests.

At times woodpeckers predation has reduced beetle populations as much as 98 per cent. Low winter temperature (−15° F for adults and −30° F for larvae) also kills most of the insects present in the bark above snow line. This is especially effective if the cold spells occur early, before the insects become cold hardy.

During the late 1940's large-scale insecticidal control operations were conducted in Colorado by the U. S. Forest Service. The treatment used consisted of spraying the lower 30 to 35 feet of the infested tree boles with orthodichlorobenze. Large quantities of this "ortho" insecticide were used in this operation, but at the present time ethylene dibromide is preferred. The dibromide is used either as an emulsion (3 pounds insecticide per 5 gallons of spray) or an oil solution (1½ pounds per 5 gallons fuel oil). The emulsion formulation is used whenever water is available in the treating area. Two-man crews are used. One operates a stirrup pump, and the other directs the spray on the tree. In the northern Rockies, where the trees are larger and the infestations occur higher on the boles, spraying standing trees is not practical. In this region some efforts have been made to reduce beetle populations by concentrating the insects in trap logs and then destroying the beetles and broods when the logs are milled.

Massey, C. L., and N. D. Wygant. 1954. *U.S.D.A. Circ.* 944.

EASTERN SPRUCE BEETLE (*Dendroctonus piceaperda* Hopk.)
SITKA SPRUCE BEETLE (*D. obesus* Mann.)
ALASKA SPRUCE BEETLE (*D. borealis* Hopk.)
Coleoptera, Scolytidae

These species are closely related to the Engelmann spruce beetle. Their appearance, gallery patterns, and habits are similar to their notorious relative. None has caused as much damage although several serious outbreaks of the eastern spruce beetle have developed following tree weakening by spruce budworm defoliation. The characteristics for these species are presented in Table 10.4. **LIFE CYCLE.** One generation per year occurs for the first two species with both adults and mature larvae overwintering. *D. borealis* occurs in colder regions; therefore it may require longer than a year to complete its development from egg to adult. **CONTROL.** Applied control methods have never been needed for these bark beetles. As for the Engelmann spruce beetle, more intensive forest management would tend to prevent trouble from developing.

Balch, R. E. 1942. *J. Forestry* 40:621–629.
Gobeil, A. R. 1941. *J. Forestry* 39:632–640.

DOUGLAS-FIR BEETLE

Dendroctonus pseudotsugae Hopk., Coleoptera, Scolytidae

This serious pest of Douglas-fir usually breeds in wind-thrown or fire-damaged trees; therefore, when catastrophic fires or severe windstorms occur, large quantities of suitable host material are produced. The resulting large beetle populations that develop infest and kill large quantities of the surrounding, uninjured, healthy trees. The attacks often occur first on the upper parts of the tree boles.

HOSTS. Douglas-fir is the principal host. Sometimes injured or felled western larch and big cone spruce also are infested.

RANGE. Throughout the west, wherever its hosts grow.

GALLERY PATTERN. *Egg galleries:* usually 2 or more feet long; rather straight to slightly sinuous; longitudinally placed in inner bark; about ¼ inch wide. *Eggs:* placed together in grooves on alternate sides of egg galleries; eggs are placed in rows, with each group containing 5 to 12. *Larval galleries:* individual; always in the inner bark. *Pupal cells:* in inner bark.

INSECT APPEARANCE. *Beetles:* ⅛ to ¼ inch long; reddish to blackish brown; lines along posterior of wing covers (elytra) deeply impressed. *Larvae:* typical. *Pupae:* elytra pads smooth, abdomen with prominent lateral spines; front of head faintly grooved.

LIFE CYCLE. This species hibernates over the winter both as adults and as full grown larvae. The beetles emerge and attack new hosts from early spring until summer. Usually there is only 1 generation per year, but there may be 2 in warm climates.

CONTROL. The only suitable control for this insect is to practice intensive forest management. The fire injured or wind-thrown timber must be salvaged rapidly so as to avoid developing large beetle populations. If this can not be done, it is unlikely that other intensive direct control methods can be carried out satisfactorily. Insecticidal sprays have not yet been found effective on the thick-barked Douglas-fir.

Bedard, U. D. 1950. *U.S.D.A. Circ.* 817.
Evenden, J. C., and K. H. Wright. 1955. *U.S.D.A. Forest Pest Leaflet* 5.
Smyth, A. V. 1959. *J. Forestry* 57:278–280.

EASTERN LARCH BEETLE

Dendroctonus simplex Lec., Coleoptera, Scolytidae

This beetle breeds in dying, felled and injured living trees. It is not very aggressive and usually infests trees repeatedly defoliated by the larch sawfly.

HOST AND RANGE. Eastern larch throughout eastern Canada and northeastern United States west to the prairies. GALLERY PATTERN. *Egg galleries:* long; slightly winding; often branched; bored longitudinal with the grain of the wood.

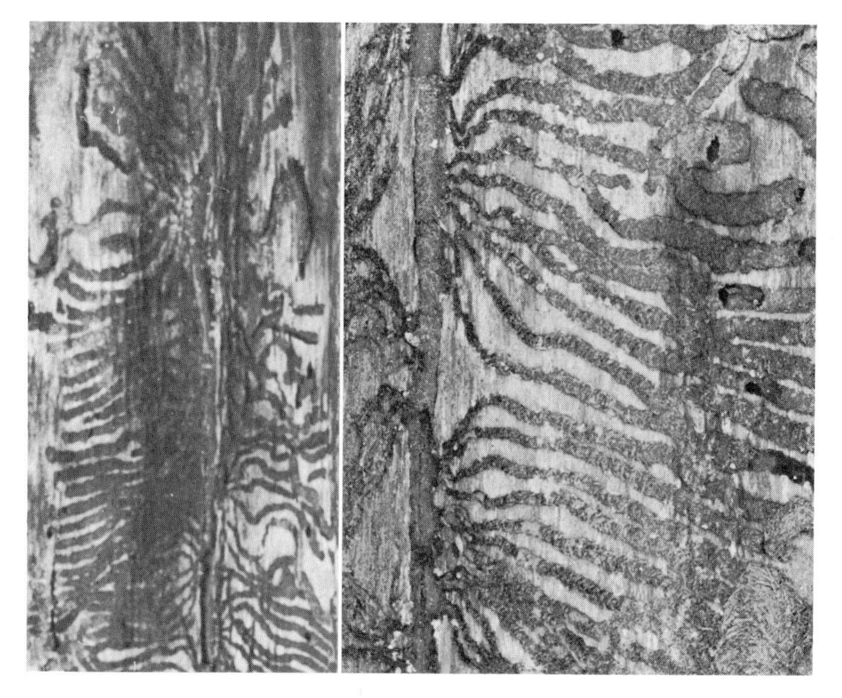

FIG. 10.13. Long longitudinal egg galleries and long larval galleries of the Douglas-fir beetle (*Dendroctonus pseudotsugae*) (X⅓ and X¾).

Eggs: in grooves; 3 to 6 eggs placed together in each. *Larval galleries:* short and always exposed in the inner bark. INSECT APPEARANCE. *Beetles:* ⅛ to $\frac{3}{16}$ inch long; dark reddish brown; lines on posterior of elytra are deeply impressed. *Larvae* and *pupae:* similar to those of the Douglas-fir beetle. LIFE CYCLE. Adults hibernate over winter beneath the bark in the host trees. These beetles emerge and attack new trees in the spring and early summer. The resulting broods mature by fall. There is only 1 generation per year. CONTROL. Applied control has never been needed for this insect, but, should conditions ever warrant, the penetrating insecticidal sprays used for the Engelmann bark beetle should be effective.

IPS ENGRAVER BEETLES

Ips spp. Coleoptera, Scolytidae

Ips engravers usually are the commonest inner bark borers in most pine and spruce forests. They are not as aggressive as species of *Dendroctonus,* for they usually infest only dying trees, severely injured trees, or green slash. Sometimes, however, epidemics develop during which many living trees are killed. Frequently these tree killing beetle populations develop in trees weakened by drought, fire, windstorm, or in the larger slash, green bolts or logs left in the woods. In large western pines the *Ips* attacks in living trees often occur only in the tops. Forest stands near pulp wood yards and sawmills also commonly are damaged by these bark beetles after they become abundant by breeding in the bolts and logs.

The many species of *Ips* have rather similar habits; therefore, they will be considered together. The more common species can be tentatively identified by referring to Table 10.5, but in order to obtain more positive identification the publications cited at the end of the general discussion on bark beetles should be consulted.

TABLE 10.5 THE COMMON SPECIES OF IPS ENGRAVER BEETLES

1. Occur in eastern North America 2
1. Occur in western North America 9
2. Beetles have 6 teeth* on each side of concavity at posterior end of wing covers (elytral concavity); host pines . . . . COARSE WRITING ENGRAVER (*I. calligraphus* Germ.)
2. Three to 5 teeth* on each side of elytral concavity 3
3. Three or 4 teeth on each side of elytral concavity; largest tooth has 2 points and sometimes appears as 2 closely united teeth 4
3. Five teeth on each side of elytral concavity 8
4. Beetles small, 2 to 3 millimeters long; 4 teeth on each side of elytral concavity; hosts pines . . . . SMALL SOUTHERN ENGRAVER (*I. avulsus* Eichh.)
4. Beetles 3 to 5½ millimeters long 5
5. Beetles with 3 teeth on each side of elytral concavity; 3 to 3½ millimeters long; hosts pine and hemlock . . . . . . SMALL NORTHERN ENGRAVER (*I. longidens* Sw.)
5. Beetles with 4 teeth on each side of elytral concavity 6
6. Beetles 3 to 4½ millimeters long 7
6. Beetles 4.7 to 5.5 millimeters long; hosts spruces . . . . NORTHERN SPRUCE ENGRAVER (*I. perturbatus* (Eichh.)
7. Pronotum distinctly longer than wide; hosts pines and spruces . . . . EASTERN PINE ENGRAVER (*I. pini* (Say)
7. Pronotum scarcely longer than wide; hosts spruce, hemlock, and fir . . . . *I. borealis* Sw.
8. Beetles 2 to 4 millimeters long; occur in southeastern United States; hosts pine . . . . SOUTHERN PINE ENGRAVER (*I. grandicollis* (Eichh.)
8. Beetles 4 to 5 millimeters long; occur in eastern Canada; hosts spruce and white pine . . . . *I. chagnoni* Sw.
9. Beetles with 6 teeth* on each side of concavity at posterior end of wing covers (elytral concavity) 10
9. Beetles with 3 to 5 teeth* on each side of elytral concavity 11
10. Two larger central teeth close together and often appear as 2 cusps of one tooth . . . . LARGE SOUTHWESTERN ENGRAVER (*I. knausi* Sw.)
10. All teeth separated . . . . WESTERN 6-SPINED ENGRAVER (*I. ponderosae* Sw.)
11. With 3 teeth on each side of elytral concavity 12

11. With 4 or 5 teeth on each side of elytral concavity 14
12. Beetles 3 to 4 millimeters long; antennal club with nearly straight lines (sutures); hosts pines . . . . SMALL WESTERN ENGRAVER (*I. latidens* Lec.)
12. Beetles 4 to 5 millimeters long; sutures on antennal club curved or arched . . 13
13. Infest sitka spruce . . . . SITKA SPRUCE ENGRAVER (*I. concinnus* Mann.)
13. Infests pines . . . . MONTEREY PINE ENGRAVER (*I. radiatae* Hopk.)
14. With 4 teeth on each side of elytral concavity; largest tooth (3rd) sometimes appears as 2 teeth closely united having 2 points 15
14. With 5 teeth on each side of elytral concavity 23
15. In the northern Rockies and the Pacific Coastal States and provinces 16
15. In the southern Rockies 22
16. Beetles large 6 to 7 millimeters long; 3rd tooth on each side of elytral concavity long, compressed, and notched at tip so as to have 2 points; hosts pines . . . . LARGE WESTERN ENGRAVER (*I. emarginatus* Lec.)
16. Beetles 3½ to 6 millimeters long 17
17. Infest pines and/or larch 18
17. Infest spruces 20
18. Pronotum distinctly longer than wide; hosts pines . . . . CALIFORNIA PINE ENGRAVER (*I. plastographus* Lec.)
18. Pronotum about or only slightly longer than wide 19
19. Infest pines; spaces between lines (striae) near base of elytra without punctures . . . . OREGON PINE ENGRAVER [*I. oregoni* (Eichh.)]
19. Infest spruces; spaces between striae near base of elytra with a few punctures . . . . *I. interruptus* (Mann.)
20. Spaces between lines (striae) on back near base of elytra not punctured . . . . NORTHERN SPRUCE ENGRAVER [*I. perturbatus* (Eichh.)]
20. Spaces between striae on back of elytra with a few punctures 21
21. Pair of projection on front of head above mouth . . . *I. tridens* (Mann.) and others
21. Front of head does not have a pair of projections above . . . *I. interruptus* (Mann.)
22. Beetles large, 5 to 8 millimeters long; 3rd tooth on each side of elytral concavity with 2 points; hosts pines . . . LARGE SOUTHWESTERN ENGRAVER (*I. knausi* Sw.)
22. Beetles 4 to 6 millimeters long; host pines and larch . . . . *I. integer* (Eichh.)
23. Occurs throughout the West except in the central and southern Rockies 24
23. Occurs in the southern Rockies 25
24. Beetles 4 to 5 millimeters long; posterior of pronotum with fine sparse punctures; hosts pines . . . . CALIFORNIA 5-SPINED ENGRAVER [*I. confusus* (Lec.)]
24. Beetles 5 to 6 millimeters long; posterior of pronotum with many coarse punctures; elytral concavity with dense long hairs; hosts pines and sitka spruces . . . . *I. vancouveri* Sw.
25. Beetles less than 4 millimeters long . . . . *I. cloudcrofti* Sw.
25. Beetles more than 4 millimeters long 26
26. Beetles 4 to 5 millimeters long . . . ARIZONA 5-SPINE ENGRAVER (*I. lecontei* Sw.)
26. Beetle about 6 millimeters long . . . . LARGE SOUTHWESTERN ENGRAVER (*I. knausi* Sw.)

* A small tooth which commonly is present near the upper edge of the concavity on each elytron should not be overlooked. The shelf-like projection on the lower edge is not counted as a tooth.

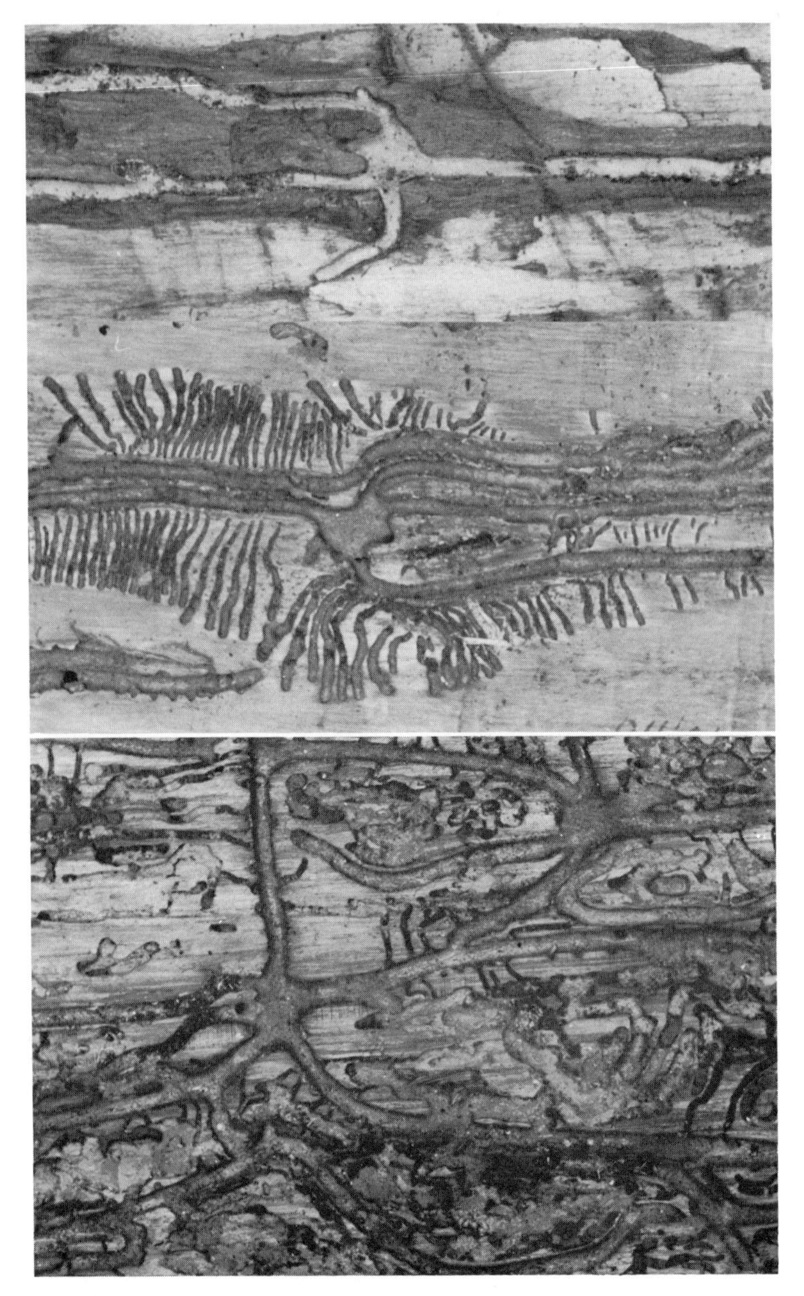

FIG. 10.14. *Ips* gallery system in the inner bark at various stages of development. Nuptial chambers, the egg galleries, and the larval galleries are shown (X1).

FIG. 10.15. A different type of egg gallery made by *I. avulsus.* Nuptial chamber shown in center is usually concealed within the inner bark. Also note the short larval galleries and the pupal chambers (X⅔).

HOSTS AND RANGE. Pines, spruces, and occasionally larch and firs, wherever the host trees grow.

GALLERY PATTERN. *Nuptial chambers:* enlarged chambers in the inner bark adjacent to each entrance hole. *Egg galleries:* in the inner bark, narrow, elongate, tunnels; 3 to 5 usually originate and radiate out from each nuptial chamber; sometimes 2 egg galleries originating from the same nuptial chamber are bored adjacent to each other and extend longitudinally along the tree bole. *Eggs:* placed in individual niches cut in sides of egg galleries. *Larval galleries:* individual; visible at the inner bark-xylem interface. *Pupal chambers:* also located in the inner bark.

INSECT APPEARANCE. *Beetles:* 1/16 to 5/16 inch long; brown to black; front of pronotum rough and projects over head; head not visible when insect is observed from above; no ridge on posterior pronotal margin; posterior of elytra concave, teeth-like projections around edge of concavity and a shelf-like projection at lower posterior edge; front tibia widened toward tip with teeth along outer edge; antennae funicle with 5 segments, club usually without lines (sutures) on inner side. *Larvae:* typical; see general bark beetle discussion. *Pupae:* resemble adults with pronotum extending over head but are whitish and have wing pads instead of wings and elytra.

LIFE CYCLE. There are 1 to 5 generations per year depending on the length and intensity of the warm summer season. Winter commonly is spent in either the adult or larval stages. Both young tan (callow) and old black adults may overwinter in the trees in which reared. Sometimes, however, the beetles pass the winter in special short-to-long winter tunnels made in the inner bark or the outer sapwood and, in the North, *I. pini* (Say), *I. grandicollis* (Erich.) and others may winter in the ground litter. When the beetles emerge in spring at about the time plant growth starts, they attack new hosts. All species of the genus are polygamous. The males make the initial entrances and excavate the nuptial chambers. They soon are joined by the females that bore the individual egg galleries. A generation is produced about every 2 months until cold weather stops their activity.

CONTROL. Preventative control methods consist of managing forests so that the trees are kept growing well. When pulpwood is cut, the remaining slash usually is so small that it dries rapidly and few beetles mature. With saw-log operations, however, the larger slash produced is more suitable breeding material, and large populations of beetles may be produced. If this breeding occurs and the logging is abruptly stopped during the summer, there is no fresh slash available; consequently, the emerging beetles attack and kill living trees. In order to avoid this trouble the logging either should be continued or stopped gradually. Of course the brood-containing slash could be burned or sprayed, but these methods would be expensive. Direct control operations generally are not used against *Ips* because the outbreaks develop and disappear rapidly. In intensively managed forests rapid salvage methods similar to those described for the southern pine beetle should be used. During outbreaks in pulpwood-size stands, thinning operations could be undertaken to provide trap logs for the purpose of absorbing the emerging beetles. This method has been described in the general discussion of bark beetle control.

Insecticidal sprays have not been used much for subduing *Ips* outbreaks primarily because they develop and decline rapidly. Should insecticidal control ever appear warranted, however, the sprays described for use against *Dendroctonus* bark beetles are effective. Of these, benzene hexachloride is preferred.

Anderson, R. F. 1948. *J. Econ. Entomol.* 41:596–602.
Clark, J. 1953. *Can. Dept. Agri. Bimonthly Prog. Rept.* 9.
Clemens, W. A. 1916. *Cornell Univ. Agri. Exp. Sta. Bull.* 383.
Hetrick, L. A. 1940. *Entomol. Soc. Wash.* 42:208–210.
————. 1942. *J. Econ. Entomol.* 35:181–183.
Lyon, R. L. 1959. *U.S.D.A., Pacific Southwest Forest and Range Exp. Sta. Misc. Paper* 33.
Struble, J. R., and R. C. Hall. 1955. *U.S.D.A. Circ.* 964.

SCOLYTUS ENGRAVER BEETLES

Scolytus spp., Coleoptera, Scolytidae

Although *Scolytus* beetles are less important as forest pests than either *Dendroctonus* or *Ips*, a few have been troublesome. The identifying characteristics of the more important species are presented in Table 10.6.

TABLE 10.6 THE MORE IMPORTANT SPECIES OF SCOLYTUS ENGRAVER BEETLES

1. Infest conifers 2
1. Infest broad-leaved (angiosperm) trees 4
2. Egg galleries extend transversely across grain of wood; beetles with very faint lines (striae) on elytra, punctures between these lines about as large as those in the striae; hosts chiefly true firs (page 236) . . . . FIR ENGRAVER
2. Egg galleries extend longitudinally along bole; beetles with a prominent spine on 2nd abdominal sternite 3
3. Two egg galleries originate from each entrance, one extends upward and the other downward; hosts spruces, larch, and firs; occurs in eastern Canada and northeastern United States west to northern Rockies . . . . SPRUCE ENGRAVER [*S. piceae* (Sw.)]
3. Single egg gallery placed longitudinally along boles 1 to 3 inches long; nuptial chamber at lower end connected to entrance hole with short tunnel placed at 45° angle to main gallery; host chiefly Douglas-fir; occurs in Pacific Coast States and British Columbia (page 237) . . . . DOUGLAS-FIR ENGRAVER
4. Infests Rosaceae hosts (apple, pear, Prunus, etc.) . . . . SHOT-HOLE ENGRAVER (*S. rugulosus* Ratz.)
4. Infests hosts other than Rosaceae; occur in eastern United States 5
5. Infests elms; beetle with one spine on 2nd abdominal sternite (page 238) . . . . SMALLER EUROPEAN ELM ENGRAVER
5. Infests hickories; beetles with 4 spines on abdominal sternites (page 238) . . . . HICKORY ENGRAVER

HOSTS. Both conifers and angiosperms are attacked, but pines are infested only rarely.

GALLERY PATTERN. *Egg galleries:* short, 1 to 6 inches long; narrow; 1 or 2 originate from each entrance or enlarged nuptial chamber. *Egg niches* and *Larval galleries:* individual. All galleries are cut deeply in both the sapwood and the inner bark: therefore, they engrave the wood surface even more than the *Ips* engravers. They are always monogamous.

INSECT APPEARANCE. *Beetles:* underside of abdomen ascends abruptly or is concaved behind the thorax; antennal basal segment (scape) very short; fore tibia with outer angle of tip a prominent spine. *Larvae* and *pupae:* typical scolytids with no distinguishing characteristics.

Black, M. W. 1934. *U.S.D.A. Tech. Bull.* 431.

FIR ENGRAVER

Scolytus ventralis Lec., Coleoptera, Scolytidae

This bark beetle has been very destructive in California where about 15 per cent of the white fir has been killed and 25 per cent more damaged. Pole size to mature timber is attacked with the upper third of the boles most commonly infested. The attacks are not concentrated to the same degree as those for many of the more aggressive bark beetles; therefore, patches of bark and tree tops are killed more frequently. When these killed bark areas heal over, pitch pockets are formed in the wood. Pitch tubes never develop at the points of attack, but liquid resin frequently exudes from the entrances and coats the tree bark below.

HOSTS. The chief hosts are white fir, grand fir, and red fir, but all species of true firs, Douglas-fir, Engelmann spruce, and mountain hemlock may be attacked.

RANGE. Wherever the hosts grow in the West.

GALLERY PATTERN. *Egg galleries:* narrow; extend across the grain of the wood; 2 usually extend in opposite directions from a central enlarged (nuptial) chamber; each is 2 to 4 inches long. *Eggs:* in individual niches. *Larval galleries:* individual; extend out at right angles from the egg galleries. All galleries groove the sap wood.

BEETLES. Typical; about ⅛ inch long; 2nd sternite of males have a small median tooth.

LIFE CYCLE. There are ½ to 1½ generations per year depending on the length and intensity of the summer season. One female constructs both egg galleries, which extend in opposite directions from each nuptial chamber. Both adults and larvae hibernate over winter. Pupation occurs in the spring and early summer so that the attacking beetles are most common during July and August.

CONTROL. Applied control measures have never been used, but as soon as the Western forests are brought under more intensive management, rapid salvage methods probably will be used to prevent large populations from developing and for reducing the losses. Removal of the weakened decadent trees would be another good preventative method.

Stevens, R. E. 1956. *U.S.D.A. Forest Pest Leaflet* 13.
Struble, G. R. 1957. *U.S.D.A. Prod. Research Rept.* 11.

FIG. 10.16. Gallery systems of *Scolytus* spp. showing nuptial chamber, egg galleries, and larval galleries. ~~Right:~~ fir engraver (*S. ventralis*) (X⅔); ~~left:~~ Douglas-fir engraver (*S. unispinosus*) (X1). left Right

DOUGLAS-FIR ENGRAVER

Scolytus unispinosus Lec., Coleoptera, Scolytidae

This bark beetle is usually only of secondary importance, for it attacks mostly weakened, injured, dying, or recently killed trees. Sometimes, however, it is a primary enemy of sapling size trees.

HOSTS AND RANGE. Douglas-fir and big cone spruce in the West, wherever the hosts grow.

GALLERY PATTERN. *Egg galleries:* narrow; rather straight; bored longitudinally with the grain of the wood; 1 to 3 inches long; nuptial chamber is at the lower end of egg gallery and is connected to the entrance hole by means of a short (3⁄16 inch long) tunnel bored at a 45° angle to the egg gallery. *Eggs:* in individual niches. *Larval galleries:* individual; start out at right angles to the egg gallery and then turn either up or down. All galleries score the sapwood.

BEETLES. Typical; small; about ⅛ inch long; 2nd abdominal sternite almost perpendicular and is armed with a stout spine.

LIFE CYCLE. Larvae hibernate over winter. Pupation and beetle emergence occurs in the spring and early summer. There may be 1 or 2 generations per year depending on the temperatures and length of the summer.

CONTROL. Applied methods never have been used to control this insect. Modifications of the methods described for the more aggressive species of *Dendroctonus* probably would be effective if infestations ever become serious.

HICKORY ENGRAVER

Scolytus quadrispinosus Say, Coleoptera, Scolytidae

This insect has killed much hickory in various parts of the East but is most troublesome only during droughts.

HOSTS AND RANGE. Only hickories, throughout the eastern half of the United States.

GALLERY PATTERN. *Egg galleries:* parallel with the grain of wood; short, 1 to 3 inches long. *Eggs:* in individual niches. *Larval galleries:* individual; they first extend out at right angles to the parent gallery and then turn and run parallel with the grain of the wood. *Pupal cells:* in bark.

BEETLES. Typical; ⅛ to ¼ inch long; abdominal sternites of males each with 4 spines; these lacking on females.

LIFE CYCLE. Winter is passed as mature larvae. Development is completed the following spring, and the resulting adults emerge to infest new host trees. Before doing so, however, they often feed on leaf petioles and the bark of twigs. There may be 1 or 2 generations per year depending on climatic conditions.

CONTROL. Applied control methods have been used to kill this insect mostly in shade and orchard (pecan) trees. Felling and burning the infested trees has been the method most frequently used. Insecticides used for killing other bark beetles have not been tried against this pest, but probably they would be effective. In valuable hickory stands it would be desirable to use rapid salvage combined with burning.

Beal, J. A., and C. L. Massey. 1945. *Duke Univ. School of Forestry Bull.* 10. Pp. 67–70.
Blackman, M. W. 1924. *J. Econ. Entomol.* 17:460–470.

SMALLER EUROPEAN ENGRAVER

Scolytus multistriatus (Marsh.), Coleoptera, Scolytidae

This bark beetle is the principal vector of the Dutch elm disease (*Ceratostomella ulmi*)—a disease introduced from Europe in infected burl

FIG. 10.17. Gallery pattern made by the hickory engraver (*Scolytus quadrispinosus*) (X⅓). Both egg and larval galleries are shown.

logs about 1930. It is not a very aggressive insect, for it breeds chiefly in fresh logs and dying trees. Nevertheless, it is effective in carrying the pathogen from diseased trees to healthy trees. This transmission occurs when the adults breed in the dying diseased trees and then feed on the bark in the twig crotches of healthy trees. They may also transmit the disease by breeding in the dying lower branches of healthy elms.

HOSTS AND RANGE. This introduced bark beetle occurs only in elms throughout most of the Northeast from Massachusetts south to Virginia and west to Wisconsin.

GALLERY PATTERN. *Egg galleries:* narrow; rather straight; short, 1 to 6 inches long; always extended with the grain of the wood. *Eggs:* in individual niches. *Larval galleries:* individual; as they extend away from the egg galleries they form a fan-shaped pattern; all galleries groove the sapwood.

BEETLES. Typical; small; about ⅛ inch long; a spine projects from the front of the 2nd abdominal sternite.

LIFE CYCLE. Winter is passed in the larval stage. The larvae mature in the spring, pupate, and emerge as adults during the summer. There are 1½ to 2 generations per year.

CONTROL. Shade trees can be protected with DDT. The formulations used are emulsions that contain 2 per cent DDT for use in hydraulic sprayers or 12 per cent DDT in oil solutions for use in mist blowers. Twenty to 30 gallons of the 2 per cent spray and only 2 to 3 gallons of the mist concentration are needed to treat an average tree about 50 feet tall. The spraying should be done in the spring before the leaves appear, at a time when the bark is dry. Thorough bark coverage is most important. Destruction of suitable breeding material has not been effective in reducing the amount of disease because of the difficulty of doing a thorough job and because the beetles can fly several miles.

Forest Insect Division, Bureau of Entomology. 1953. *U.S.D.A. Leaflet* 185 (revised).
Whitten, R. R. 1956. *U.S.D.A. Central States Forest Exp. Sta. Misc. Release* 10.

CONIFER-INFESTING BARK BEETLE GENERA OF LESSER IMPORTANCE

PSEUDOHYLESINUS BARK BEETLES

Pseudohylesinus spp., Coleoptera, Scolytidae

These bark beetles have not been considered serious forest pests for they have usually attacked only slash, wind-thrown, or otherwise weakened trees. Recently, however, an outbreak of *P. grandis Sw.* and *P. granulatus* (Lec.) in northeastern Washington has resulted in the death of more than half a billion board feet of silver fir. Some species infest mostly the tops, whereas others attack lower on the tree boles.

HOSTS. *P. nebulosus* (Lec.), infests chiefly Douglas-fir poles and saplings; *P. grandis Sw.* and *P. granulatus* (Lec.) attack grand fir, silver fir, and Douglas-fir; and *P. sitchensis Sw.* infests sitka spruce. Other species chiefly breed in firs and hemlock. One species has been reported from pine.

RANGE. Western North America, wherever the hosts grow.

GALLERY PATTERN. *P. nebulosus*—*Egg gallery:* short, ½ to 3 inches long; longitudinal, with the grain of the wood; whole gallery pattern very similar to that produced by *Scolytus unispinosus,* but differs in that there is no well-defined nuptial chamber visible on the inner-bark surface. *P. grandis, P. granulatus,* and *P. sitchensis*—egg galleries are bored transversely with a nuptial chamber present near the middle, from which the 2 egg galleries extend in opposite directions. *Egg niches* and *larval galleries:* individual. *Pupal chambers:* mostly in the bark.

BEETLES. Small to medium size, ⅛ to ¼ inch long; head visible from above; body covered with scales; color dull and variegated; base of each elytron with fine teeth; antennae with dense plumose hairs; funicle with 7 segments; club segmented; fore coxae placed close together.

LIFE CYCLE. The adults spend the winter in the host trees or in moss. They emerge in the spring and attack new hosts. There are usually 2 generations per year. These insects are always monogamous.

CONTROL. Applied controls have seldom been used. When outbreaks occur, however, the various salvage and insecticidal methods used for other species of bark beetles should be effective.

Blackman, M. W. 1942. *U.S.D.A. Misc. Publ.* 461.

EASTERN BALSAM BARK BEETLE

***Pityokteines sparsus* (Lec.), Coleoptera, Scolytidae**

Species of *Pityokteines* have never caused serious trouble for they commonly breed only in dying or recently cut trees. **HOSTS.** *Abies* spp. and Douglas-fir. **RANGE.** *P. sparsus* occurs in northeastern North America, whereas several other species occur in various parts of the West. **GALLERY PATTERN.** *Egg galleries:* 3 to 5 narrow tunnels radiate out from a central nuptial chamber. *Egg niches* and *larval galleries:* individual. **BEETLES.** See Table 10.3. Front of head is usually clothed with dense yellow hairs. **LARVAE AND PUPAE.** Typical. **LIFE CYCLE.** Typical; see page 208; polygamous. **CONTROL.** Applied methods have never been needed to kill these bark beetles.

WESTERN BALSAM BARK BEETLE

***Dryocoetes confusus* Swaine, Coleoptera, Scolytidae**

There are many *Dryocoetes* species, but only the one listed here is a tree killer. Even this species has not received much attention, because its hosts are not very valuable. **HOSTS.** True firs. Other species attack firs, Douglas-fir, spruces, or larches. A few even infest broad-leaved deciduous trees. **RANGE.** Rocky Mountains. Other species occur wherever the hosts grow. **GALLERY PATTERN.** *Egg galleries:* 3 to 6 or more narrow galleries radiate out from a central nuptial chamber. *Egg niches* and *larval galleries:* individual. **BEETLES.** Two to 5 millimeters long; cylindrical; resemble *Ips* but lack the posterior toothed elytral convavity; females have dense, short, yellowish hairs on front of head; see Table 10.3. **LARVAE AND PUPAE.** Typical. **LIFE CYCLE.** Young adults winter in the outer bark. These emerge in the spring and attack new hosts. There are 1 to 2 generations per year. They are polygamous. **CONTROL.** In the past the low value of the western true firs has not stimulated the development of control measures for this species. Very likely, however, the general methods used for the Engelmann spruce beetle would be effective.

FOUR-EYED SPRUCE BARK BEETLE

***Polygraphus rufipennis* (Kirby), Coleoptera, Scolytidae**

This species is not very injurious. It is commonly found infesting dying or recently cut trees. **HOSTS.** Spruces and larch; occasionally attacks pine. **RANGE.** Wherever the hosts grow. **GALLERY PATTERN.** *Egg galleries:* several (3 to 5) narrow tunnels radiate from a central nuptial chamber. *Egg niches* and *larval galleries:* individual. **BEETLES.** About ⅛ inch long; resemble *Dendroctonus;*

head visible from above; posterior of elytra not concave; differ by having 4 compound eyes; color reddish to black. **LIFE CYCLE.** Typical; see page 208; polygamous. **CONTROL.** Applied control has never been needed for this beetle.

CEDAR, CYPRESS, JUNIPER, AND REDWOOD BARK BEETLES

Phloeosinus spp., Coleoptera, Scolytidae

The *Phloeosinus* bark beetles are not very aggressive; therefore, they seldom are the primary cause of tree mortality. Usually they breed in the limbs and boles of weakened, broken, dying, or felled trees. Sometimes, however, they become abundant and attack apparently healthy trees. The newly emerged beetles commonly bore into and kill the twigs on vigorous growing trees. These remain hanging on the trees and appear unsightly. Sometimes the beetles also eat the leaves and cause partial defoliation.

HOSTS. Members of the families Cupressaceae and Taxodiaceae such as cedars, junipers, redwoods, and cypress. One species breeds in pines and 2 in spruces.

RANGE. Wherever the hosts grow. Thirty-five of the 45 North American species occur in the West.

GALLERY PATTERN. *Egg galleries:* short, 1 to 3 inches long; straight and bored longitudinally with the grain of the wood; often the sapwood is grooved rather deeply; nuptial chamber sometimes is present at lower end; egg galleries usually

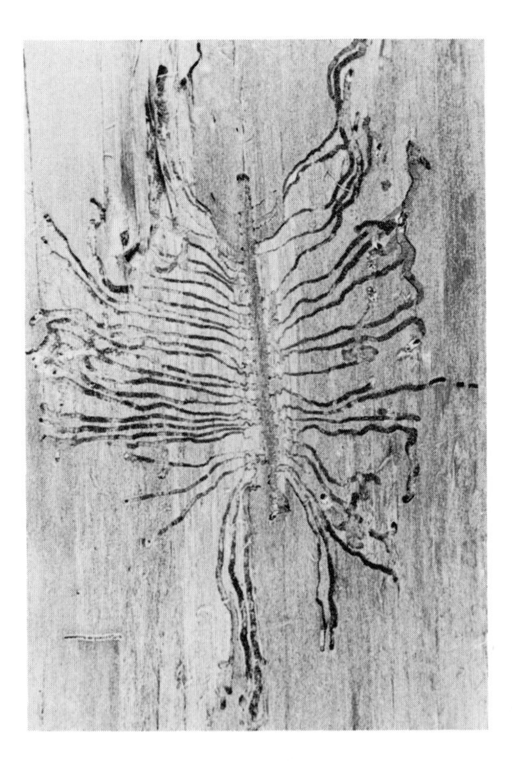

FIG. 10.18. Gallery pattern made by the *Phloeosinus* bark beetles (X1). Nuptial chamber, egg gallery, and larval galleries are shown.

kept free of borings. *Eggs:* in individual niches; placed close together. *Larval galleries:* individual; extend out at right angles from the parent galleries.

BEETLES. Small, 1½ to 4 millimeters long; head visible from above; antennal club conical much longer than wide, segmented; funicle 5 segmented; eyes deeply notched; posterior of elytra of males with small toothed projections on alternate spaces between the lines (striae).

LIFE CYCLE. These insects hibernate over winter both as adults and as larvae. The beetles emerge during the spring and summer and first feed on twigs as described previously, before they attack new hosts. There are 1 to 2 generations per year. They are monogamous.

CONTROL. Applied control has never been needed in forests. For ornamentals, the insecticidal methods suggested for the more aggressive *Dendroctonus* species might be tried should conditions ever warrant action.

Blackman, M. W. 1942. *Proc. U. S. Natl. Museum* 92 (3154). 397–474.

HARDWOOD-INFESTING BARK BEETLES

Coleoptera, Scolytidae

None of the bark beetles that infest hardwoods are as common or as injurious as some of those that attack conifers. At times, however, they may be observed attacking various trees; therefore, a few of these will be considered here. These beetles can be tentatively identified on the basis of host infested and general appearance of the beetles. The *Scolytus* species have been discussed previously. See page 235.

NATIVE ELM BARK BEETLE

Hylurgopinus rufipes (Eich.), Coleoptera, Scolytidae

This species, like *Scolytus multistriatus,* is of importance as a shade tree pest because it transmits the Dutch elm disease, *Ceratostomella ulmi.* HOSTS. Only dying and recently felled elms. RANGE. Eastern North America, wherever the American elm grows. GALLERY PATTERN. *Egg galleries:* narrow; rather short and straight; extend transversely across the grain of the wood; 2 branches extend in opposite directions from each entrance hole. *Eggs:* in individual niches. *Larval galleries:* individual, extend with the grain of the trees. BEETLES. Resemble *Dendroctonus* spp.; head visible from above and posterior of elytra not concave; differ by having 7 segmented antennal funicle. LIFE CYCLE. Typical; see page 208; 1 to 2 generations per year. CONTROL. To protect shade elms from becoming diseased, the method as described for *S. multistriatus* should be used.

Kaston, B. J. 1939. *Conn. Agri. Exp. Sta. Bull.* 420. (See references following discussion of *S. multistriatus.*)

OAK BARK BEETLES

Pseudopityophthorus spp., Coleoptera, Scolytidae

These insects are of little importance as forest pests, but they may be encountered infesting dying tree limbs. Recently *P. minutissimus* (Zimm.) has been reported to be a vector for the oak wilt fungus. **HOSTS.** Oaks are preferred, but beech and birch are also commonly infested. Some species also infest other species of Fagaceae and a few attack other hosts such as hickory and dogwood. **RANGE.** Wherever the hosts grow. **GALLERY PATTERN.** *Egg galleries:* 2 short (½ to 1 inch long), straight narrow tunnels extend out across the grain of the wood in opposite directions from each entrance. *Egg niches* and *larval galleries:* individual. **BEETLES.** Small, less than 1/16 inch long; slender; head not visible from above; posterior of elytra not concave; mouthparts densely clothed with stiff hairs. **LIFE CYCLE.** Typical. There probably are 2 to 5 generations per year. **CONTROL.** Applied methods have never been needed.

ASH BARK BEETLES

***Leperisinus* spp., Coleoptera, Scolytidae**

These bark beetles attack only weakened, dying, and recently felled trees; therefore, they are not serious tree pests. **HOSTS.** *Fraxinus* spp. **RANGE.** Wherever the hosts grow. **GALLERY PATTERN.** Similar to that made by Scolytus. *Egg galleries:* from each entrance 2 short, straight tunnels extend in opposite directions across the grain of wood. *Egg niches* and *larval galleries:* individual. **BEETLES.** Less than ⅛ inch long; body covered with scales that are in light and dark patches giving the beetles a varigated appearance. **LIFE CYCLE.** The adults mostly hibernate over winter in short galleries. There are usually 2 or 3 generations per year. **CONTROL.** Applied methods have never been needed to control these insects.

ALDER BARK BEETLE

***Alniphagus aspericollis* (Le Conte), Coleoptera Scolytidae**

This species usually attacks only weakened, dying, or recently felled trees. **HOSTS.** Western species of *Alnus.* **RANGE.** Pacific coastal region. **GALLERY PATTERN.** *Egg galleries:* short, 2 to 5 inches long, narrow, straight, and extend with the grain of the wood. *Egg niches* and *larval galleries:* individual. **BEETLES.** About ⅛ inch long; similar to *Dendroctonus*, but differs by having 7 segmented antennal funicle. **LIFE CYCLE.** Typical with 2 generations per year. **CONTROL.** Applied methods have never been needed.

FLAT-HEADED INNER-BARK BORERS (METALLIC BEETLES)

Coleoptera, Buprestidae

Only a few of the many species of flat-headed inner-bark borers have been troublesome forest pests, and even these usually only attack weakened trees. The infested trees usually do not die rapidly but continue to live for many years. External symptoms consist of thinning foliage, branch dying, and, in smooth-barked trees such as birch, the bark surfaces may become irregular. Buprestids, like other inner-bark borers, cause a girdling type of damage, but generally they are not virulent tree killers. Some of the flat-headed borers that are more injurious as wood borers also bore in the inner bark for a time before they enter the wood; therefore, the beetle and larval characteristics of the various common genera of all tree-boring buprestids are presented in Tables 10.7 and 10.8.

TABLE 10.7 COMMON GENERA OF FLAT-HEADED BORER LARVAE*

1. Last abdominal segment with a pair of spines that project backwards; prothoracic back plate with a single longitudinal median line; maximum width of galleries about 1/8 inch; infest hardwoods (page 248) . . . . *Agrilus*

1. Larvae without spines on last abdominal segment; galleries broader than 1/8 inch wide 2

2. Prothoracic plates roughened 3

2. Prothoracic plates not roughened 7

3. Prothoracic plates with definite margins; roughness on plates tends to form ridges; markings dark and distinct 4

3. Prothoracic plates with indefinite margins; roughness on plates consists of small points; markings light 5

4. Back of pronotum (pronotal plate) with a Y-shaped mark (page 306) . . *Chalcophora*

4. Pronotal plate with a V-shaped mark; in hardwoods . . . . *Chalcophorella*

5. Lower prothoracic plate (prosternal plate) with a median groove extending from the back margin forward 2/3 to 3/4 the length of the plate (page 302) . . . *Buprestis*

5. Prosternal plate with a median groove which extends backward from the front margin 6

6. Prothoracic back plate (pronotal plate) oblong; prosternal plate narrow; with median groove extending complete length of plate; prothorax about as long as the following 3 segments combined; conifers hosts chiefly (page 251) . . . . *Melanophila*

6. Prothoracic plates circular; prosternal plate never completely bisected by median line; prothorax about as long as following 3½ segments (page 253) . . . . *Chrysobothris*

7. Prothoracic back plate (pronotal plate) with 2 dark lines forming a Y or V; prosternal plate with median line extending length of plate 8

7. Pronotal plate with a single median longitudinal line or groove that may widen at the anterior end 11

8. Pronotum with a V-shaped groove; prosternum with median line not bisecting plate; metathorax with 2 pairs of small tubercles . . . . *Anthaxis*

8. Pronotum with lines forming a V or Y shape; prosternal median line bisecting plate; tubercles absent on metathorax 9

9. Front (stem) end of pronotal V- or Y-shaped lines widened by numerous fine lines 10

9. Pronotal V-shaped lines not widened at pointed end by fine lines (p. 254) . . . . *Poecilonota*

10. Widened stem areas of Y mark on pronotal plate forms a depressed diamond-shaped area; hosts cypress, redwood, or cedar (page 305) . . . . *Trachykele*

10. Pronotal plate with V-shaped lines; stem area broad with fine lines; hosts pines, firs, poplar, and cypress (page 306) . . . . *Dicerca*
11. First abdominal segment as wide as or narrower than 2nd; prothorax about as long as the following 2½ segments combined . . . . *Chrysophana*
11. First abdominal segment wider than 2nd; prothorax about as or not as long as the following 2 segments combined . . . . *Acmaeodera*

* Adapted from Burke, H. E. 1917. *U.S.D.A. Bur. Entomol.* 437.

TABLE 10.8 COMMON GENERA OF BUPRESTID WOOD-BORER BEETLES*

1. Posterior coxae (coxal plates) not much widened near mid-part of body 2
1. Each posterior coxal plate greatly widened near place of attachment with rest of leg 4
2. Posterior pronotal margin straight 3
2. Posterior pronotal margin irregular; prothorax not grooved to receive antennae; 1st segment of hind tarsi as long as 3 following segments (page 248) . . . *Agrilus*
3. Mesonotum (scutellum) indistinct . . . . *Acmaeodera*
3. Scutellum distinct . . . . *Ptosima*
4. Posterior third of elytral margin with coarse teeth . . . . *Chalcophorella*
4. Posterior third of elytral margin smooth or with fine teeth 5
5. Beetles longer than ¾ inch; pronotum and elytra roughly sculptured so they appear like pressed leather (page 306) . . . . *Chalcophora*
5. Beetles shorter than ¾ inch or, if longer, the elytra are either smooth or contain many longitudinal lines 6
6. Eyes widely spaced on top of head; inner margins of the 2 eyes nearly parallel 7
6. Eyes close together on top of head; inner margins of eyes not parallel; 1st segment of hind tarsi much longer than 2nd (page 253) . . . . *Chrysobothris*
7. Posterior projection of prosternum (prosternal spine) not widened much behind legs; tip of projection rounded 8
7. Prosternal spine widened behind legs so as to form an acute point on each side; front edge of lower lip membranous (page 251) . . . . *Melanophila*
8. Front plate-like edge of lower lip (mentum) membranous, often sunken and sometimes paler in color (page 302) . . . . *Buprestis*
8. Front edge of mentum hard 9
9. Elytra without longitudinal lines (striae) (page 305) . . . . *Trachykele*
9. Elytra with striae; elytral tips often narrowed, prolonged and truncate or only truncate with 2 spines on tip of each 10
10. Small sclerite between inner corners of elytral bases (scutellum) small, round or oval, (page 306) . . . . *Dicerca*
10. Scutellum trapezoidal, much broader than long; pronotum with a smooth median line (page 354) . . . . *Poecilonota*

* Table adapted from Knull, J. N. 1950. *U.S.D.A. Misc. Publ.* 657:187–97 and Franklin, R. T., and H. O. Lund. 1956. *Ga. Agri. Exp. Sta. Tech. Bull.* N. S. 3.

GALLERY PATTERN. The adult beetles never re-enter the hosts after they have emerged; therefore, only larval galleries are present in the inner bark. Borings never are ejected to the outside, because the larvae do not make openings through the bark; consequently, the larval galleries are always tightly packed with fine borings. These often appear to be packed in curved layers.

INSECT APPEARANCE. *Larvae:* body elongate, soft, whitish, segmented, somewhat flattened and worm-like; head flattened, invaginated somewhat into the prothorax; prothorax enlarged, much widened and flattened, with well-developed hardened plates on both upper and lower surfaces; these are of about equal size and marked with lines or grooves; legless. *Pupae:* resemble the adults, but are whitish and have wing cases instead of wings and elytra. *Beetles:* body hard, somewhat flattened and boat shaped; color irridescent, brown to black but sometimes bright green or blue; antennal segments resemble saw teeth (serrate); legs with anterior coxae globular, femora attached to the sides of the trochanters; prosternum has a process that extends back and fits into a groove in mesosternum. *Eggs:* whitish to translucent; oval to ovoid; laid on the

FIG. 10.19. Typical flat-headed borer (*Buprestidae*) larvae showing the enlarged prothorax with the lined back (pronatal) plate. Above: (X2); below: (X6).

bark of the host trees. The common name, metallic wood borers, is derived from the metallic sheen of the beetles, whereas the enlarged, flattened prothoracic segment resembles an enlarged head; therefore, the larvae are called flat-headed borers.

LIFE CYCLE. Winter is passed as either partially grown or mature larvae. The latter pupate during the spring, and the adults appear during the spring or summer. The beetles frequently like hot weather; therefore, they are most active during sunny weather and are attracted to those portions of the tree boles exposed to the sun. The eggs are glued in bark crevices or beneath bark scales. After the larvae hatch, in a few days to a week, they bore directly into the inner bark where they tunnel and feed until mature. Then they form small pupal chambers, usually in the outer layers of the wood, in which they remain in a doubled up position until the following spring. The length of time required to complete 1 generation varies from 1 to several years depending on the physiological condition of the host trees.

HOST CONDITION. Many bupresteds attack chiefly weakened trees, and in these hosts larval development often is retarded. The larvae may survive and grow slowly for several years but mature only if and when the host dies. Apparently complete cessation of the radial growth must occur before the insects can complete their development but the actual mechanism involved is unknown. Here again is an example of where the success of insect development is dependent on the physiological condition of the host trees.

CONTROL. Certain inner-bark flat-headed borers have caused concern at various times, but none is primarily responsible for any major forest protection problem. Inasmuch as the attacks are associated so frequently with weakened host condition, the insect problems usually can be solved best by using better silvicultural methods to keep the trees growing vigorously.

TAXONOMIC REFERENCES

Burke, H. E. 1917. *U.S.D.A. Bull.* 437.
Franklin, R. T., and H. O. Lund. 1956. *Ga. Agri. Exp. Sta. Tech. Bull.* N.S. 3.
Knull, J. N. 1925. *Ohio State Univ. Studies* 2:1–71.
———. 1950. *U.S.D.A. Misc. Publ.* 657:187–197.

AGRILUS FLAT-HEADED BORERS

Agrilus spp., Coleoptera, Buprestidae

Bronze birch borer (*A. anxius* Gory.), the two-lined chestnut borer [*A. bilineatus* (Web.)], and the bronze poplar borer [*A. liragus* (B. and B.)] are common tree-infesting species.

When dying trees are heavily infested with *Agrilus* the insects commonly are thought to be the cause of tree death. Most of those who have worked on this problem, however, think that these borers are in the dying trees because the host material is suitable and are not the primary cause of tree decadence. Sometimes, however, they do kill trees such as the European varieties of white birch. Droughts also predispose trees to attack by causing temporary partial to complete cessation of radial growth. A serious problem affecting the yellow birch in eastern Canada is known as "Birch dieback." For a time it was thought that the bronze birch borer was the cause of this trouble, but now the insect appears to be of only secondary importance. The actual cause of the trouble is still unknown.

HOSTS. Species of *Agrilus* infest many hosts including *Ostrya,* hickories, butternut, oaks, boxelder, beech, *Crataegus,* poplars, willows, maples, dogwood, honey locust, black locust, and hackberry. The most important of these infest birches (*A. anxius*), oaks (*A. bilineatus*), and *Populus* (*A. liragus*).

RANGES. *A. anxius* and *A. liragus* occur throughout northern North America and south in the Rockies wherever the host birches and poplars grow. *A. bilineatus* occurs throughout eastern United States west to Colorado.

GALLERY PATTERN. Winding; maximum width about ⅛ inch; always tightly packed with fine borings; never any openings through the bark until the beetles emerge. The bronze poplar borer makes a compact zigzag type of gallery in host trees that are still growing radially. At intervals these galleries depart from the inner-bark region and dip into the sapwood. Molting occurs during these short departures into the wood.

INSECT APPEARANCE. *Larvae:* see Table 10.7; rather long and narrow; prothorax distinctly larger and broader than the following segments; prothoracic plates well developed; last abdominal segment with a pair of backward projecting spines. *Beetles:* generic characteristics are presented in Table 10.8; the bronze birch borer and the bronze poplar borers are olivaceous bronze, and *A. bilineatus* is bluish black with a whitish line along each side of the prothorax and elytra.

LIFE CYCLE. Winter is passed as young or mature (prepupae) larvae. The latter pupate in the spring and in early summer the beetles emerge. In suitable host material there is one generation per year, but in trees still growing, larval growth is repressed. These insects may live as larvae for several years and probably mature only if and when the hosts die.

CONTROL. The only practical type of applied control under forest

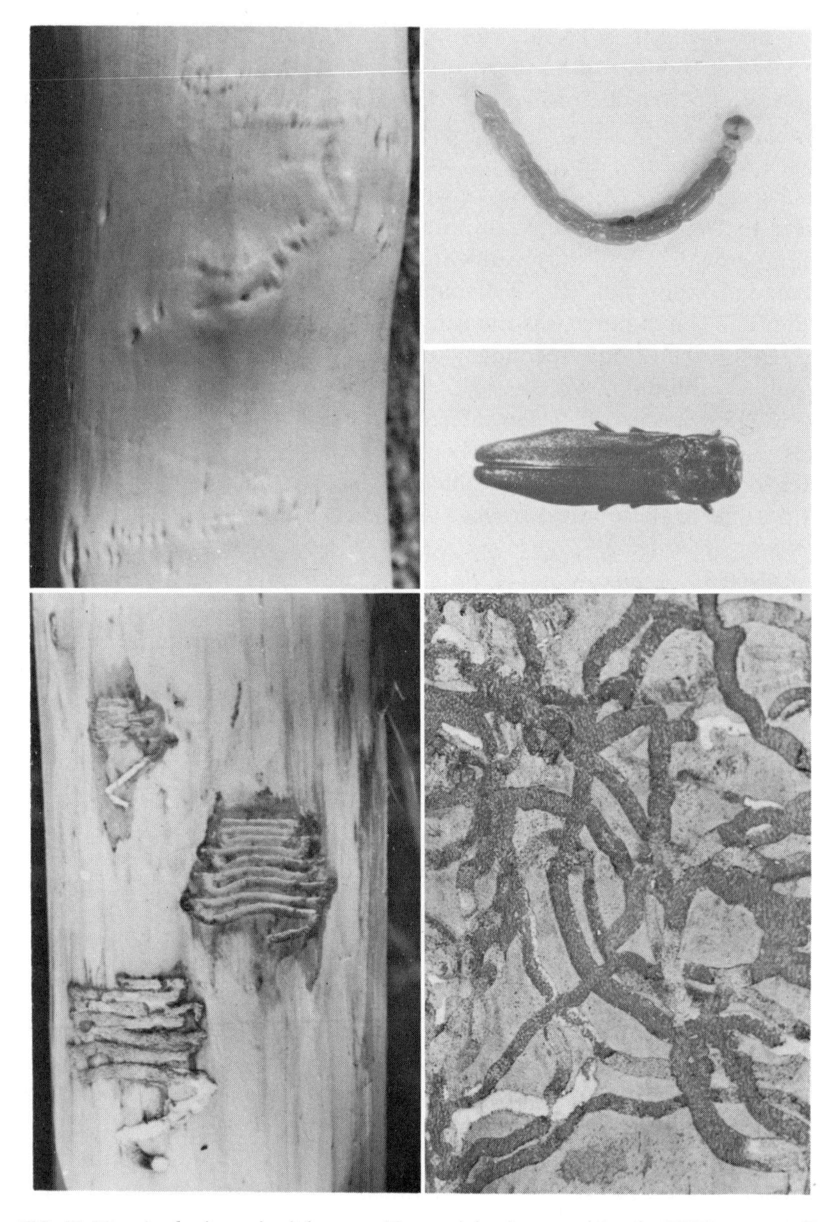

FIG. 10.20. *Agrilus* inner-bark borers. Upper right: larva and beetle (X3¼); upper left: distorted woody growth covering old larval galleries (X½); lower left: gallery pattern made by a larva of *A. liragus* in a living tree (X½); the areas between the different parts of the gallery are where the larva bored into the wood for the purpose of molting; lower right: *A. anxius* larval galleries (X1).

conditions consist of using silvicultural methods to keep the trees growing vigorously. Special care must be exercised when selectively logging yellow birch. Severe soil disturbance must be avoided and enough trees left to shade the soil around the uncut trees. Drying and heating of the surface soil causes excessive root mortality and subsequent tree decadence. This vulnerability to exposure dictates that the stands should be either cut completely (clear cut) or cut in groups.

Anderson, R. F. 1944. *J. Econ. Entomol.* 37:588–596.
Balch, R. E., and J. S. Prebble. 1940. *Forestry Chronicle* 16:179–201.
Chapman, R. N. 1915. *J. Agri. Research* 3:282–297.
Chittenden, F. N. 1909. *U.S.D.A. Bur. Entomol. Circ.* 24 (revised).
Hall, R. C. 1933. *Univ. Mich. School Forest and Conserv. Bull.* 3.
Hawboldt, L. S. 1952. *Nova Scotia Dept. Lands and Forests Bull.* 6.
Spaulding, P., and H. J. MacAloney. 1931. *J. Forestry* 29:1134–1149.

MELANOPHILA FLAT-HEADED BORERS

Melanophila spp., Coleoptera, Buprestidae

The eastern hemlock borer [*M. fulvoguttata* (Harr.)], the California flat-headed borer (*M. californica* Van. D.), the pine flat-headed borer (*M. gentilis* Lec.), and the flat-headed fir borer [*M. drummondi* (Kby.)], are common species.

The *Melanophila* borers are not very aggressive, for they generally infest only dying, recently felled, or very slow-growing trees. It is possible, however, that at times they may be more injurious. During the 1930's much hemlock timber died in the Lake States which was heavily infested with the eastern hemlock borer. Studies showed, however, that this mortality was not caused by the borer, since successful attacks occurred only in trees in which nearly all the roots were dead. Some of the western species are known as "fire bugs" because they are strongly attracted to forest, oil, and other fires. In burning forests they often try to land and oviposit on the hot, smoldering timber. They also often annoy the fire fighters by biting and crawling on them.

HOSTS AND RANGES. *Eastern hemlock borer:* hemlock and spruce in eastern Canada and northeastern United States. *California flat-headed borer:* pines in the far West and northern Rockies. The other two species listed above occur throughout the West, with *M. drummondi* also occurring in northeastern North America. *M. gentilis* infests pines and *M. drummondi* attacks spruces, firs, and pines. Other species infest pines, firs, and spruces wherever the hosts occur. Two species have been reported to infest hardwoods.

GALLERY PATTERN. Larger tunnels are always broader than ¼ inch; fine borings always tightly packed and frequently show curved packing

FIG. 10.21. Eastern hemlock borer (*Melanophila fulvoguttata*) beetle (X5) and larval gallery (X3).

lines; never any visible openings through the bark until beetles emerge; small tortuous galleries in slow-growing trees (*M. californica* and *M. fulvoguttata*) become resin soaked and covered with scar tissues.

INSECT APPEARANCE. *Larvae:* typical flatheads with characteristics as outlined in Table 10.7. *Beetle:* ¼ to 7/16 inch long; *M. gentilis;* bright bluish green, others are greenish bronze or black; *M. fulvoguttata* sometimes has yellowish spots on elytra; other generic characteristics are outlined in Table 10.8.

LIFE CYCLE. Winter is passed as larvae. Those which are mature (prepupae) pupate in the late spring and soon emerge as adults. In suitable host material there is 1 generation per year, but when they infest trees that are still growing, the larvae grow very slowly and may live for several years. Apparently only when the radial tree growth stops are they able to grow rapidly and mature.

CONTROL. Applied control consists of either clear cutting the stands or selectively cutting by groups and by removing the slow-growing and overmature trees. When eastern hemlock stands are selectively cut in the usual manner, they suffer severe post logging decadence. This occurs because the resulting soil exposure causes most of the roots to die from desication and heat.

Graham, S. A. 1943. *Univ. Mich. School of Forestry and Cons. Bull.* 10.
Secrest, H. C., H. J. MacAloney, and R. C. Lorenz. 1941. *J. Forestry* 39:3–12.
Sloop, K. D. 1937. *Univ. Cal. Publ. in Entomol.* 7:1–19.
West, A. S. 1941. *J. Econ. Entomol.* 34:43–45.
———. 1947. *Can. J. Research* 25 D:97–118.

CHRYSOBOTHRIS FLAT-HEADED BORERS

Chrysobothris spp., Coleoptera, Buprestidae

The flat-headed apple tree borer [*Chrysobothris femorata* (Oliv.)] and the Pacific flat-headed borer (*C. mali* Horn) are the most common species of this genus.

These insects have not been serious forest pests but have been troublesome to recently transplanted orchard and shade trees. The shock of transplanting causes retardation or temporary cessation of radial growth. Such trees are very attractive to the beetles and most suitable for subsequent larval development. When thin-barked trees are infested, the bark over the galleries often cracks and peels or sometimes becomes depressed. Sap and borings may exude from these places, but in thicker barked trees the injury always is concealed.

HOSTS. The two species listed previously infest almost all deciduous fruit and shade trees. A few species infest conifers, and one even breeds in cedar.

RANGES. *C. mali* occurs throughout the West and *C. femorata* is distributed throughout much of North America.

GALLERY PATTERN. Broad, larger ones ¼ inch or wider; irregular, shallow mines in both the inner bark and outer sap wood; in smaller thin-barked trees the galleries are deeper; borings tightly packed and often show the characteristic curved packing lines; pupal cell in the outer sapwood; beetle exit hole is oval.

INSECT APPEARANCE. *Larvae:* typical; with the distinguishing characteristics as presented in Table 10.7. *Beetles:* broad and flat; about ½ inch long; color dark brown on back with indistinct spots and bands of gray; bright metallic greenish blue beneath elytra; bronze underneath; whole insect has a brassy sheen; other generic characteristics are presented in Table 10.8. *C. femorata* has front of prosternum straight and basal half of large spine on front femur is serrate. *C. mali* has front of prosternum lobed and fore femur spine is smooth.

LIFE CYCLE. Both species overwinter as mature larvae (prepupae). Pupation occurs in the spring, and the adults emerge from late spring until summer. The adults prefer very warm situations; consequently, they are most frequently observed running over or resting on the sunny portions of the trees. The eggs are laid in the bark cracks. Usually there is only 1 generation per year, but in locations where the warm sea-

FIG. 10.22. Flat-headed apple tree borer (*Chrysobothris femorata*) beetle (X2) and larval gallery, (X2). Note the curved packing lines in the borings.

son is short it may take a year and a half or 2 years to complete their life cycle.

CONTROL. Applied control has never been needed in forests. For preventing attacks to shade and orchard trees the main parts of the boles can be wrapped with paper. The effectiveness of the newer insecticides has not yet been demonstrated, but probably some would work. If only a few trees are infested, the damaged bark areas can be opened and the larvae removed. The resulting wounds should be treated with tree paint.

Brooks, F. E. 1922. *U.S.D.A. Farmers' Bull.* 1065.
Burke, H. E. 1929. *U.S.D.A. Tech. Bull.* 83.
Fisher, W. S. 1942. *U.S.D.A. Misc. Publ.* 470.

POPLAR AND WILLOW FLAT-HEADED BORERS

Poecilonota spp., Coleoptera, Buprestidae

P. cyanipes (Say) is the most common species, but there are several others occurring in the West. The larvae bore and feed on the inner bark usually beneath and around wounds caused by other agents. Sometimes the mines extend into the sapwood. They have not been serious pests.

HOSTS. Willows and poplars.

RANGES. *P. cyanipes* occurs throughout northeastern North America west to the Rockies. Other species occur in the West.

INSECT APPEARANCE. *Larvae:* typical; identifying characteristics are presented in Table 10.7. *Beetles:* about ½ inch long; color irridescent bronze with sharp coppery-colored elytra tips; posterior of prosternum projection rounded; mesonotum (scutellum) large and trapezoidal shaped.

LIFE CYCLE. Two years.

CONTROL. Applied controls have never been needed in forests. For shade and ornamental trees the controls suggested for *Chrysobothris* are effective.

ROUND-HEADED INNER-BARK BORERS (LONG-HORNED BEETLES)

Coleoptera, Cerambycidae

Only a few cerambycids restrict their tunneling to the inner bark, and these are not very injurious. Many others, however, that are chiefly wood borers also bore in the inner bark for some time before they enter the wood. Therefore, these species together with the general characteristics for all round-headed borers are discussed in more detail in Chapter 11. A few species that restrict their boring only to the inner bark are discussed briefly below. Applied controls have never been needed to reduce the numbers of any of these round-headed inner-bark borers.

GALLERY PATTERN. The adult beetles do not re-enter the host after they have emerged; therefore, the boring damage is always caused only by the larvae. The irregular larval galleries are filled with coarse borings (frass), which rapidly turns brown.

ROUND-HEADED FIR AND LARCH BORERS

***Tetropium* spp., Coleoptera, Cerambycidae**

These inner-bark feeding cerambycids usually infest only felled trees, but a western species, *T. abietis* Fall, has been suspected of killing trees. **HOSTS.** Conifers, especially the true firs and larches. **RANGE.** Wherever the hosts grow. **BEETLES.** ½ to ¾ inch long; velvety brown beetles. **LARVAE.** Typical; body segmented, soft, whitish; head wider than long. See Tables 11.4 and 11.5. **LIFE CYCLE.** Little is known about the biology of these species. Probably they complete 1 generation per year. See Table 11.7.

ROUND-HEADED RIBBED PINE BORER

***Stenocorus lineatus* Oliv., Coleoptera, Cerambycidae**

This borer is a common inhabitant in the inner bark of felled logs and dying trees, but it is of no economic importance. The larvae produce much fibrous

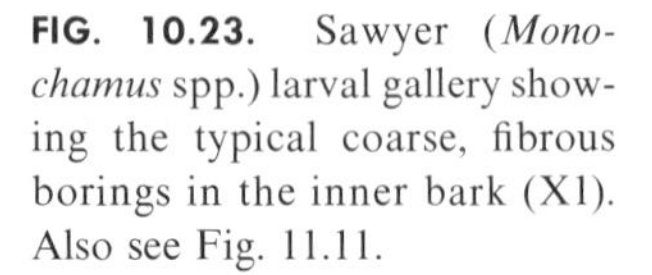

FIG. 10.23. Sawyer (*Monochamus* spp.) larval gallery showing the typical coarse, fibrous borings in the inner bark (X1). Also see Fig. 11.11.

brown borings, and when mature they construct characteristic nest-like pupal chambers in the inner bark. **HOSTS.** Pines and other conifer species, especially those cut or those that die during the late fall or winter. **RANGE.** Throughout North America wherever the hosts grow. **BEETLES.** ½ to ¾ inch long; dark gray with 3 lighter cross bands; prothorax cylindrical with a spine on each side; antennae short, about as long as the head and half the prothorax; each elytron with 6 raised longitudinal lines. **LARVAE.** Body segmented, soft, whitish, and very flattened; head wider than long; mandibles deeply notched at the tips. See Tables 11.4 and 11.5. **LIFE CYCLE.** The young beetles hibernate over winter in nest-like pupal chambers. They emerge very early in the spring and, after mating, the females lay their eggs in bark crevices of felled or dying trees. The resulting larvae bore irregular galleries in the inner bark and fill these with much coarse dark brown borings. Fibrous nest-like pupal cells are constructed late in the summer. In these the insects transform to adults.

ROUND-HEADED PINE BARK BORERS

Acanthocinus spp., Coleoptera, Cerambycidae

The boring characteristics and the nest-like pupal chambers are similar to those for the ribbed pine borer. **HOSTS.** Dying or recently felled pines, spruces, and firs. **BEETLES.** Elongated, somewhat flattened, ⅜ to 1 inch long; color grayish, brown mottled, or striped with black; antennae longer than whole body, with basal segments fringed with hairs underneath; females have long ovipositors **LARVAE.** Body very flattened; head longer than wide; mandibles pointed. **LIFE**

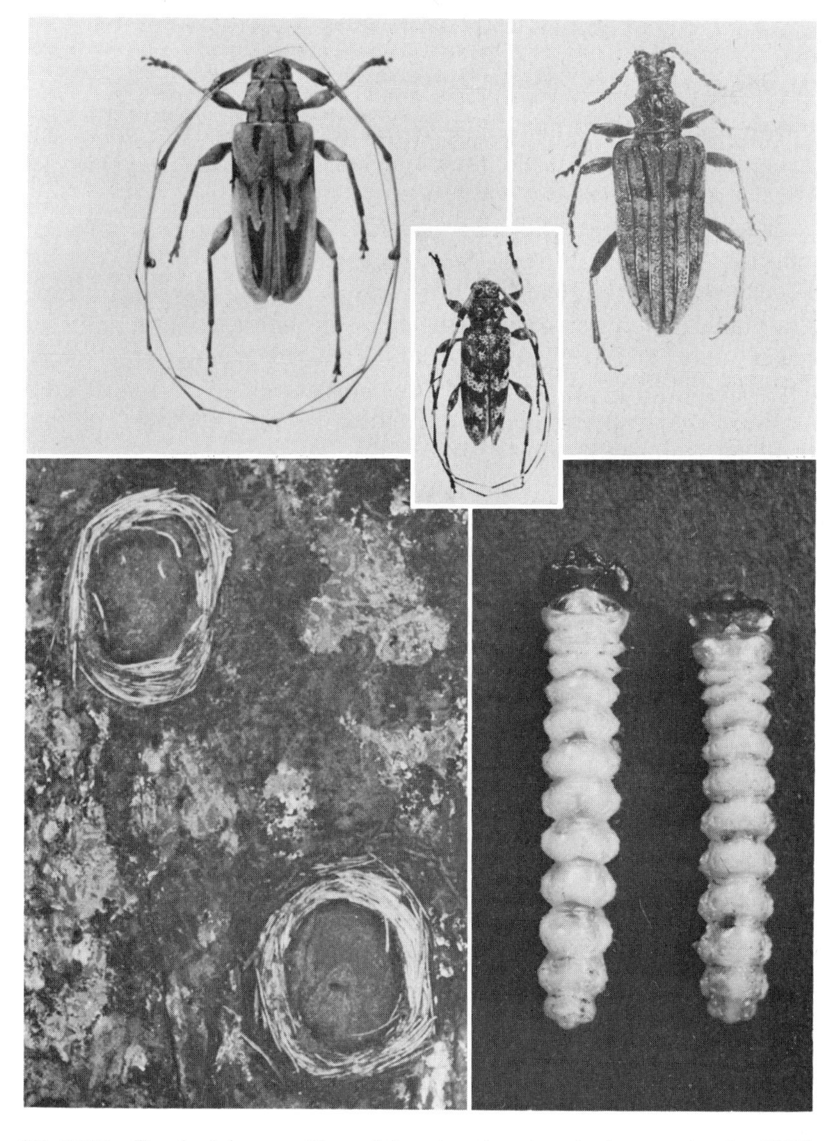

FIG. 10.24. Pine bark borers. Upper left and center: *Acanthocinus* spp. beetles (X1½); upper right: *Stenocorus lineatus* beetle (X2½); lower left: nest-like pupal chambers made by the larvae of both the above genera (X⅔); lower right: larvae of *S. lineatus* (X2½).

CYCLE. The beetles may be present anytime during the summer. After mating the females lay their eggs either in niches they gnaw or in bark beetle holes. There is only 1 generation per year. The nest-like pupal cell in the inner bark is similar to that formed by the ribbed pine borer.

WEEVIL INNER-BARK BORERS

Coleoptera, Curculionidae

Most weevil borers that attack the main tree boles cause no serious damage because they usually infest only dying or felled trees. One species, the root collar weevil, which is a pest of living trees, is considered in Chapter 17 along with the root feeders. Several others are most important as terminal borers. See Chapter 13.

GALLERY. Elongate, irregular to winding larval tunnels; often packed with granular borings; when living trees are infested the borings are ejected; pupal cells oval; some species (*Pissodes*) may construct small, fibrous, wood-chip cocoons. The adults never re-enter the hosts after they have emerged, but they often feed on the inner bark after cutting small holes through the outer bark.

INSECT APPEARANCE. *Larvae:* similar to bark beetle larvae except that they grow larger; head globular tan or brown, with mouthparts directed downward; body segmented, whitish, and cylindrical with long axis curved; legless. *Weevils:* beetles having the front of head prolonged into a long beak, which is curved downward; small chewing mouthparts at tip of beak; clubbed antennae attached to the beak.

LIFE CYCLE. Weevils commonly hibernate through the winter as adults. They appear in the spring or summer and feed for a time on inner bark by cutting small holes through the bark. The females cut other small circular holes in the bark in which the eggs are deposited. Usually there is only 1 generation per year.

POPLAR AND WILLOW WEEVIL

***Cryptorhynchus lapathi* (L.), Coleoptera, Curculionidae**

This European species has become established in the United States, where it is a serious pest of recently transplanted willows and poplars.

HOSTS. Sapling-sized trees of willows, poplars, alders, and birch.

RANGE. Eastern United States south to Virginia and in Washington and Idaho.

GALLERY PATTERN. Only larval galleries present in the inner bark; these usually extend transversely around the tree trunk, but they may be irregular and even zigzag; brown to black granular borings are extruded through small openings cut in the bark; mature larvae bore 1 to several inches into the wood to form pupal chambers; borings granular when in inner bark but coarse and fibrous when boring in wood.

INSECT APPEARANCE. *Larvae:* typical. *Weevils:* typical; ⅓ to ⅖ inch long; elongate; densely clothed with black and pale-colored scales intermixed with erect large, black bristles; body dark colored except posterior fourth of elytra, which is a pale yellow; beak as long as head and thorax and, when not feeding, is almost concealed in a groove on the underside of the thorax.

LIFE CYCLE. The weevils, which have never been observed to fly, appear during late July and August. They feed on the inner bark of year-old shoots for about 10 days before mating. The females then cut small round holes in the older bark and in each of these deposit 1 to 4 eggs. Egg laying continues until fall. The resulting larvae grow and mature by early summer the following year. There is only 1 generation per year.

CONTROL. Applied controls have not been used for this pest in recent years. It is very likely, however, that a spray treatment similar to that used for the black turpentine beetle would be effective.

Matheson, R. 1917. *Cornell Univ. Agri. Exp. Sta. Bull.* 388.

PINE BARK WEEVILS

Pissodes spp., Coleoptera, Curculionidae

Although these inner-bark boring weevils are of no economic importance, they are commonly present in logs and the boles of dying trees. They are most prevalent during the cooler spring and fall seasons when the bark beetles and sawyers are less active.

HOSTS AND RANGE. Pines, spruces, and firs, wherever the hosts grow.

GALLERY PATTERN. Only larval galleries present; rather uniform, long, straight to winding with tightly packed granular borings; pupal chambers about ½ inch long, constructed in the surface of the sapwood and are completely enveloped with tightly packed, coarse, fibrous shreds of wood (chip cocoons). See fig. 13.1.

INSECT APPEARANCE. *Larvae:* typical; see page 258. *Weevils:* typical; front of head prolonged into a slender beak which is grooved along each side to receive basal antennal segments; body color reddish brown to black; 3/16 to 5/16 inch long; sparsely to densely clothed with slender to broad scales which form spots; mandibles each with 3 teeth; antennae elbowed, with a compact club, place of antennal attachment on middle of beak and not visible when insect viewed from above; ventral abdominal segments of unequal length; 3rd and 4th much shorter than either 2nd or 5th.

LIFE CYCLE. Similar to that for the terminal feeding *Pissodes* weevils such as the white pine weevils.

CONTROL. These insects do not cause any damage.

ELM BARK WEEVILS

Magdalis spp., *M. barbita* (Say), and *M. armicollis* (Say), Coleoptera, Curculionidae

These weevils are not serious tree pests, for they generally infest only dying trees, branches, or green, recently felled trees.

HOSTS. The two species listed above infest only elms. Other species infest hickories, willows, chestnuts, butternuts, and pines.

RANGE. Wherever the hosts grow.

GALLERY PATTERN. Larval tunnels are about 1½ inches long; usually bored longitudinally in the inner bark; borings granular; oval pupal chamber also occurs in the inner bark.

INSECT APPEARANCE. *Larvae:* typical; see page 258. *Weevils:* small, ⅛ to 3/16 inch long; body somewhat wedge shaped black to bluish, greenish to reddish brown; (*M. barbita* is black, and *M. armicollis* is reddish brown); beak slender; antennae attached near tip of beak; hind angles of pronotum extend over bases of elytra; abdominal segments underneath of unequal length with segments 1 and 2 long and 3, 4, and 5 short.

LIFE CYCLE. The weevils hibernate over winter. They emerge in late spring and then feed for a time on elm foliage. The females lay their eggs in small circular holes in the bark. The larvae mature in the late summer and transform to adults. There is only 1 generation per year.

CONTROL. Applied methods have never been needed.

Felt, E. P. 1905. *N. Y. State Museum Bull.* 8:73–75.

INNER-BARK BORING CATERPILLARS

Lepidoptera, Aegeriidae, and Pyralidae

Caterpillar inner-bark borers are not very injurious either as forest or shade tree pests.

GALLERY PATTERN. Irregular tunnels through the inner bark. These may contain both globular excrement pellets and sometimes a small amount of silk. The external evidence of attack consists of exudations of resin, sap, and/or borings. Only the larvae cause the boring damage.

INSECT APPEARANCE. *Caterpillars:* typical; head globular, brownish; body naked, cylindrical, whitish to pinkish; 3 pairs of jointed, short, peg-like thoracic legs and 5 pairs of fleshy abdominal legs; latter have 1 or 2 rows of minute hooks (crochets) on ends, which are not of different intermixed lengths. *Moths:* clear-winged moths—wasp-like; colored tan to blue-black with a metallic iridescence; body frequently marked with yellow, orange, or red spots or bands; wings usually only partly covered with scales, which suggests the common name; depending on species, the wingspread varies from ½ to 1¾ inches. Pyralid moths, small and drably colored, have no good distinguishing characteristics.

PITCH MOTHS (PITCH MASS BORERS)

Vespamima spp., Lepidoptera, Aegeriidae, and *Dioryctria zimmermani* (Grote), Lepidoptera, Pyralidae

The caterpillars of pitch moths bore in the inner bark of conifers and cause a copious resin flow. They seldom kill trees but cause damage in the form of wood defects known as pitch pockets and pitch seams. These may be very large—up to 60 feet long and 6 or more inches deep—and subsequently become embedded in the wood as the trees heal over the injuries. *Dioryctria* spp. also kills the tops of large trees. Nevertheless, the total amount of damage these insects cause is not great.

HOSTS AND RANGE. Pines, Douglas-fir, and sometimes larches throughout North America. The damage has been most prevalent in the northern Rockies.

GALLERY PATTERNS. Several larvae live together forming winding and irregular tunnels 2 to 6 inches long. New galleries are frequently adjacent to older attacks or other wounds. Large masses of gummy pitch mixed with borings appear at the entrances. The moths, of course, are not borers.

INSECT APPEARANCE. *Caterpillars:* ¾ to 1½ inches long when full grown; color whitish, yellowish, or greenish; head tan or brown; 5 pairs prolegs. *Moths:* *Vespamima* spp.—clear-winged moths that resemble

FIG. 10.25. Pitch moth (*Vespamima* spp.) damage showing the resinous exudate (X½).

wasps; forewings transparent with dark borders; wingspread 1 to 1½ inches; body black with some abdominal segments bordered with yellow or red; sometimes with red spots on thorax or orange tuft at posterior end. *Dioryctria* spp.—about ½ inch long; wingspread 1¼ to 1½ inches; wings completely covered with scales, dark gray irregularly marked with reddish bands; hindwings pale yellowish white.

LIFE CYCLE. The moths appear during midsummer. After mating, the females lay their eggs adjacent to tree wounds. Two to four years are required for the *Vespamima* larvae to mature, but *Dioryctria* completes 1 generation per year. Pupation occurs within the resinous mass, but just before the moths emerge the pupae work their way to an outside edge. The caterpillars hibernate over winter.

CONTROL. These insects commonly repeatedly reinfest the same trees; therefore, it has been suggested that these "brood" trees should be removed. If ornamental trees are infested, the larvae can be removed and the wounds treated with a wound dressing. The insecticidal sprays recommended for treating deciduous trees infested with boring caterpillars probably would also be effective against these pitch mass borers.

Brunner, J. 1914 and 1915. *U.S.D.A. Bull.* 111, 255, and 295.

INNER-BARK BORING CATERPILLARS OF DECIDUOUS BROAD-LEAVED TREES

Lepidoptera, Aegeriidae

The various common species can be tentatively identified on the basis of hosts. These are as follows: maple callus borer [*Sylvora acerni* (Clem.)], *Acer* spp; dogwood borer [*Thamnosphecia scitula* (Harr.)], *Cornus* spp; ash borer [*Podosesia syringae fraxini* (Lug.)], *Fraxinus* spp; persimmon borer (*Sannina uroceriformis* Wlk.), *Diospyros* spp.; the hornet moth [*Aegeria apiformis* (Clerck)] poplar (*Populus* spp.), willow (*Salix* spp.), and the peach borers [*Sanninoidea exitiosa* (Say)], and [*Synanthedon pictipes* (G. and R.)], *Prunus* spp. Some of these, especially the last three, probably would be classified more correctly as wood borers, but inasmuch as they all infest smaller trees they are included here.

These inner-bark boring caterpillars are chiefly orchard and shade tree pests. The external evidence of attack consists of exudations of borings and/or gum from infested trees. Inside the galleries will be found the typical caterpillars and sometimes strands of silk and/or excrement pellets.

HOSTS AND RANGE. Various tree species as indicated previously, wherever the hosts grow.

GALLERY PATTERNS. Irregular tunnels in the inner bark and sapwood with borings (plus gum for *Prunus* spp.) ejected to the outside through holes in the bark.

INSECT APPEARANCE. *Caterpillars:* typical; ½ to 1 inch long when full grown; *Moths:* ½ to 1⅔ inches wingspread; wasp-like insects; color tan to blue-black with metallic iridescence; frequently marked with yellow, orange, or red spots or bands; wings usually only partly covered with scales, which suggests the common name "clear-winged moths."

LIFE CYCLE. The larvae hibernate over winter. Pupation occurs in the spring, usually in the old galleries or among the extruded gum and debris. Moth emergence occurs in late spring and early summer. After mating, the females deposit their eggs in cracks and crevices, usually around old wounds. These hatch within a week or two, and the resulting larvae bore

FIG. 10.26. Enlarged gall-like distorted branch growth produced by a peach tree borer (*Synanthedon pictipes*) (X⅔).

directly into the inner bark. Generally there is only 1 generation per year, but for some species there may be 2.

CONTROL. Applied control has never been needed to protect forest trees, but for orchard or ornamental trees various insecticides are being used. One spray consists of 3½ pounds of actual DDT (7 pounds 50 per cent wettable powder) per 100 gallons of water applied 1 to 4 times per season, depending on severity of the infestation. Applications must be made when the moths first appear and every 3 or 4 weeks thereafter. Malathion (1 pound actual insecticide per 100 gallons spray) can also be used in the same way. Shade trees can be treated by opening the places where borings are being ejected and injecting a fumigant such as carbon disulphide or calcium cyanide. Immediately after treating, the holes should be closed with putty or other similar plastic material. The caterpillars also can be removed mechanically and the wounds treated with any good tree wound dressing.

Snapp, O. I., and J. R. Thompson. 1943. *U.S.D.A. Tech. Bull.* 854.
Snapp, O. I. 1954. *U.S.D.A. Farmers' Bull.* 1861 (revised).

DIPTEROUS INNER-BARK BORERS

CAMBIUM MAGGOTS

Agromyza spp., Diptera, Agromyzidae

The maggots of these insects bore in the cambium and cause streak-like defects to appear in the wood, which are named "pith-flecks." Therefore, the damage, which is most frequently observed as a wood defect, never interferes seriously with tree growth but disfigures the wood so that it is less useful for fine cabinet work. See also page 343.

HOSTS AND RANGE. Various species infest birch, cherry, maple, and possibly other tree species. In the West a similar type of damage occurs to pines and other conifers. These insects probably occur wherever the hosts grow in North America.

GALLERY PATTERN. Small, less than 1/16 inch wide; extremely long; originate in twigs and extend down through the cambium of the bole to the tree bases. The resulting flecks in the wood are more commonly observed than the galleries in the cambium.

INSECT APPEARANCE. *Maggots:* about 1 inch long when full grown; very slender 1/32 inch in diameter; whitish; cylindrical; head indistinct and pointed, with pair of black, toothed mouth hooks; pair of spiracles on prothorax and another pair on last abdominal segment. *Puparia:* typical; brown, oval, cylindrical, and segmented; about 3/16 inch long.

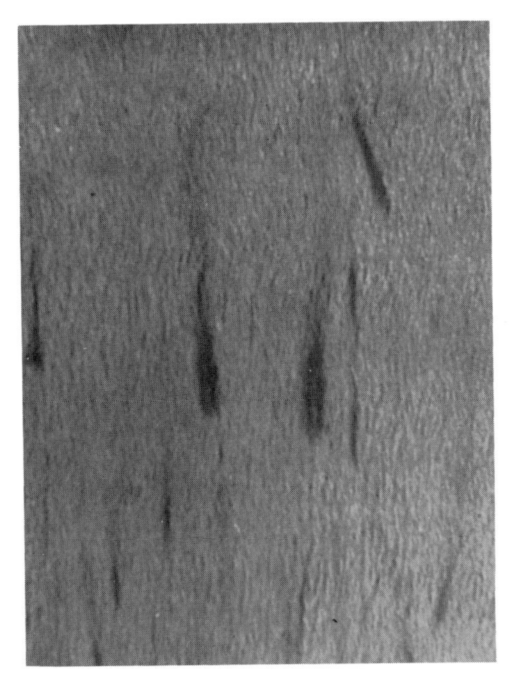

FIG. 10.27. Flecks in wood caused by a cambium miner (*Agromyza* spp.) (X2).

Flies: small ⅛ to $\frac{3}{16}$ inch long; typical flies having 2 membranous wings but without any good distinguishing characteristics.

LIFE CYCLE. The maggots mature and emerge from the trees during the summer. They immediately enter the soil where they pupate. In this stage they hibernate over winter. The flies emerge in the spring, mate, and the females lay their eggs on the lower branches of living trees. The resulting maggots bore into the cambium and there proceed toward the base of the trees. There is probably only 1 generation per year.

CONTROL. These insects have not caused much damage, and no applied control methods have been developed.

Brown, H. P. 1913. *U.S.D.A. Forest Serv. Circ.* 215.
Greene, C. T. 1914. *J. Agri. Research* 1:471–474.

HEMLOCK BARK MAGGOT

Cheilosia alaskensis Hunter, Diptera, Syrphidae

This cambium borer causes the formation of small, brown to black, resinous scarred areas, which are about an inch in diameter. These become imbedded in the wood and cause the lumber or other products cut from heavily infested timber to be severely degraded.

HOSTS AND RANGE. Western hemlock is the host of *C. alaskensis,* but a related species, *C. hoodiana* Bigot, infests white and grand fir. These insects occur in the Pacific coastal region from Alaska to Oregon. The fir-infesting species also occurs in the southern Rockies.

INSECT APPEARANCE. ***Maggots:*** ⅝ to ¾ inch long when full grown; whitish; body cylindrical; covered with microscopic hairs; head indistinct with mouth hooks; whip-like tail at posterior end; fleshy legs present on most segments. ***Flies:*** typical, small, 2-winged flies; wingspread about ¾ inch. ***Puparia:*** typical, segmented, brown, barrel-shaped structure.

LIFE CYCLE. The maggots hibernate in the galleries over winter. They pupate during the spring in the outer parts of the resin masses. After mating, the females deposit their eggs in wounds, and the resulting maggots bore into the cambium where they form the typical damage. As the larvae feed they always have their "tails" extending through the resin mass to the outside air. Three to five years are required for these insects to complete one generation.

CONTROL. There is no practical control for these insects.

CHAPTER 11

Wood-Boring Insects

Many insect species and a few other invertebrate animals are wood borers. Some of these obtain both sustenance and shelter from the wood, whereas others use the wood only as a place in which to live. Certain species attack only living trees; others are found chiefly in freshly felled or dying trees; a few infest only dry seasoned wood, and still others attack only old moist wood. Those that attack trees and fresh logs frequently bore and live in the inner bark for a variable period of time before they enter the wood; therefore, these also can be considered to be inner-bark borers. Nevertheless, since these insects usually are more injurious as wood borers, they are considered here. Some that attack only freshly killed or felled trees can survive and develop slowly in dried wood; therefore, these species often continue boring in wood that has been dried and manufactured. The various types of damage are outlined in Table 11.1. See also page 343 for additional information dealing with wood discolorations associated with insect borers.

Most insects found boring or living in wood are beetle larvae. The numerous species belong to many families, but only a few of these contain most of the important wood borers. These can be identified by referring to Table 11.2.

TABLE 11.1 COMMON TYPES OF WOOD-BORING DAMAGE

1. Galleries more or less filled with powdery or granular boring dust 2
1. Galleries either free of borings, or they are coarse and fibrous; soil sometimes present 7
2. Galleries and beetle emergence holes up to 1/8 inch in diameter; larvae with globular heads; body cylindrical, curved somewhat in long axis; with 3 pairs small, jointed, thoracic legs (page 307) . . . . SMALL POWDER POST BEETLES 3

2. Some galleries and adult exit holes usually larger than 1/8 inch in diameter . . . 5
3. Powdery borings contain small oval to cylindrical pellets which can be seen with the aid of a hand lens; infest both sapwood and heartwood of both conifers and hardwoods; larvae usually hairy, head conspicuous, with mouthparts located close to prothorax (page 311) . . . . ANOBIIDAE
3. Borings powdery; pellets absent; only sapwood of ring porous hardwoods mostly attacked; larval mouthparts directed forward 4
4. Borings tend to clump; larger galleries and beetle exit holes usually larger than 1/16 inch in diameter (1/16 to 1/8 inch); infest hardwoods and conifers; larvae with posterior spiracles not much larger than others (page 309) . . . . . BOSTRICHIDAE
4. Galleries always less than 1/16 inch in diameter; borings always flour-like; only in sapwood of hardwoods; larvae with posterior pair of spiracles much larger than the others (page 307) . . . . LYCTIDAE
5. Galleries circular in shape; up to 1/4 inch in diameter; in wood of both hardwoods and conifers; larvae with a single pointed hardened spine extending backward from the last abdominal segment; body cylindrical and elongate; head globular with mouthparts directed downward; no eyes; thoracic legs fleshy and not jointed (page 333) . . . . HORNTAILS
5. Larval galleries oval in shape; beetle exit holes sometime circular; usually larger than 1/4 inch; larvae whitish, elongate, cylindrical to flattened; head flattened with mouthparts directed forward . . . . LARGE POWDER POST BEETLES 6
6. Galleries are flattened oval shaped; borings fine and tightly packed in galleries; frequently packed in layers so that curved arc-like lines appear; larval prothorax flattened and distinctly wider than abdominal segments with distinct similar-sized plate-like areas on back (pronotum) and underneath (prosternum); pronotal plate always with a distinct longitudinal median line, V- or Y-shaped marking or groove; always legless; spiracles cresent shaped (page 301) . . . . FLAT-HEADED BORERS
6. Galleries broadly oval shaped; borings usually granular; larval prosternal plates less well developed than pronotal plates; spiracles oval in shape; with or without jointed thoracic legs (page 297) . . . . ROUND-HEADED BORERS
7. Damage occurs only to wood exposed in sea or brackish water; galleries of some always with lime coated walls (page 339) . . . . MARINE BORERS 8
7. Terrestrial wood borers 9
8. Marine borers that honeycomb the surface of the wood by extending burrows only into the outer shell of the solid wood; holes 1/16 to 1/2 inch in diameter; animals with seven pairs of jointed legs and 2 pairs of antennae; head united with thorax; 1/4 to 1/2 inch long (page 342) . . . . WOOD LICE
8. Tunnels circular, larger ones 3/8 to 7/8 inch in diameter; lined with a whitish hard lime coating; body unsegmented, very elongate worm-like without a distinct head, with a small pair of shell-like structures at the anterior end (page 339) . . . . SHIP WORMS
9. Galleries less than 1/8 inch in diameter (pinholes) 10
9. Larger galleries more than 1/8 inch in diameter 11
10. Pinholes rather uniform in size; walls usually dark stained; small cylindrical beetles with elytra (wing covers); antennae elbowed with a head-like club (capitate); larvae legless and resemble bark beetles (page 312) . . . . AMBROSIA BEETLES

10. Tunnels vary considerably in size; walls usually not stained; larvae with 3 pairs jointed thoracic legs; body elongate (page 316) . . . . TIMBER WORMS
11. Galleries irregular in shape and interconnecting so as to "honeycomb" the wood; soil often present but galleries generally free of wood borings 12
11. Galleries oval to circular in shape; if the wood is rather "honeycombed" fibrous borings may be present; soil never present in galleries (grub holes) 14
12. Small oval excrement pellets (less than 1/32 inch long) present; these with rounded ends and smooth concave depressions on sides; whitish, thick-waisted, ant-like insects; most are wingless but sometimes black-winged forms are present; antennae with rounded bead-like segments (page 324) . . . . DRY WOOD TERMITES
12. Pellets never present 13
13. Walls of galleries roughly coated with a whitish plaster-like material; soil sometimes packed in galleries; small earthen tubes about (3/16 inch in diameter) frequently present on outside of infested wood and on adjacent masonry; insects similar to those for dry wood termites (page 318) . . . . SUBTERANIAN TERMITES
13. "Honey combing" damage without the characteristics listed in 13 above; piles of borings outside entrances are coarse and fibrous; adults insects are large black ants; bodies constricted between thorax and abdomen with the two parts connected with a small node-like segment; winged or wingless; live in colonies (page 334) . . . . CARPENTER ANTS
14. Galleries circular; walls smooth, appear gouged out; crosswalls of macerated wood at 1/2- to 3/4-inch intervals sometimes present; circular exit hole always present; borers in dry posts and timbers; larvae whitish, grub-like, 1/2 to 3/4 inch long when full grown; with small head and weakly developed mouthparts (page 377) . . . . CARPENTER BEES
14. Galleries generally oval in shape; do not have characteristics listed above; borers in living or dead trees, wet or dry wood 15
15. Larvae legless or with 3 pairs of small peg-like thoracic legs; elongate, whitish; head flattened with mouthparts directed forward; excrement never globular pellets (page 271) . . . . ROUND-HEADED WOOD BORERS
15. Sometimes a light silken web may be present in galleries; larvae are whitish, cylindrical caterpillars with 3 pairs of short peg-like thoracic legs and 5 pairs of fleshy abdominal legs (prolegs) armed with groups of minute hooks (crochets); excrement pellets globular (page 325) . . . . CARPENTER WORMS

TABLE 11.2 LARVAE OF COMMON FAMILIES OF BEETLES FOUND IN SOUND WOOD

1. Legs absent 2
1. With 3 pairs of jointed thoracic legs present 5
2. Head flattened with mouthparts projecting forward; body elongate, cylindrical or flattened 4
2. Head rather globular with mouthparts located near the prothorax; body cylindrical, usually curved somewhat in the long axis 3
3. Parent beetles also usually present in galleries; beetles small; body cylindrical; antennae each with head-like club (capitate) (page 312) . . . . (ambrosia beetles) SCOLYTIDAE and PLATYPODIDAE

3. Adult weevils never present in galleries, but occasionally may be found in pupal chambers . . . . (weevils) CURCULIONIDAE
4. Prothoracic plate-like area on back (pronotal plate) about equal in size to that on underside (prosternal plate); prothorax enlarged, much widened and flattened; spiracles cresent-shaped (page 301) . . . (flat-headed wood borers) BUPRESTIDAE
4. Pronotal plate larger than the prosternal plate; prothorax sometimes somewhat flattened and widened; spiracles oval (page 271) . . . . (round-headed wood borers) CERAMBYCIDAE
5. Posterior of abdominal, segment bears a hard median toothed ridge or a single long projection (page 316) . . . . (timber worms) LYMEXYLIDAE
5. Posterior of abdomen without a hard median toothed ridge or a single long projection, but sometimes a pair of projecting structures (cerci) may be present on the back of the last abdominal segment 6
6. Cerci jointed . . . . HISTERIDAE
6. Cerci, if present, not jointed 7
7. Head and body whitish and soft; parasites on wood borers . . . . (cocoon-forming beetles) BOTHRIDERIDAE
7. Head hardened and colored darker than body 8
8. Posterior abdominal segment with 2 or more pairs of spines (cerci) (page 316) . . . . (timberworm, *Strongylium* spp.) TENEBRIONIDAE
8. Cerci absent or only one pair present 9
9. Legs well developed; longer than half the thickness of thorax 10
9. Legs poorly developed; length less than half the thickness of thorax 12
10. Posterior edge of mouth parts on underside of head attached behind where the mandibles are attached (retracted) 11
10. Mouthparts not retracted (page 211) . . . . (checkered beetles) CLERIDAE
11. Sac-like depression between pair of recurved spines (cerci) on posterior abdominal tergite; body cylindrical and elongate; predators on ambrosia beetle larvae . . . . (cylindrical bark beetles) COLYDIIDAE
11. Not as in 11 above, but cerci present (page 211) . . . . . . (ostomids) OSTOMIDAE
12. Head flattened with mouthparts directed forward; body elongate, cylindrical to somewhat flattened (page 271) . . . (round-headed wood borers) CERAMBYCIDAE
12. Head rather globular; body cylindrical and curved somewhat in long axis . . . 13
13. Legs each with only one segment plus a claw (page 316) . . . . (timberworms) BRENTIDAE
13. Legs with 3 or 4 segments plus a claw . . . . (small powder post beetles) 14
14. Posterior abdominal spiracle much larger than others, mouthparts directed forward (page 307) . . . . LYCTIDAE
14. Posterior spiracle about as large as others 15
15. Body usually hairy; head conspicuous; mouthparts located close to prothorax (page 311) . . . . ANOBIIDAE
15. Body not hairy; head retracted into prothorax, often inconspicuous; mouthparts directed forward (page 309) . . . . BOSTRICHIDAE

LARVAL TAXONOMY

Böving, A. G., and F. C. Craighead. 1931. *Larvae of Coleoptera.* Brooklyn Entomol. Soc., Brooklyn, N. Y. 351 pp.
Craighead, F. C. 1950. *U.S.D.A. Misc. Publ.* 657:153–343.
Peterson, A. 1948 and 1951. *Larvae of insects.* Edwards Bros., Inc., Ann Arbor, Michigan. Part I: 315 pp. Part II: 416 pp.

GENERAL REFERENCES

Chamberlin, W. J. 1949. *Insects affecting forest products and other materials.* Oregon State College Cooperative Association, Corvallis, Oregon. 159 pp.
Craighead, F. C. 1950. *U.S.D.A. Misc. Publ.* 657:153–343.

ROUND-HEADED WOOD BORERS (LONG-HORNED OR LONGICORN BEETLES)

Coleoptera, Cerambycidae

Adult cerambycids are commonly known as long-horned beetles because they have long antennae. The other name, round-headed wood borer, probably originated from the fact that some species make circular emergence holes. The larvae do not have round heads, and they always bore oval tunnels.

Both the larvae and beetles of cerambycids are commonly observed by foresters. Although some are pests of living trees, few are tree killers. Smaller trees sometimes are girdled and killed, whereas heavy infestations in larger trees weaken them so that they may be broken easily by wind. Those that bore in dry, dead wood are serious pests, but those that require moist dead wood generally are beneficial, for they help disintegrate the wood debris present in forests. Round-headed wood borers that infest living trees are mostly restricted to the broad-leaved deciduous hosts. In these hosts their grub holes often result in such heavy losses that many infested logs are not even worth removing from the woods. Shade trees also are frequently attacked and damaged. Those that infest freshly felled or killed timber often cause much loss following catastrophic timber destruction by wind storm, fire, or attacks by other insects. If this killed timber cannot be salvaged immediately it soon becomes infested and damaged by round-headed and other borers. Only the larvae cause the wood-boring damage. The adults never reenter the hosts after they have emerged.

GALLERY PATTERN. The oval larval galleries usually wind irregularly through the inner bark and/or wood. The larger tunnels of most species are $\frac{3}{16}$ to $\frac{5}{16}$ inch by $\frac{3}{8}$ to $\frac{3}{4}$ inch in cross section; but the holes bored by large species, such as the ponderous borer, may be 1 by 2 inches across. Either granular and/or fibrous borings are produced. Often, however, those that make only granular borings throughout most of their larval lives construct a fibrous plug adjacent to their pupal chambers. Some keep their galleries clear of borings by ejecting them through small holes in the bark. Others pack the frass and borings in the galleries. The females of some species cut small niches through the bark in which the eggs are deposited.

INSECT APPEARANCE. *Larvae:* body segmented, whitish, elongate, robust to slender, soft and fleshy, naked (not hairy), cylindrical to somewhat flattened; head flattened, usually partly imbedded in prothorax; mouthparts directed forward; main jaws (mandibles) with blunt, rounded or pointed tips; without grinding (molar) areas; prothorax is largest of the thoracic segments and has an indistinct plate on back (pronotum) that is larger and more distinct than that underneath (prosternum); pronotum without V- or Y-shaped markings, but some have a median longitudinal groove; legs small, short, peg-like, and widely spaced occur on many species, but some are legless; abdominal segments telescope and have prominent, fleshy, ambulatory swellings (ampullae) on the

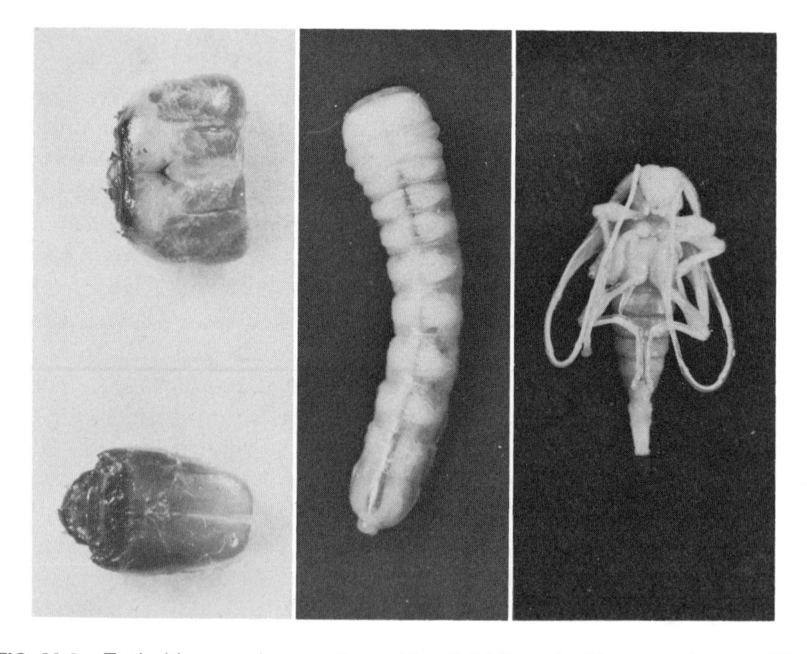

FIG. 11.1. Typical larva and pupa of round-headed (*Cerambycidae*) wood borers (X2). Left: the elongate and the broad types of larval heads (X5).

FIG. 11.2. The adult of a parasite (*Megarhyssa lunator*) that attacks wood-boring larvae. The long ovipositor is inserted into the wood for the purpose of placing eggs near the host grubs (X⅔).

back, sides, and underneath on the first 6 or 7 abdominal segments; spiracles oval shaped. ***Pupae:*** somewhat resemble adults except they are whitish colored and have wing cases instead of fully developed elytra. The appendages are never tightly appressed to the body. ***Beetles:*** cylindrical; oblong, elongate to somewhat flattened; sometimes beautifully colored; antennae usually thread-like, long, extending at least beyond the middle of the pronotum, but usually they are much longer, often being as long or longer than body; legs long with 5 segmented tarsi, which sometimes appear 4 segmented; 4th segment usually is small and closely united to the 5th; 3rd segment bilobed except on *Parandra* spp. ***Eggs:*** whitish; commonly elongate tapering to points at each end (fusiform).

LIFE CYCLE. The length of the life cycle varies from 1 to many years, depending on the species and host condition. For those having larvae that bore extensively in the rich inner-bark region, the 1-year life cycle prevails; whereas those that bore chiefly in the wood usually require 2 to 4 years to reach maturity. The beetles may appear from early spring to late fall, depending on the species, and live as adults for only a few weeks. Some long-horned beetles feed on the tender bark of twigs, others eat leaves, a few eat fungi, and some do not feed. The oviposition habits vary according to the species. Some females lay their eggs beneath bark scales or in bark crevices. Others place their eggs in the inner bark after they first cut niches through the dry outer bark. Some even use holes made by other insects. Winter usually is passed either as partially grown or as mature (prepupae) larvae. A few, however, hibernate as young adults in pupal chambers. There are so many species of round-headed wood borers that only a few can be considered here in detail. For some of the others the identifying and biological characteristics are summarized in Tables 11.3, 11.4, 11.5, 11.6, and 11.7. Additional information is presented below for the more important species.

Craighead, F. C. 1915. *U.S.D.A. Bur. Entomol. Rept.* 107.

————. 1923. *Can. Dept. Agri. Bull.* 27 (new series).

————. 1950. *U.S.D.A. Misc. Publ.* 657:228–271.

Chamberlin, W. J. 1949. *Insects affecting forest products and other materials.* Oregon State Coll. Coop. Asso., Corvallis, Oregon. Pp. 61–75.

Dillon, L. S., and E. S. Dillon. 1941. *Reading Public Museum and Art Gallery Sci. Publ.* 1.

Keen, F. P. 1952. *U.S.D.A. Misc. Publ.* 273 (revised). 173–176, 191–198.

Knull, J. N. 1946. *Ohio Biological Survey Bull.* 39.

TABLE 11.3 COMMON GENERA OF ROUND-HEADED BORER LARVAE THAT INFEST THE MAIN BOLES AND LARGER BRANCHES OF BROAD-LEAVED TREES, LOGS, OR WOOD*

1. Larvae bore in living trees 2
1. Larvae bore in recently dead trees, dry or moist dead wood; larvae always with heads wider than long† 15
2. Borings mostly packed in galleries 3
2. Borings mostly ejected from galleries 9
3. Galleries chiefly in or beneath the bark but sometimes also in the wood 4
3. Galleries mostly in the wood; usually so abundant that wood is honeycombed; larvae with heads always wider than long†; with 3 pairs thoracic legs 8
4. Galleries mostly in the dry outer bark of oaks and chestnut, but these may extend to the inner bark; larvae with heads wider than long;† legs present 5
4. Infest species of hardwoods other than oak and chestnut; galleries beneath the bark and also in wood 6
5. Borings mostly granular, but there may be a fibrous plug near pupal chambers; 1 pair distinct simple eyes; tips of mandibles rounded (Table 11.6) . . . . OAK BARK BORER (*Romaleum cortiphagus* Craighead)
5. Borings mostly fibrous, some of which may be ejected; larvae with indistinct simple eyes; mandibles pointed at tips . . . . CHESTNUT BARK BORER [*Leptura nitens* (Forst.)]
6. Infest various hardwood but not ash or poplar; borings mostly fibrous; larvae with head longer than wide†; legless; segmental body swellings (ampullae) roughened (page 293) . . . . *Saperda* spp.

6. Infest ash or poplar; borings mostly granular except near pupal chambers; larvae with heads wider than long†; with or without legs 7
7. Infest ash; larvae with 3 pairs small jointed legs (Table 11.6) (page 282) . . . . BROWN ASH BORER (*Tylonotus bimaculatus* Hald.)
7. Infest base of poplars; galleries often so numerous that the wood is honeycombed; larvae legless (page 290) . . . . POPLAR BUTT BORER (*Xylotrechus obliteratus* Lec.)
8. Large oval galleries; larger ones wider than ¾ inch; mostly packed with fibrous borings; larval pronotum smooth; prosternum with 2 groups of small dot-like points; each segmental body swelling with 2 impressed cross lines (Table 11.6) (page 282) . . . . HARDWOOD STUMP BORER (*Stenodontes dasystomus* Say)
8. Galleries less than ¾ inch wide; borings mostly granular; larval pronotum roughened with small points (Table 11.6) . . . POLE BORER [*Parandra brunnea* (F.)]
9. Boring fibrous; larvae with heads longer than wide†; legless 10
9. Borings mostly granular but there may be fibrous plugs near pupal chambers; larvae with heads wider than long†; 3 pairs thoracic legs 13
10. In poplars, willows, basswood or Rosaceae species 11
10. In oaks, beeches, hickories, elms, ironwoods and sycamore; larval segmental body swellings (ampullae) with small tubercles (page 282) . . . . TRUNK BORERS (*Goes* spp.)
11. In bases of cottonwood and willows throughout Central and Southern United States; larval pronotum brownish; segmental body swellings (ampullae) with small rounded tubercles (page 290) . . COTTONWOOD BORER [*Plectrodera scalator* (F.)]
11. Occurs throughout range of hosts; larval ampullae with small points . . . . SAPERDA 12
12. Infests *Populus* spp. (page 290) . . . . ASPEN BORER (*Saperda calcarata* Say)
12. Infests species other than poplar (page 293) . . . . *Saperda* spp.
13. In oaks; larvae with segmental body swellings (ampullae) covered with fine lines; legs each 4 segmented including claw (page 282) . . . . RED OAK BORER (*Romaleum rufulum* Hald.)
13. In host trees other than oaks; legs each 3 segmented including claw; ampullae finely granular 14
14. In hard maples; posterior of pronotum with network of fine lines; mesonotum with a V-shaped impression (Table 11.6) . . . . SUGAR MAPLE BORER [*Glycobius speciosus* (Say)]
14. In black locust (page 288) . . . . . . LOCUST BORER [*Magacyllene robiniae* (Forst.)]
15. Galleries in old, dead, moist wood; often so numerous so that the wood is honeycombed larvae always with 3 pairs thoracic legs 16
15. Galleries beneath the bark of recently dead or felled trees, but the larvae often continue to bore in the wood of logs and dried lumber 18
16. Large oval galleries; larger ones wider than ¾ inch; borings fibrous; larval body swellings (ampullae) each with 2 transverse impressed lines 17
16. Galleries smaller than ¾ inches wide; borings mostly granular; larval pronotum roughened with small points (Table 11.6) . . POLE BORER [*Parandra brunnea* (F.)]
17. Larvae with 2 flat ridge-like teeth on front of head (Table 11.6) . . . . BROWN PRIONID [*Othosoma brunneum* (Forst.)]
17. Larvae do not have ridge like teeth on front of head; prosternum with 2 groups

of 5 to 12 small dot-like elevations (Table 11.6) (page 282) . . . . HARDWOOD STUMP BORER (*Stenodontes dasystomus* Say)

18. Borings mostly granular 19
18. Borings fibrous 26
19. Borings packed in galleries 20
19. Borings ejected through a small hole in bark; larvae with 3 pairs legs 24
20. Galleries in wood loosely packed with borings but borings ejected from galleries beneath bark; larvae with 3 pairs legs; under margin of head roughened (page 293) . . . . BANDED HICKORY BORER [*Chion cinctus* (Drury)]
20. All galleries tightly packed with borings 21
21. Galleries chiefly in wood; larvae with 3 pairs thoracic legs; boring in dry seasoned wood as well as in that freshly dead 22
21. Galleries chiefly beneath bark; larvae legless (Table 11.6) . . . . RUSTIC BORER (*Xylotrechus colonus* F.)
22. Larval legs small; each with 2 or 3 segments including claw 23
22. Larval legs distinct; each with 4 segments including claw (page 299) . . . . IVORY-MARKED BEETLE (*Eburia* spp.)
23. Larvae with smooth white triangular area on prosternum (page 299) . . . . FLAT OAK BORER [*Smodicum cucujiforme* (Say)]
23. Larvae not as in 23 above; body finely granulated (page 300) . . . . ASH BORERS (*Neoclytus* spp.)
24. Borings ejected from galleries beneath bark but are loosely packed in wood; larvae with 1 pair simple eyes (ocelli) (page 293) . . . . BRANDED HICKORY BORER [*Chion cinctus* (Drury)]
24. Borings mostly ejected from all galleries; larval legs each with 3 or 4 segments including claw; 2 or 3 pairs ocelli 25
25. Larval legs each with 4 segments including claw; 2 pairs ocelli (Table 11.6) (page 293) . . . . SPINED BARK BORER [*Elaphidion mucronatum* (Say)]
25. Larval legs each with 3 segments including claw; 3 prs. ocelli (page 288) . . . . PAINTED HICKORY BORER (*Megacyllene caryae* Gahan)
26. Larvae with pronotum brownish pubescent; small tubercles on body . . . . *Leptostylus* spp.
26. Pronotum with small hard points; hosts, elms, basswood or hickory and others (see Nos. 6 and 11) (page 293) . . . . *Saperda* spp.

* Derived from Craighead, F. C. 1950. *U.S.D.A. Misc. Publ.* 657:231–271.

† Heads must be removed to determine the relation of head length to width.

TABLE 11.4 COMMON GENERA AND SPECIES OF ROUND-HEADED BORER LARVAE THAT INFEST THE MAIN BOLES AND LARGER BRANCHES OF CONIFERS*

1. Infest species of Pinaceae (pines, spruces, firs, larch, and hemlock) 6
1. Infest species of Cupressaceae and Taxaceae (cedars, junipers, cypress, and red woods); larval heads always wider than long†; mandible tips rounded (except *Atimia*); small thoracic legs present 2

2. Borings mostly ejected from galleries; granular except near pupal chambers . . 3
2. Borings mostly packed in galleries; either granular and/or coarse and fibrous . . 4
3. Galleries mostly beneath the bark, but deeply scoring the wood leaving sharp ridges; pupal chambers deep in wood; larvae flattened; underside of heads with 4 small tubercles; eyes absent (page 299) . . . . BLACK-HORNED BORERS (*Callidium* spp.)
3. Larvae bore extensively in both the inner bark and wood; larvae slender; head with many curved hairs on sides; 1 pair simple eyes (page 293) . . . . CYPRESS BORER (*Oeme rigida* Say)
4. Borings mixed granular and fibrous; galleries mostly in inner bark, but mature larvae enter wood to pupate; 2 spines on last abdominal segment (page 293) . . . . SMALL CEDAR BORER (*Atimia confusa* Say)
4. Borings mostly granular, but fibrous wads may be present near pupal chambers; galleries mostly in inner bark but sapwood is deeply scared; pupal cells in sapwood 5
5. Larvae with 2 dark spots underneath last abdominal segment; infest cypress (page 293) . . . . CYPRESS BARK BORER (*Physocnemum andreae* Hald.)
5. Infest chiefly cedars and junipers; larvae with many curved hairs on sides of heads (page 293) . . . . LARGE CEDAR BORER [*Semanotus ligneus* (F.)]
6. Most of borings ejected from galleries, but some may be present near pupal chambers 7
6. Borings mostly packed in galleries 10
7. Borings contain much coarse fibrous material; galleries mostly in inner bark but somewhat scoring wood; mature larvae bore a U-shaped gallery in the wood; larvae with heads longer than wide†; legless; back of segmental body swellings with small rounded tubercles (page 293) . . . . SAWYERS (*Monochamus* spp.)
7. Borings mostly granular but may be fibrous near pupal chambers; galleries in inner bark and sapwood; larvae with heads wider than long†; 3 pairs thoracic legs 8
8. Galleries mostly beneath the bark but deeply score sapwood; pupal chambers deep in wood; larvae flattened; under side of heads with 4 small tubercles; eyes absent (page 293) . . . . BLACK-HORNED BORERS (*Callidium* spp.)
8. Galleries under bark and also in the sapwood and heartwood 9
9. Larvae slender; mandible tips rounded, gouge-like; sides of heads with many curved hairs; 1 pair simple eyes (ocelli) (page 293) . . . . *Oeme* spp.
9. Tips of mandibles pointed; 3 pairs of ocelli; western species (page 293) . . . . *Arhopalus* spp.
10. Galleries only in dry seasoned or old moist wood; larvae with heads always wider than long† 14
10. Galleries beneath the bark, but sometimes some boring may occur in the wood; larval head varies 11
11. Larvae with heads longer than wide†; legless 13
11. Larvae with heads wider than long†; with 3 pairs of thoracic legs 12
12. Galleries only in inner bark; pupal chambers nest-like, made of coarse fibers and are ¾ to 1 inch in diameter; never infest living trees; larvae flattened with heads wider than thorax; mandibles notched at tips (page 255) . . . . ROUND-HEADED RIBBED PINE BORER (*Stenocorus lineatus* Oliv.)

12. Galleries mostly in inner bark, but pupal cells may be either in the sapwood or the bark; sometimes infest living trees; larvae with 2 small spines on back of last abdominal segment; sides of heads with long curved hairs; upper lip (labrum) wider than long (page 255) . . . . ROUND-HEADED FIR AND LARCH BORERS (*Tetropium* spp.)
13. Galleries only in inner bark; larvae rather flattened; pronotum and segmental abdominal segments (ampullae) pubescent (page 256) . . . . ROUND-HEADED PINE BARK BORERS (*Acanthocinus* spp.)
13. Galleries mostly in inner bark, but the mature larvae bore U-shaped galleries into the wood; larval ampullae with small tubercles (page 293) . . . . SAWYERS (*Monochamus* spp.)
14. Larvae infest dry seasoned wood; borings loose and granular; larvae with V-shaped impression on metanotum; with 3 pairs of thoracic legs; mandibles with a rounded gouge-like cutting edge; 3 pairs ocelli (page 299) . . . . OLD HOUSE BORER [*Hylotrupes bajulus* (L).]
14. Larvae infest only moist dead wood or the heartwood of living trees in which they gain entrance through wounds; larval mandibles pointed with oblique cutting edges (except *Xylotrechus*) 15
15. Large oval galleries with larger ones larger than ¾ inch wide; borings mostly coarse and fibrous; larvae large being over 2 inches long when full grown with a row of tubercles or ridges on front of head 16
15. Oval galleries less ¾ inch wide; borings mostly granular but may have coarse shreds intermixed; larval heads without tubercles or ridges 18
16. Larvae with 2 flat ridge-like teeth on front of head; segmental body swellings (ampullae) each with 2 transverse impressed lines (Table 11.7) . . . . BROWN PRIONID (*Orthosoma brunneum* Forst.)
16. Larvae with a row of 4 sharp or rounded tubercles across front of head 17
17. Tubercles on front of head blunt and rounded; occurs only in West (page 297) . . . . PONDEROUS BORER (*Ergates spiculatus* Lec.)
17. Tubercles on front of head sharp edged (Table 11.7) . . . . HAIRY PINE BORER (*Tragosoma harrisi* Lec.)
18. Galleries numerous often interconnecting so as to honeycomb the wood; larvae with 3 pairs well-developed legs; each 5 segmented including claw; back of pronotum and segmental body swellings roughened (Table 11.7) (page 282) . . . . POLE BORER [*Parandra brunnea* (F.)]
18. Galleries solitary, not interconnecting 19
19. Larvae legless (Table 11.7) . . . . RUSTIC BORER (*Xylotrechus sagittatus* germ.)
19. Larvae with 3 pairs legs; 2 small spines on back of last abdominal segment; sides of head with long curved hairs; upper lip (labrum) as long as or longer than wide (Table 11.7) . . . . BLACK SPRUCE and PINE BORERS (*Asemum* spp.)

* Derived from Craighead, F. C. 1950. *U.S.D.A. Misc. Publ.* 657:231–271.

† The heads must be removed in order to determine the relation between width and length.

TABLE 11.5 COMMON GENERA AND SPECIES OF LONG-HORNED WOOD-BORING BEETLES

1. Prothorax with a ridge along each side; on some species these ridges extend out to produce spines 6
1. Prothorax not margined as indicated in 1 above but some species have a single spine on each side 2
2. An oblique groove extends along tibia of each front leg 13
2. Front tibia not grooved 3
3. Prothorax with a spine on each side 11
3. Prothorax not spined on sides, but there may be 2 or more small tubercles or teeth on back 4
4. With 2 pairs of compound eyes (page 255) . . . . . FIR AND LARCH BORERS *Tetropium* spp.
4. With 1 pair eyes 5
5. Inner margin of eyes curved inwards (emarginated), so that the eyes partly surround bases of antennae 28
5. Eyes usually emarginated, but bases of antennae are not within area formed by indentation 20
6. Antennae short, not extending beyond base of elytra; 3rd tarsal segments not bilobed or notched; 4th tarsal segments distinct; beetles brown 3/8 to 3/4 inch long; prothorax rather square without spines or teeth on sides (Table 11.6) (page 282) . . . . . POLE BORER *Parandra brunnea* F.
6. Antennae long; 3rd tarsal segment bilobed with small 4th segment attached in the notch 7
7. Prothorax with 4 or more small spines on each side 8
7. Prothorax with 3 or fewer teeth on each side 9
8. Third antennae segment much longer than the others (page 297) . . . . . PONDEROUS BORER (*Ergates spiculatus* Lec.)
8. Third antennal segment not much longer than others; beetles 1 to 1 1/2 inches long; reddish brown and somewhat flattened (Table 11.6) (page 282) . . . . . HARDWOOD STUMP BORER (*Stendontes dasystomus* Say)
9. Prothorax with 1 spine on each side; beetles 3/4 to 1 1/2 inches long; underside of body hairy; elytra with several raised lines; small spine at tip of each elytron (Table 11.7) . . . . . HAIRY PINE BORERS (*Tragosoma* spp.)
9. Prothorax with 2 or 3 spines on each side 10
10. Antennae with 11 segments, thread-like; beetles brown, flattened and large, 3/4 to 1 1/2 inches long (Table 11.6) . . BROWN PRIONID (*Orthosoma brunneum* Forst.)
10. Antennae with 12 segments; these often appear toothed or resemble overlapping scales (page 400) . . . . . ROOT BORERS (*Prionus* spp.)
11. Antennae shorter than combined length of head and prothorax; segments 5 to 11 thicker than basal 4 segments; elytra strongly ridged (page 255) . . . . . ROUND-HEADED RIBBED PINE BORER (*Stenocorus lineatus* Oliv.)
11. Antennae about as long or longer than body; 2 small spines at tip of each elytron 12

12. Brown, elongate beetles; ⅝ to 1¼ inches long; an oblique or crescent-shaped yellowish spot near middle of each elytron (page 293) . . . . BANDED HICKORY BORER [*Chion cinctus* (Drury)]
12. Light brown, robust, elongate beetles; ⅝ to 1 inch long; pair of shining white spots at base and another similar pair near middle of each elytron; 2 pairs (1 pair small) raised smooth spots on pronotum (page 299) . . . . IVORY-MARKED BEETLE [*Eburia quadrigeminata* (Say)]
13. Prothorax with a spine on each side 14
13. Prothorax not spined 17
14. Large stout beetle, 1 to 1¼ inches long; color black with whitish pubescence forming lines which produce black rectangular marks on elytra (page 290) . . . . COTTONWOOD BORER (*Plectrodera scalator* F.)
14. Beetles not colored as in 14 above 15
15. Basal antennal segment (scape) with a distant scar-like spot (cicatrix) at tip . . 16
15. Scape without a cicatrix; body color black mottled with gray to brown pubescence; females with long ovipositers; beetles ½ to 1 inch long; antennae very long (page 256) . . . . PINE BARK BORERS (*Acanthocinus* spp.)
16. Antennae always longer than body; color black, mottled gray or brownish gray (page 293) . . . . SAWYERS (*Monochamus* spp.)
16. Antennae about as long as body on females and slightly longer on males; color brownish; elytra may be banded (page 282) . . . . . . TRUNK BORERS (*Goes* spp.)
17. Each of the 2 claws on each leg has 2 teeth; beetles ⅜ to ⅝ inch long; color yellowish (page 385) . . . . TWIG BORER (*Oberea* spp.)
17. Two simple claws on each tarsus 18
18. Two claws on each tarsus gradually diverge to form a fork 19
18. Two claws on each tarsus diverge widely from the base; middle tibia also grooved (page 384) . . . . TWIG GIRDLERS (*Oncideres* spp.)
19. Body short and broad, about 2 times longer than broad; femora club shaped . . . . *Leptostylus* spp.
19. Body elongate, about 3 times longer than broad; femora slender (pages 290 and 293) . . . . *Saperda* spp.
20. Eyes hairy; beetles about ½ inch long; light to dark brown; pronotum and elytra finely granulated (Table 11.7) . . . . BLACK PINE AND SPRUCE BORERS (*Asemum* spp.)
20. Eyes not hairy 21
21. Eyes coarsely granulated 22
21. Eyes finely granulated 23
22. Inner margin of eyes deeply curved in (emarginated); body narrow and flattened; tan to light brown; beetles ¼ to ⅜ inch long; femora of legs club shaped, being slender near body (page 299) . . FLAT OAK BORERS (*Smodicum cucujiforme* (Say)
22. Eyes somewhat emarginate; body robust elongate; color light to dark brown; front of head deeply grooved; beetles dark brown; about 1 inch long (page 293) . . . . *Arhophalus* spp.
23. Femora of legs rather slender near body but expands greatly on outer ¾ of length so they appear club shaped 25
23. Legs not club shaped 24

24. Head narrower than front edge of pronotum; pronotum with 3 to 5 tubercles or slightly raised irregular smooth spots; beetles ¼ to 1½ inches long; color dark brown to bluish; some have orange or red areas; sides of elytra parallel (page 293) . . . . LARGE CEDAR BARK BORERS (*Semanotus* spp.)
24. Pronotum narrowed toward head so that head is wider; pronotum without tubercles, posterior corners extended into points; sides of elytra narrowed from base to posterior (wedge shaped) . . . . *Leptura* spp.
25. Pronotum with distinct tubercles on back 26
25. Pronotum not as above 27
26. Beetles dark brown or black; sometimes with 2 indistinct light lines crossing elytra; pronotum with 3 low tubercles (page 299) . . . . OLD HOUSE BORER [*Hylotrupes bajulus* (L.)]
26. Color reddish brown with arched whitish mark on each elytra (CYPRESS BARK BORER) or bluish black with 3 short raised whitish lines on basal half of each elytra; beetles ⅜ to ¾ inch long (page 293) . . . . ELM BARK BORER (*Physocnemum* spp.)
27. Body dark bluish black; beetles about ½ inch long; pronotum with shallow depressed areas on each side of middle (page 299) . . . . BLACK-HORNED BORERS (*Callidium* spp.)
27. Body colored variously tan to brown; one species has dark blue elytra; beetles ¼ to ½ inch long . . . . *Phymatodes* spp.
28. Eyes coarsely granulated; elytra not marked with cross bands 29
28. Eyes finely granulated; elytra often marked with gray, yellow, or orange cross bands 34
29. Short to long spines on sides or tips of several of the basal antennal segments; femora not club shaped; brown beetles sparsely covered with a grayish pubescence 32
29. Antennae without spines; femora of at least the last 2 pairs of legs club shaped (slender near base and enlarged toward tibia) 30
30. Insects with bodies rather densely covered with gray pubescence but with small bare shiny spots forming irregular designs; elytral tips truncate or notched; beetles small and stout, ¼ to ⅜ inch long (page 293) . . . . SMALL CEDAR BORERS (*Atimia* spp.)
30. Insect bodies without the characteristics listed in 30 above; with sparse pubescence especially on upper sides 31
31. Elytra color brown with a tan oval spot near middle and another smaller tan spot at tip of each; pronotum with a median line and 2 small smooth shining spots; most antennal segments have a groove on outer sides; beetles about ½ inch long (Table 11.6) . . . . ASH BORER (*Tylonotus bimaculatus*) Hald.
31. Color uniformly light brown or tan; beetles elongate; ½ to ¾ inch long; head with a median depressed line (page 293) . . . . CYPRESS BORER [*Oeme rigida* (Say)]
32. Small spines on sides of some antennal segments; elongate slender tan or light brown beetles; ½ to ¾ inch long (page 293) . . . . CYPRESS BORER [*Oeme rigida* (Say)]
32. Spines on tips of antennal segments; beetles dark brown irregularly covered with grayish pubescence 33
33. Large robust beetles, ¾ to 1 inch long and 3/16 to 5/16 inch wide; pronotum with-

out smooth raised spots as below (page 282) . . . . RED OAK BORER [*Romaleum rufulum* (Hald.)]

33. Smaller brown beetles, usually ½ to ¾ inch long and up to 3/10 inch wide; pronotum with a smooth ridge along mid-line plus 2 to 4 other raised smooth spots; with irregular patches of white pubescence; antennae, elytra, and femora with long spines (Table 11.6) . . . . SPINED BARK BORER (*Elaphidion* spp.)
34. Some antennal segments with spines on tips 35
34. Antennae not spined 36
35. Pronotum marked with 4 yellowish narrow cross lines and elytra crossed with 6 or 7 wavy narrow yellowish cross lines (page 288) . . . . *Megacyllene* spp.
35. Pronotum marked with 2 yellowish stripes on each side, but these never cross back; posterior fourth of elytra yellow with a thin black cross stripe and a black spot near tip of each elytron; base elytra with 3 yellow cross band with center one W-shaped; beetles ⅞ to 1 inch long (Table 11.6) (page 292) . . . . SUGAR MAPLE BORER (*Glycobius speciosus*)
36. Ridge on front of head forming a V or W or 2 or 3 raised lines; beetles ⅜ to ¾ inch long; light to dark brown with gray or yellow cross lines or along edges of elytra (Table 11.6) . . . . RUSTIC BORERS (*Xylotrechus* spp.)
36. Head does not have characteristics as noted in 36 above; elytra with 4 light gray to yellowish crossbands with first 2 almost forming circles; one species with prothorax reddish; tips elytra with spines or rounded (page 300) . . . . ASH BORERS (*Neoclytus spp.)*

ROUND-HEADED BORERS THAT INFEST LIVING TREES

Coleoptera, Cerambycidae

Living angiosperm trees are more frequently and more heavily attacked by cerambycid borers than are the conifers. In addition to those considered below are *Tylonotus bimaculatus* Hald. (living ash and privet), *Xylotrechus aceris* Fisher (living red maple), and *Tetropium* spp. (living larches and firs). *Parandra brunnea* F. (the pole borer) and *Stenodontes dasystomus* (Say) (the hardwood stump borer) infest the wood of living trees only when they can gain entrance through wounds.

The general biological characteristics of these are presented in Tables 11.6 and 11.7.

TRUNK AND BRANCH BORERS

Goes spp. and *Romaleum* spp., Coleoptera, Cerambycidae

Common species of trunk and branch borers are the White oak borer [*Goes tigrinus* (Deg.)], the oak branch borer [*G. debilis* (Lec.)], the living hickory borer [*G. pulcher* (Hald.)], the living beech borer [*G. pulverulentus* (Hald.)], the oak sapling borer [*G. tesselatus* (Hald.)], and the red oak borer [*Romaleum rufulum* (Hald.)]. Of these, *G. tigrinus* and *R. rufulum*

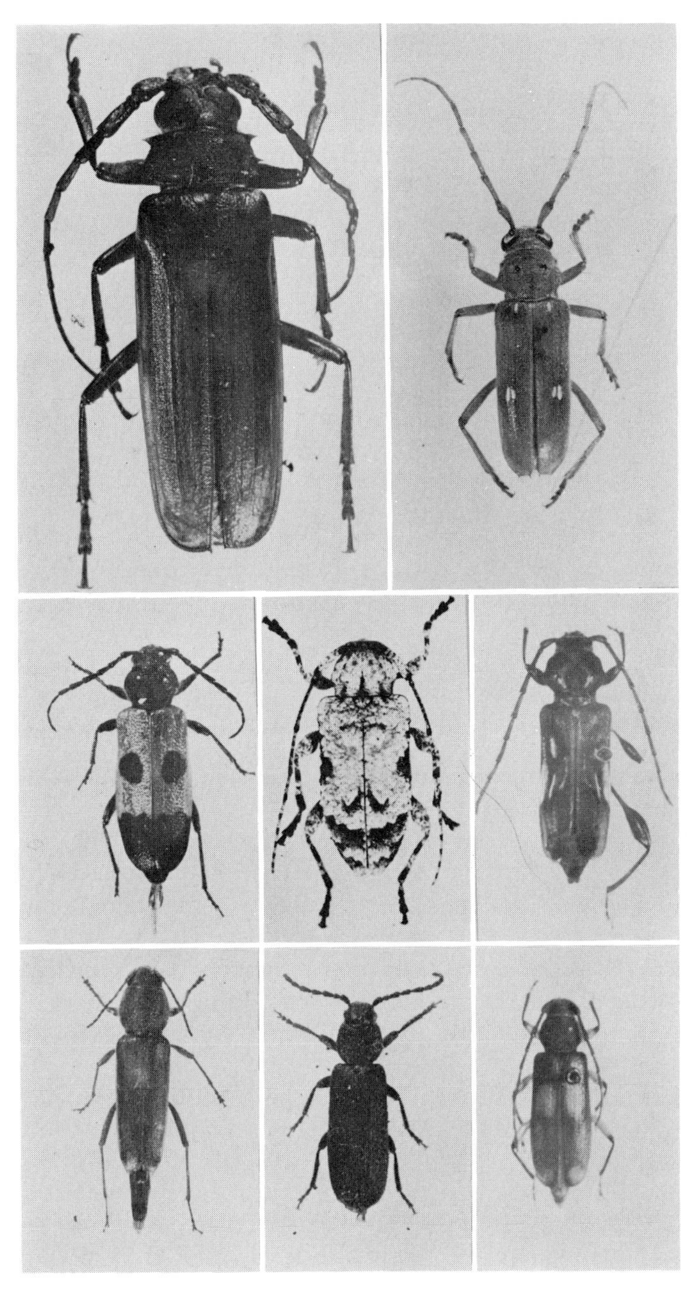

FIG. 11.3. Some genera of round-headed wood borers of lesser importance. Upper left: *Orthosoma;* upper right: *Eburia;* center left: *Semanotus;* center: *Leptostylus;* center right: *Psysocnemum;* lower left: *Xylotrechus;* lower center: *Asemum;* lower right: *Tylonotus;* (all about X1⅔).

FIG. 11.4. Pole borer (*Parandra brunnea*) beetle (X1⅔) and damage (X⅔).

together with the carpenter worms are responsible for much of the grub hole damage that occurs in living oaks. Often 10 to 40 per cent of the lumber cut from black oaks is damaged because of the holes made by these insects. The larvae of *G. debilis* attack the branches and, therefore, cause little serious damage. *G. tesselatus* larvae bore in the bases of small trees.

INJURY. Oval galleries; larger ones commonly ¼ to ½ inch in cross section; wind irregularly through the wood; borings mostly ejected from the galleries; *Romaleum* produces granular borings; borings for the *Goes* spp. are fibrous.

HOSTS. Living trees that are mostly species of oaks. *G. pulcher* infests only hickories, and *G. pulverulentus* attacks various hardwoods from oak to sycamore.

RANGE. Throughout eastern United States wherever the hosts grow.

INSECT APPEARANCE. The general generic characteristics of both the adults and larvae are presented in Tables 11.3 and 11.5. Some of the more specific beetle characteristics are as follows. *Goes* spp. have a small to stout spine on each side of the prothorax. These are absent on species of *Romaleum.* **White oak borer:** robust; about 1 inch long; color

TABLE 11.6 BIOLOGICAL CHARACTERISTICS OF THE COMMON GENERA OF ROUND-HEADED BORERS THAT INFEST DECIDUOUS BROAD-LEAVED TREES

Genus	Host Conditions				Location in Host		Characteristics of Borings				Time of Beetles' Appearance			Length of Life Cycle	
	Living Trees	Freshly Killed Trees	Moist Dead Wood	Dry Dead Wood	Inner Bark	Wood	Granular*	Fibrous	Ejected from Galleries	Packed in Galleries	Spring	Summer	Fall	1 Year	2 to Many Years
Goes	x	—	—	—	x	x	—	x	x	—	x	—	—	—	x
Plectrodera	x	—	—	—	x	x	—	x	x	—	—	x	—	—	x
Saperda	x	x	—	—	x	x	—	x	x	x	—	x	—	—	x
Glycobius	x	—	—	—	x	x	x	—	x	—	—	x	—	—	x
Megacyllene	x	x	—	—	x	x	x	—	x	—	x	—	x	x	—
Romaleum	x	—	—	—	x	x	x	—	x	x	x	—	—	—	x
Tylonotus	x	—	—	—	x	x	x	—	—	x	—	x	—	—	x
Chion	—	x	—	x‡	x	x	x	—	x	x	—	x	—	—	x
Elaphidion	—	x	—	x‡	x	x	x	—	x	—	—	x	—	—	x
Neoclytus	—	x	—	x‡	x	x	x	—	—	x	x	x	x	x	x
Xylotrechus	x†	x	—	—	x	x	x	—	—	x	—	x	—	x	—
Eburia	x†	—	—	x	—	x	x	—	—	x	—	x	—	—	x
Smodicum	—	—	—	x	—	x	x	—	—	x	—	x	—	x	x
Orthosoma	—	—	x	—	—	x	—	x	—	x	—	x	—	—	x
Parandra	x†	—	x	—	—	x	x	—	—	x	—	x	—	—	x
Stenodontes	x†	—	x	—	—	x	—	x	—	x	—	x	—	—	x

* A fibrous plug may be present adjacent to pupal chambers.
† Enter through wounds and bore only in wood.
‡ Females oviposit only on bark of green logs or dying trees, but the larvae may continue to live and bore in the dead host material for a year or longer.

TABLE 11.7 BIOLOGICAL CHARACTERISTICS OF THE COMMON GENERA OF ROUND-HEADED WOOD BORERS THAT INFEST CONIFERS

Genus	Main Hosts		Host Condition				Location in Host		Borings				Beetle Appearance			Length of Life Cycle	
	Pine, Spruce, Fir, Etc.	Cy-press, Cedar, Etc.	Living Trees	Freshly Killed Trees‡	Moist Dead Wood	Dry Dead Wood	Inner Bark	Wood	Gran-ular*	Fi-brous	Ejected from Gal-leries	Packed in Gal-leries	Spring	Sum-mer	Fall	1 Year	2 to Many Years
Acanthocinus	x	—	—	x	—	—	x	—	—	x	—	x	—	x	—	x	—
Monochamus	x	—	—	x	—	—	x	x	—	x	x	x	x	x	—	x	—
Callidium	x	x	—	x	—	x‡	x†	x	x	—	x	—	x	—	—	x	—
Hylotrupes	x	—	—	—	—	x	—	x	x	—	—	x	—	x	x	—	x
Xylotrechus	x	—	—	x	—	—	—	x	x	—	—	x	—	x	—	x	—
Stenocorus	x	—	—	x	—	—	x	—	—	x	—	x	x	—	—	x	—
Arhopalus	x	—	—	x	—	—	x	x	x	—	x	—	—	x	—	x	—
Asemum	x	—	—	—	x	—	—	x	x	x	—	x	—	x	—	—	x
Tetropium	x	—	x	x	—	—	x	—	x	—	—	x	—	x	—	x	—
Ergates	x	—	—	x	x	—	—	x	—	x	—	x	—	x	—	—	x
Orthosoma	x	—	—	—	x	—	—	x	—	x	—	x	—	x	—	—	x
Parandra	x	—	x§	—	x	—	—	x	x	—	—	x	—	x	—	—	x
Tragosoma	x	—	—	—	x	—	—	x	—	x	—	x	—	x	—	—	x
Oeme	x	x	—	x	—	—	x	x	x	—	x	—	x	—	x	x	—
Physocnemum	—	x	—	x	—	—	x†	x	x	—	—	x	—	x	—	x	—
Semanotus	—	x	—	x	—	—	x†	x	x	—	—	x	x	—	—	x	—
Atimia	—	x	—	x	—	—	x	x	—	x	—	x	x	—	x	x	—

* A fibrous plug may be present near the pupal chambers.
† Sapwood deeply scored.
‡ Females oviposit only on bark of green logs or dying trees, but the larvae may continue to live and bore in the dead host material for some time.
§ Enter living trees only through wounds.

FIG. 11.5. Trunk and branch borers. *Goes tigrinus* above and *Romaleum rufulum* below (X1¾). The grub hole damage was caused by *Goes* (X¾).

brown irregularly covered with white pubescence; basal third of elytra with small raised points. **Oak branch borer:** small and slender; about ⅝ inch long; color brown with yellowish pubescence on head, thorax, and tip third of elytra; gray pubescence on basal ⅔ of elytra; 2 irregular brown bands cross elytra. **Oak sapling borer:** ¾ to 1 inch long; color brown, densely clothed with gray and yellowish pubescence, the latter forms small, yellowish spots over elytra. **Living beech borer:** ¾ to ⅞ inch long; color brown, uniformly covered with short whitish hairs; elytra indistinctly barred across the base and the middle with darker pubescence; **Living hickory borer:** ¾ to ⅞ inch long; dark brown covered with yellowish pubescence with a darker cross band near the middle of elytra and another across the base. **Red oak borer:** robust ⅞ to 1 inch long; color brown with yellowish pubescence over body but denser on head and elytra; tip of elytra prolonged to produce 2 small spines; no spines on sides of prothorax.

LIFE CYCLE. The beetles emerge in the spring. After mating, the females deposit their eggs beneath bark scales. The resulting larvae bore into and feed on the inner bark for a short time before entering the wood. Most of the boring occurs in the wood. Two or three years are required to develop from egg to adult.

CONTROL. At the present time there is no known way to reduce the damage these borers cause to forest grown trees. Infested shade trees can be treated by injecting a fumigant such as carbon disulphide or calcium cyanide into the galleries via the openings through which the borings are ejected. Immediately following treatment the openings should be plugged with putty or other similar plastic material.

LOCUST BORER

Megacyllene robiniae **(Forst.), Coleoptera, Cerambycidae**

Black locust has been grown extensively for soil erosion control because this species will grow on poor sites and because it produces desirable rot-resistant wood. These stands often are seriously damaged by the locust borer. A closely related species, the painted hickory borer (*M. caryae* Gahan) infests freshly cut hickory and sometimes other hardwood species.

INJURY. Irregular oval grub-holes—larger ones 3/16 x 1/2 inch in cross section—that wind through both the sapwood and heartwood. Sometimes they are so abundant that the wood is "honeycombed." The borings are mostly extruded to the outside.

HOST AND RANGE. Living black locust throughout most of the United States, wherever its host grows except in the Pacific Coastal States.

INSECT APPEARANCE. The generic characteristics of both the beetle and the larvae are presented in Tables 11.3 and 11.5. Additional characteristics of the beetles are as follows: about 3/4 inch long; colored black, marked with narrow cross bands of yellow, 4 of these cross prothorax and 7 cross the wing covers (elytra), 3rd and 4th elytral cross bands form two W-shaped figures; antennae of male slender, shorter than elytra; pale pubescence on each side (episternum) of 3rd thoracic segment not divided or only narrowly divided at middle. *M. caryae* differs in that male has stout antennae that are longer than elytra; and the pale pubescence on each side (episternum) of the 3rd thoracic segment is broadly divided to form two spots. *Larvae:* legs 3 segmented; 3 simple eyes (ocelli); abdominal body swellings finely granulated; spiracles broadly oval.

FIG. 11.6. Locust borer (*Megacyllene robiniae*) beetle (X2) and damage (X¾).

LIFE CYCLE. The adults appear in the fall from late August to October and feed on the flowers of goldenrod. After mating the females deposit their eggs in the bark crevices of the host trees, usually around wounds. The eggs hatch in about a week, and the resulting larvae bore into the inner bark. Here they live until the following spring. Early the next summer they bore into the wood and pupation occurs mostly during August. There is 1 generation per year.

CONTROL. Vigorously growing trees produce unfavorable conditions for the locust borer. Even though these trees may be attacked, few of the larvae survive long enough to cause much damage; therefore, whenever possible, black locust should be grown on the best sites available. In addition, any other practical cultural practices that will stimulate growth should be used. On poor sites tree growth can be stimulated by first allowing the trees to grow for 5 or 6 years until heavily attacked and then cutting the stand. The resulting sprouts grow so vigorously that conditions are unfavorable for the larvae. The old stand should be cut in the winter and the stems burned. After one year of sprout growth the sprouts should be thinned to one per stump.

Borer-resistant varieties ("shipment locust" and "highbee locust") have been reported, but the extent of their resistance under poor growing conditions is unknown to the writer.

Spraying the tree boles with emulsions or wettable powder formulations of 5 per cent DDT, or ½ per cent gamma isomer of benzene hexachloride just before beetle emergence may be effective, but results of such tests have not been reported.

Hall, R. C. 1942. *U.S.D.A. Circ.* 626.

POPLAR BORERS

Saperda spp., *Plectrodera* spp., and *Xylotrechus* spp., Coleoptera, Cerambycidae

The more important poplar borers are the aspen borer (*Saperda calcarata* Say), the cottonwood borer [*Plectrodera scalator* (F.)], and the poplar-butt borer [*Xylotrechus obliteratus* (Lec.)]. Of these, the aspen borer probably is the most injurious.

INJURY. The aspen borer larvae bore in both the inner bark and the wood of the main boles. They make winding, oval tunnels up to 3⁄16 x ½ inch in cross section. Most of the fibrous borings are ejected from the galleries. The short life span of 25 to 35 years for aspen on the poor sites and only 40 to 50 years on average sites apparently is due in part to the combined action of these insect attacks, bark cankers, and wood rot. In Michigan, Graham and Harrison (1954) found that almost all bark canker infections occurred in insect injuries.

The other two species attack chiefly the tree bases. Of these the cottonwood borer extrudes much of its borings, whereas the poplar butt borer packs granular borings in the galleries.

HOSTS. Living trees of *Populus* spp. Willows are also attacked by *Plectrodera, Xylotrechus,* and *S. concolor.*

RANGE. Most of these species occur wherever the hosts grow except *Plectrodera.* It appears to be limited to the southern half of the eastern United States.

INSECT APPEARANCE. The general generic characteristics are presented in Tables 11.3 and 11.5. Additional characteristics are as follows: *Aspen borer beetles:* tip of each wing cover (elytra) extended into a small spine; body ¾ to 1 1⁄10 inches long; dense pubescences colors beetles a mottled, greenish, yellowish gray; body covered with minute dark dots where pubescence is absent; antennae as long or longer than body; no spines on sides of prothorax. **Cottonwood borer bettles:** 1¼ to 1½ inches long; color black, overlaid with lines of cream-colored scales forming irregular black patches; basal 2-antennal segments and legs appear either rusty or grayish; prothorax with a stout spine on each side; antennae as long or longer than body. **Poplar-butt borer beetles:** ⅜ to ¾ inch long; body colored dark with yellow cross bands at anterior and posterior margins of prothorax; three other yellow bands cross elytra, the first is oblique, the middle band curved, and the third straight.

LIFE CYCLE. The Saperda beetles appear during the summer and immediately start feeding on the bark of young twigs. After mating, the females cut oblong holes in the bark and deposit 1 or 2 eggs in each. These eggs hatch in about 3 weeks. The resulting larvae bore in the inner bark until the following spring and then enter the wood. Much fibrous borings (frass) are extruded through the enlarged entrance hole. Pupation occurs during the summer near the lower end of the gallery and the adults emerge through the holes used for ejecting the borings. Three or four years are required for the development of 1 generation. There are differences in the habits of the other species as follows: *S.*

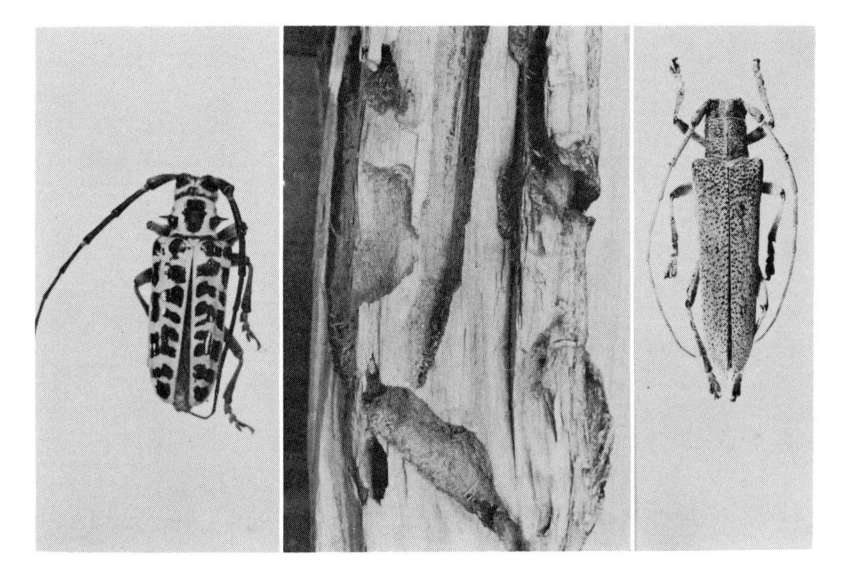

FIG. 11.7. Poplar borers. Right: aspen borer (*Saperda calcarata*) beetle (X1½); left: the cotton-wood borer (*Plectrodera scalator*) beetle (X1). The damage was caused by the aspen borer (X¾).

concolor and *Xylotrechus* do not extrude much frass. *Xylotrechus* beetles fly late in the summer, deposit their eggs in the bark crevices or on the exposed wood, and the larval borings are of a granular consistency. *Plectodera* usually completes 1 generation in only 2 years.

CONTROL. Aspen should be grown in dense stands and that located on the poorer sites should be harvested for pulpwood as soon as stand deterioration starts. The borers in shade poplars or willows can be killed by injecting a fumigant such as carbon disulphide or calcium cyanide in the openings and then containing the fumigant by closing the holes with a material such as putty. Thereafter every effort should be made to increase the growth rate of the trees.

Graham, S. A., and R. R. Harrison. 1954. *J. Forestry* 52:741–743.
Hofer, G. 1920. *U.S.D.A. Farmers' Bull.* 1154.
Milliken, F. B. 1916. *U.S.D.A. Bull. (Prof. Paper)* 424.

ROUND-HEADED WOOD BORERS THAT ATTACK DYING OR RECENTLY FELLED TREES

Coleoptera, Cerambycidae

Numerous species of cerambycids infest the boles and larger limbs of dying and recently felled trees, but only a few of these insects can be

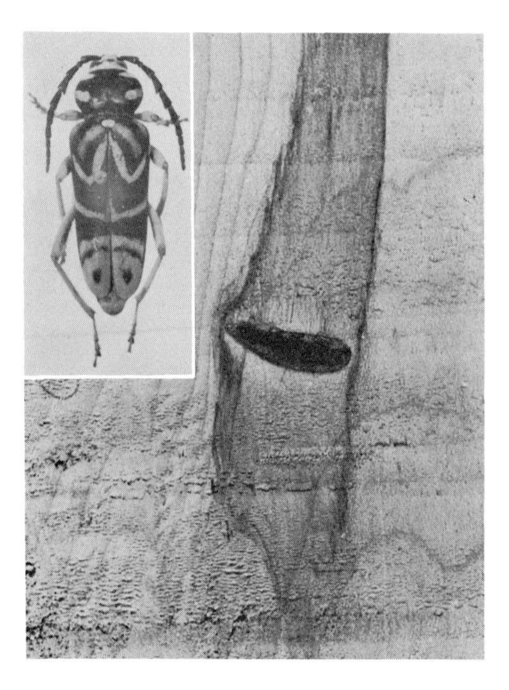

FIG. 11.8. The sugar maple borer (*Glycobius speciosus*) beetle (X1½) and damage (X1).

FIG. 11.9. Round-headed apple tree borer beetle (*Saperda candida*) (X2¼).

considered here in detail. Others include several species of *Saperda* that are common borers in hardwood logs. These are the round-headed apple tree borer, *S. candida* F. (Rosaceae spp.); the hickory saperda, *S. discoidea* F. (hickories and butternut); the elm borer, *S. tridentata* Oliv. (elms); and the basswood borer, *S. vestita* Say (basswood). There are also a few other species that attack the twigs and limbs of poplars and thorn apple. *Megacyllene caryae* Gahan commonly infests hickory, but sometimes it is also found in ash, osage orange, and hackberry. The banded hickory borer [*Chion cinctus* (Drury)] and the spined bark borer [*Elaphidion mucronatum* (Say)] larvae often are found in various hardwood species. *Arhopalus* spp. infests inner bark and wood of Douglas-fir, true firs, and spruces, whereas species of *Oeme, Physocnemum, Semanotus,* and *Atimia* attack chiefly members of the Cupressaceae (cypresses, cedars, and *Sequoia*).

The general biological characteristics of these various genera are presented in Tables 11.6 and 11.7.

SAWYERS

Monochamus spp., Coleoptera, Cerambycidae

Sawyers usually are the most common cerambycid invaders of dying or freshly felled pines, spruces, and firs. They often attack shortly after the primary bark beetles have become established and, because of their large size and the loud gnawing noise they make, are often erroneously

FIG. 11.10. Several round-headed borers that infest dying or recently felled trees. Left: banded hickory borer (*Chion cinctus*) (X¾) damage and beetle (X1⅓); upper right: black-horned borer (*Callidium* spp.) beetle (X1½) and damage (X½); lower right: the cypress borer beetle (*Oeme rigida*) (X2) and its typical arrow-shaped pupal chamber (X1¼).

thought to be the cause of tree death. Sawlogs that are milled soon after being cut are not damaged by sawyers. They can be most injurious, however, when large quantities of windstorm or fire-killed saw timber accumulate and can not be salvaged quickly. Pulpwood, on the other hand, frequently remains in storage long enough to become heavily infested. The volume loss may be almost 1 per cent for a 10-inch diameter stick to as much as 1½ per cent for a 6-inch stick, when these pulpwood sticks are completely saturated with sawyer larvae.

INJURY. The larvae first bore in the inner bark, where each makes an irregular gallery covering an area of about 18 square inches. See Figure 10.23. Usually the wood is also scored. As they approach maturity each larva bores an oval hole into the wood for a distance of ¾ to 2½ inches and then returns to the surface. This forms a U-shaped tunnel. Each larva of *M. titillator* destroys about ⅖ cubic inch of wood. The tunnels in both the inner bark and wood are more or less filled with fibrous borings. Some of this frass is extruded through small openings in the bark and forms small piles on the outside. The emergence holes are circular.

HOSTS AND RANGE. Freshly cut, dying, or recently dead pines, spruces, firs, and Douglas-fir, wherever the hosts grow. Different sawyer species occur in the different regions.

INSECT APPEARANCE. The general generic characteristics for both the larvae and adult beetles are presented in Tables 11.4 and 11.5. Additional beetle characteristics of the common species are as follows: ***M. titillator* (Fab.):** ⅝ to 1 inch long; reddish brown, marbled with white and brown; tip of each wing cover projects as a small spine; occurs in the eastern and southern parts of the United States. ***M. notatus* (Drury):** ⅝ to 1¼ inches long; female with head greatly prolonged and flattened; color grayish with a few small patches of dark brown pubescence on elytra; northeastern North America west to the Great Lakes region. ***M. scutellatus* Say:** ⅜ to 1 inch long; color dull blackish; mesonotum (scutellum) with white pubescence; northern North America. ***M. oregonensis* Lec.:** ⅝ to 1 inch long; similar to *M. scutellatus* but blacker, and mesonotum has a median glaborous line extending the full length; western North America. ***M. marmorator* Kby.:** ¾ to 1 inch long; wing covers taper gradually toward tip; color is dark brown, marbled, with irregular bands of white and yellow; northeastern North America west to the Great Lakes and south to North Carolina. ***M. maculosus* Hald.:** ⅝ to 1 inch long; color dark reddish to blackish with yellowish patches of pubescence; tip of each elytron prolonged to form a tooth; western North America.

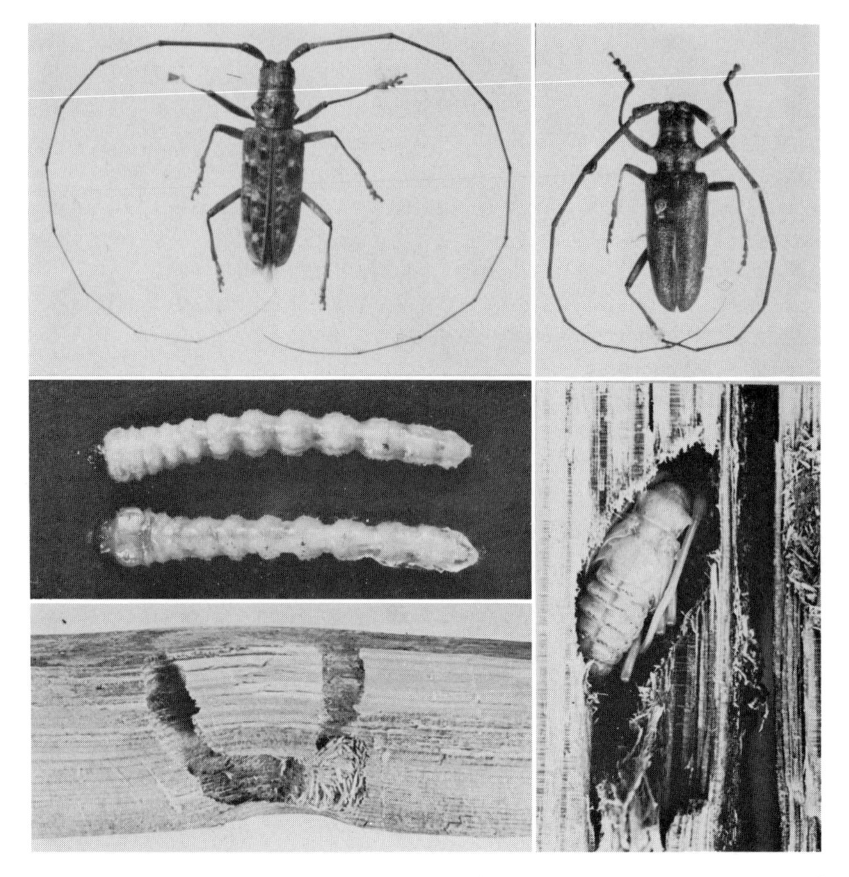

FIG. 11.11. Sawyers (*Monochamus* spp.). Upper left: the southern pine sawyer (*M. titillator*) (X1); upper right: the Oregon sawyer (*M. oregonensis*) (X1¼); below are shown larvae (X1¼), pupa (X1), and wood damage (X½). Also see Fig. 10.23.

LIFE CYCLE. The beetles emerge in the spring and summer and start feeding on the tender bark of twigs. After mating, the females gnaw elliptical niches into the inner bark and in each deposits one to several eggs. The resulting larvae first bore in the inner bark for 1 to 2 months and then enter the wood where they make the characteristic U-shaped holes. Pupation occurs the following spring or summer, and the adults emerge through a circular exit hole. Sawyers usually develop from egg to adult in 1 year, but in the South there is more than 1 generation per year, and in very cold climates the insects may require longer than 1 year to mature.

CONTROL. Rapid utilization of all dying, killed, or felled timber is the best method for preventing damage. Sawyer susceptible timber, which can not be utilized within a period of about a month after felling during the summer months, can be protected by spraying with benzene hexachloride. The bark surfaces should be thoroughly wet by the spray, which should be formulated as a No. 2 fuel oil (diesel oil) solution containing ¼ per cent by weight of the gamma isomer.

Becker, W. B., H. G. Abbott, and J. H. Rich. 1956. *J. Econ. Entomol.* 49:664–666.

PONDEROUS BORER

Ergates spiculatus **Lec., Coleoptera, Cerambycidae**

This species has caused much loss to catastrophically killed timber that could not be salvaged within a period of a couple of years.

INJURY. The larvae mine deep into the heartwood of dead, dying, and recently felled trees, logs, stumps, and poles. They bore oval galleries that may be very large (1 x 2 inches in cross section). These are filled with coarse, fibrous borings.

HOSTS. Recently felled, dead, or decaying logs, stumps, and poles of Douglas-fir and pines. Sometimes the wood of other Pinaceae conifers also is attacked.

RANGE. Throughout western North America.

INSECT APPEARANCE. Some larval and beetle characteristics are presented in Tables 11.4 and 11.5. Additional characteristics are as follows: *Larvae:* large, 2 to 3 inches long when full grown; front of head with a transverse row of 4 rounded tubercles; inner surface of mandibles with numerous lines; 3 pairs of indistinct simple eyes (ocelli). *Beetles:* 1½ to 2¼ inches long; reddish brown; pronotum with 5 or more small spines along each side; 3rd joint of antennae much longer than the others.

LIFE CYCLE. The adults emerge during the summer. After mating, the females deposit their eggs in bark crevices or in cracks of the wood. Three or more years are required for the insect to complete 1 generation.

CONTROL. Rapid utilization of all dead timber is the only practical method that will prevent damage by this borer.

ROUND-HEADED POWDER POST BEETLES

Coleoptera, Cerambycidae

A number of cerambycids infest dried wood and produce granular borings that either remain in the galleries or are ejected. In addition to the species discussed below there are the conifer-infesting black-horned

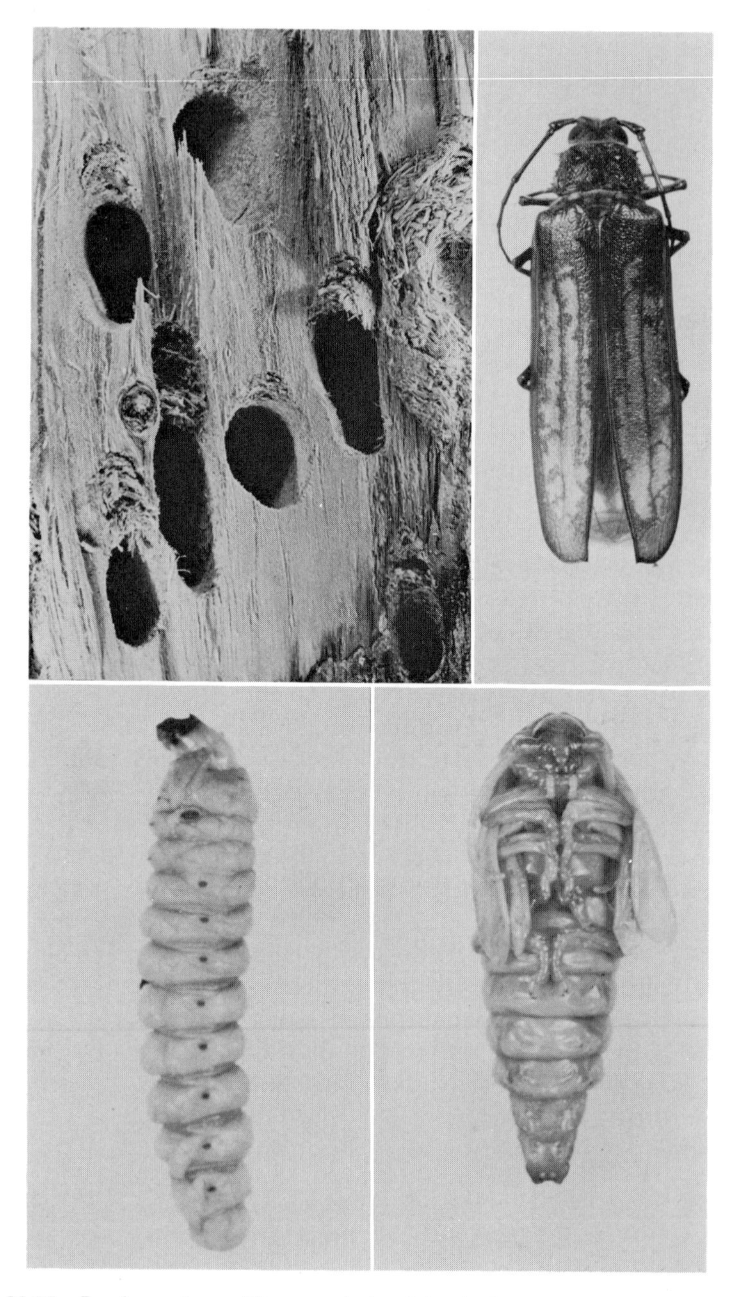

FIG. 11.12. Ponderous borer (*Ergates spiculatus*) beetle, larva, pupa, (X1) and damage (X⅓).

beetles (*Callidium* spp.) and several species that infest hardwoods. These include the flat oak borer [*Smodicum cucujiforme* (Say)] and *Eburia* spp. Although species of *Callidium, Chion, Neoclytus,* and *Elaphidion* attack only dead or recently felled trees, and only when the bark is still on the boles, the resulting larvae often continue boring after the wood has dried. The general biological characteristics for the above are presented in Tables 11.6 and 11.7.

OLD HOUSE BORER

Hylotrupes bajulus (L.), Coleoptera, Cerambycidae

INJURY. This large powder posting beetle is becoming more abundant in North America and is causing an increasing amount of damage to wooden buildings. The oval galleries (3⁄16 x ½ inch) are loosely packed with mixed, fine powdery borings and small pellets. Usually some of the frass also falls out of the infested wood.

HOSTS. Only the sapwood of dry seasoned coniferous (Pinaceae) wood.

RANGE. This insect was introduced from Europe and is now becoming abundant throughout the United States. There are several western species that also infest coniferous wood.

INSECT APPEARANCE. See Tables 11.4 and 11.5 for the general larval and beetle characteristics. **Beetles:** black; wing covers (elytra) sometimes

FIG. 11.13. Old house borer (*Hylotrupes bajulus*) beetle (X1⅓) and damage (X1).

brownish; pronotum wider than long, sides hairy, with 3 low tubercles on back; 2 light colored bands indistinctly cross elytra; head with pit between lower portion of eyes. *Larvae:* head wider than long; tip of mandibles rounded, 3 pairs of ocelli; prothorax smooth; body swellings covered with network of lines.

LIFE CYCLE. The adults appear late in the summer and, after mating, the females deposit their eggs in cracks or irregularities of the wood. Usually 3 or more years are required to complete 1 generation.

CONTROL. Infested timbers should be brush or spray treated with one of the insecticides listed for the control of *Lyctus* powder post beetles. The applications must be liberal so that the insecticides will penetrate and soak through the borings to where the larvae are feeding.

ASH BORERS

Neoclytus spp., Coleoptera, Cerambycidae

Two common borers that attack felled hardwoods are the banded ash borer, *N. caprea* (Say), and the redheaded ash borer, *N. acuminatus* (F.).

INJURY. The larvae riddle the sapwood by boring oval galleries ($\frac{3}{16}$ x $\frac{1}{2}$ inch) in which are packed granular borings.

FIG. 11.14. Banded ash borer (*Neoclytus caprea*) beetles (X1$\frac{1}{4}$) and damage (X$\frac{1}{2}$).

HOSTS AND RANGE. Ash, oak, hickory, persimmon mesquite, and hackberry, in eastern United States west to Arizona.

INSECT APPEARANCE. The general characteristics are presented in Tables 11.3 and 11.5. The beetles are dark colored and cross banded with yellow. The redheaded ash borer has a rusty colored head and prothorax.

LIFE CYCLE. Winter is passed either as larvae or as young adults in pupal chambers. The beetles emerge in early spring about the time when red maple blooms. After mating, the females deposit their eggs in bark crevices. The resulting young larvae first bore for several weeks in the inner bark and then enter the wood. *N. caprea* can complete 1 generation each year, whereas *N. acuminatus* may have 2 or more generations per year. Whenever the wood dries too rapidly, however, development from egg to adult may require several years.

CONTROL. Rapid utilization of cut material is the best method to prevent damage. Should conditions warrant, the felled logs can be protected by spraying with benzene hexachloride as suggested for preventing sawyer damage.

FLAT-HEADED WOOD BORERS (METALLIC BEETLES)

Coleoptera, Buprestidae

Buprestid beetles are commonly called metallic wood borers because they usually have an iridescent metallic luster. The larvae are named flat-headed borers because the prothorax, which appears to be the head, is large, flattened, and distinctly broader than the other body segments.

INJURY. Only the larvae do the boring. They produce oval to flattened tunnels up to ½ or ¾ inch wide, which are always tightly packed with fine borings. These characteristics are somewhat similar to those produced by some powder post beetles, but differs in that the tightly packed borings usually are arranged in concentric layers so that arc-like bands appear when the galleries are exposed. Gallery characteristics alone can not be used to identify the various species. Generally only dying or felled trees are attacked, but some species always attack living trees. Though the larvae of some species first bore in the inner bark for a variable period of time before they enter the wood, the species considered here are more injurious as wood borers. Other general characteristics of the buprestids are presented in the section dealing with inner-bark borers in Chapter 10.

HOSTS AND RANGE. Most genera contain some species that infest only broad-leaved trees and others that infest only conifers. They occur wherever the hosts grow.

INSECT APPEARANCE. The general family characteristics of the larvae are presented in Table 11.1, whereas the generic characteristics are outlined in Tables 10.7 and 10.8.

LIFE CYCLE. Development from egg to adult usually requires several years; therefore, many larvae hibernate over the winter. When these insects mature they pupate, and the adults winter over in pupal chambers. Beetle emergence usually occurs during the spring or summer. After mating the females deposit their eggs in bark crevices or in cracks in the wood.

CONTROL. Applied methods are not needed for most species. Specific recommendations are presented for the more important species considered separately.

BUPRESTIS WOOD BORERS

Buprestis spp., Coleoptera, Buprestidae

There are many species of *Buprestis* wood borers. The more important of these are the two discussed in detail here. Both conifers and hardwoods are infested by the different species.

FLAT-HEADED TURPENTINE BORER

Buprestis apricans Hbst., Coleoptera, Buprestidae

INJURY. The turpentine borer has been very injurious to pines cropped for naval stores, but at present it is not as troublesome because better turpentining practices are being used. This insect attacks only pines that have been injured so that dry wood is exposed on living trees. The larvae bore extensively in both sapwood and heartwood making irregular, oval galleries that are tightly packed with arc-like layers of resin-soaked borings (frass). Sometimes the wood in a 2- or 3-foot length of bole behind a dry turpentine face may contain several dozen larvae. This boring weakens the trees so that much wind breakage results. In addition, the lumber cut from these infested places has no value.

HOSTS AND RANGE. Exposed dry wood of living southern pines, throughout southeastern United States west to Texas.

INSECT APPEARANCE. *Larvae:* typical buprestid; body when full grown about 1½ inches long and ⅜ inch broad across prothorax. The specific larvae characteristics of the genus are presented in Table 10.7. *Beetles:* large, ¾ to 1¼ inches long by ¼ to ⅜ inch broad; somewhat flattened; color dark brown with a metallic luster; each wing cover (elytron) with 8 rows of large punctures. See Table 10.8 for the generic characteristics.

FIG. 11.15. Turpentine borer (*Buprestis apricans*) beetle, larva (X2½) and damage (X1⅓).

LIFE CYCLE. The young adult beetles pass the winter in pupal chambers about ¼ to ½ inch below the surface of the wood. Emergence occurs early in the spring during late February and early March, at about the time pawpaw blooms. The newly emerged beetles feed on the pine needles for about a month before mating and ovipositing. The females deposit their eggs in checks, cracks, or crevices of dry exposed wood. The eggs hatch in a few days, and the resulting larvae immediately bore into the wood. Here they live where each bores a gallery 10 to 30 inches long. The time required to complete a generation is 3 or more years.

CONTROL. Turpentine borer damage can be prevented by using conservative resin cropping practices. The faces must be kept covered with resin so that no part of the wood becomes exposed or dry. Excessive streaking and bleeding often cause the trees to develop dry faces. Improper deep installation of the collecting gutters may also cause the surrounding wood to become dry and exposed. Good fire protection in turpentine orchards is also a necessity because the resin-covered faces ignite and burn readily leaving charred, dry, susceptible wood exposed.

Beal, J. A. 1932. *U.S.D.A. Circ.* 226.

GOLDEN BUPRESTID

Buprestis aurulenta Linn., Coleoptera, Buprestidae

INJURY. This flat-headed borer breeds in dying trees, stumps, felled logs, and sometimes even in partially to completely seasoned wood. The dry wood is less attractive, but when it is infested the presence of the borers may not be detected until the adults emerge. It is the most common and most injurious western *Buprestis* spp. The boring damage is typical for flat heads.

HOSTS AND RANGE. Dead, recently cut, drying, or seasoned wood of pines, firs, Douglas-fir, and possible other conifers throughout western North America.

INSECT APPEARANCE. *Larvae:* typical buprestids; characteristics of the genus are presented in Table 10.7; about 1½ inches long when full grown. *Beetles:* ½ to ⅞ inch long; color yellowish to green or blue green

FIG. 11.16. Golden buprestid (*Buprestis aurulenta*) beetle (X3).

with a metallic, iridescent luster; edges of elytra copper colored; prothorax green, blue, or copper; 4 ridges along each elytron. Generic characteristics are presented in Table 10.8.

LIFE CYCLE. Usually 2 to 4 years are required to develop from egg to adult, but under adverse conditions this period may be as much as 10 years or more. Other details of the life cycle of this important species have never been published.

CONTROL. Damage can be greatly reduced by milling the logs immediately after cutting and leaving no bark (wane) on the edges of the lumber. As with most buprestids, the boring larvae produce no external evidence that the wood is infested; therefore, the damage can be detected only when the galleries are exposed or the beetles emerge. Sometimes it may be feasible to apply an insecticidal treatment as described for use against *Lyctus* powder post beetles to prevent additional damage.

WESTERN CEDAR FLAT-HEADED BORER

Trachykele blondeli Mars., Coleoptera, Buprestidae

INJURY. The insect mentioned above is the commonest wood borer in living western red cedar. It frequently makes the wood worthless for higher grade uses such as boats, shingles, and furniture. The larvae bore oval winding tunnels (⅛ x ⅜ to ½ inch in cross section) that are tightly packed with powdery borings intermixed with oval pellets. These occur in both the sapwood and heartwood of the branches and main trunk. The tunnels mostly follow the annual rings but frequently cross from one ring to another. Infestation can be determined only by felling and cutting into the tree boles or sometimes by the presence of oval exit holes, dead tops, or dead branches.

HOSTS. Healthy, living western red cedar. Several other species infest hosts such as cypress, juniper, incense cedar, Sequoia spp., firs, and hemlock.

RANGE. Western parts of North America below elevations of about 7,500 feet. Most of the other species also occur in the West, except the flat-headed cypress borer (*T. lecontei* Gory), which occurs in the Southeast.

INSECT APPEARANCE. *Larvae:* typical; the specific generic characteristics are presented in Table 10.7; about 1½ inches long when full grown. *Beetles:* ⅝ to ⅜ inch long and about ¼ inch wide; color a brilliant bronze-green, usually with 2 rows of velvety black spots on each wing cover (elytron); the generic characteristics are presented in Table 10.8.

LIFE CYCLE. At least 3 years are required for this insect to complete 1 generation, but often this period of time is much longer. Either larvae or young beetles in pupal chamber live over the winter. The beetles emerge in the spring or early summer and probably feed on the foliage of the host trees. After mating, the females lay their eggs under bark scales on branches of healthy living trees. The resulting larvae bore down through the branches and eventually enter the main boles. Here they mature several years later. Pupal chambers are excavated near the surface of the wood. They transform to adults during the late summer and fall of the year preceding their emergences.

CONTROL. There is no applied method whereby the damage from this insect can be reduced. Standards have been established by the pole industry whereby the usefulness of any pole is determined on the basis of the number of galleries in each of the exposed ends. For example, a 7-inch diameter pole 25 feet long is allowed to have 4 galleries exposed in one end or a total of 6 in both ends.

Burke, H. E. 1928. *U.S.D.A. Tech. Bull.* 48.

SCULPTURED METALLIC WOOD BORERS

Chalcophora spp., Coleoptera, Buprestidae

INJURY. The larvae of sculptured wood borers mostly bore in sapwood and heartwood of stumps and logs. Sometimes, however, they also attack living trees provided they have direct access to the wood through wounds. The damage is typical of that produced by wood-boring buprestids. They are not serious pests. HOSTS. The various *Chalcophora* species attack different hosts. Some infest only conifers, whereas others attack broad-leaved Angiosperm trees. RANGE. Wherever the hosts grow throughout North America. INSECT APPEARANCE. The generic characteristics for both the larvae and adults are presented in Tables 10.7 and 10.8. These buprestids are among the largest found in North America. The larvae may grow to lengths of 2 inches and as wide as ½ inch, whereas the beetles may be over an inch long and as much as ⅜ inch wide. LIFE CYCLE. The beetles appear in the spring and summer, at which times they are commonly observed because of their large size. After mating, the females deposit their eggs directly on the wood, and the resulting larvae immediately start boring. Several years are required to complete 1 generation. CONTROL. Serious damage seldom is caused by these borers except when catastrophic fires or windstorms kill large quantities of timber that can not be harvested within a few months. Damage to living trees is nonpreventable, but usually there is not much of this damage.

RIBBED METALLIC WOOD BORERS

Dicerca spp., Coleoptera, Buprestidae

Most of the data presented for the sculptured metallic wood borers also applies to the ribbed metallic wood borers, except that they differ in appearance. The generic characteristics for both the beetles and larvae are presented in Tables 10.7 and 10.8. The beetles are about ⅝ to ¾ inch long. Most species have numerous parallel raised lines along elytra. The elytra tips are often prolonged.

FIG. 11.17. Right: sculptured metallic wood-borer (*Chalcophora* spp.) beetle (X1⅓); left: ribbed metallic wood-borer (*Dicerca* spp.) beetle (X1⅓).

SMALL POWDER POST BEETLES

Coleoptera, Ptinoidea

The wood-boring larvae of many beetles reduce the wood to a powdery or granular residue, most of which remains in the galleries. Many of these bore in seasoned wood and commonly reinfest the same pieces until much of the wood is destroyed. Some, however, can attack only wood in trees that have died recently. Since the larger cerambycid and buprestid powder posting species have been discussed previously, only the smaller species are considered here. The general characteristics for these are presented in Table 11.1 and in the discussions that follow.

LYCTUS POWDER POST BEETLES

***Lyctus* spp., Coleoptera, Lyctidae**

INJURY. *Lyctus* beetles are the most injurious powder posting pests in North America. The larvae bore in the sapwood of seasoned hardwoods and make small tunnels up to 1/16 inch in diameter. Most of these galleries are filled with fine, powdery borings, some of which usually drop out and form small piles beneath the infested wood. Severely infested wood attacked repeatedly for many years is nearly reduced to powder but remains somewhat covered by an intact, holey, exterior shell. The success of brood development is correlated with the amount of starch in the wood—the more starch the more suitable for the insects. Infestations commonly occur in freshly seasoned wood.

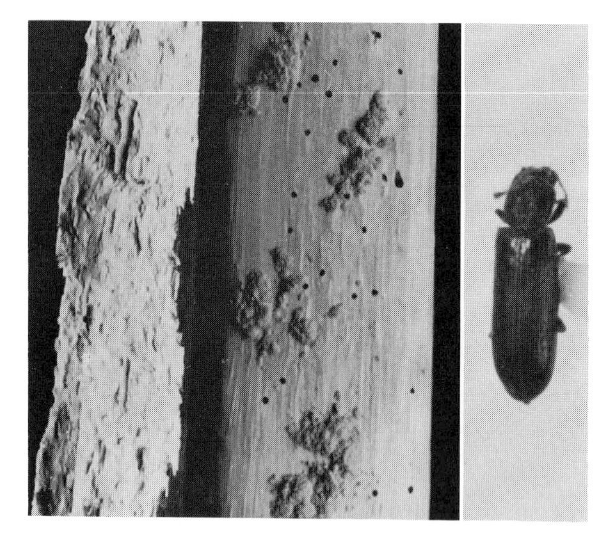

FIG. 11.18. *Lyctus* powder post beetle (X7) and damage (X⅔).

HOSTS. Only the sapwood of seasoned hardwoods with moisture contents between 6 to 30 per cent is attacked. Woods most commonly infested are those species with large pores (ring porous) such as ashes, hickories, and oaks. Sometimes, however, wood with smaller pores also is attacked.

RANGE. These insects, widely dispersed by commerce, now occur everywhere.

INSECT APPEARANCE. *Larvae:* small; up to ¼ inch long when full grown; body soft skinned, whitish, cylindrical with enlarged thoracic segments; body C-shaped with long axis of body curved toward the under side; head yellowish and globular with mouthparts directed forward; short, 3-segmented thoracic legs present; spiracles (breathing pores) on last abdominal segment larger than the others; *Beetles:* small; about ⅕ inch long; body elongate, somewhat flattened, colored brown to black; head not covered by prothorax; antennae attached in front of eyes; antennae with terminal 2 segments forming a club; tarsi with 5 segments, first one very short and somewhat fused with second; femora joined to tips of trochanters; abdomen with 5 visible ventral segments, the first is longer than second.

LIFE CYCLE. Usually there is only 1 generation per year, but under very favorable conditions some species can complete 1 generation in less than half a year. On the other hand, when conditions are unfavorable, the period of development may be 2 or more years. Winter is passed in the larval stage. The period of greatest beetle emergence occurs during the spring, but some may continue to emerge throughout the summer. Each female lays about 60 elongate eggs. These are placed inside the wood pores about ⅛ inch deep by means of flexible extendable ovipositers. The resulting larvae bore through the wood in all directions and eventually pupate in cells at the end of the burrows. The exit holes of the adults are circular.

CONTROL. Preventative measures in wood-using industries consist of locating infestations by making frequent, systematic inspections and by using adequate control methods. All susceptible wood should be inspected on entering the premises and at least once each year thereafter. Heartwood is not attacked; therefore, stock containing no sapwood should be segregated and need not be inspected. Wood refuse and unusable heavily damaged stock should be burned. Lightly infested material can be treated. Other preventative measures consist of using insecticidal dips, sprays, or surface-sealing coatings. For the latter linseed oil, wax, or any other type of finish that closes the pores in the wood is suitable. Many insecticides are effective, but two which have long residual toxic properties are 5 per cent DDT and 5 per cent pentachlorphenol. These should be dissolved in light nonstaining oils. For preventative purposes only the surface needs treatment, but when the boring larvae must be killed, the spray, brush, or dip must be applied heavily so that the liquid penetrates to where the insects are located. Penetration usually occurs rather rapidly for the powdery borings act as wicks to transport the solution.

Sometimes, when kiln facilities are available, heat can be used to stop infestations. The time required to kill all the boring larvae at different temperatures at 100 per cent relative humidity are 3, 5, and 7 hours for lumber 1, 2, and 3 inches thick respectively at 140° F; 4, 6, and 8 hours at 135° F; and 8, 10, and 12 hours at 130° F. Veneers can also be treated in packs up to 3 inches thick, but glued material such as plywood often will not stand high humidities. When lower humidities are used the length of exposures must be increased. For example, reducing the relative humidity to 60 per cent requires the time to be increased by about 20 per cent. Of course, heat-treated wood is susceptible to reinfestation as soon as it cools.

Anon. 1954. *U.S.D.A. Leaflet* 358.
Snyder, T. E. 1938. *U.S.D.A. Farmers' Bull.* 1477.

BOSTRICHID POWDER POST BEETLES

Coleoptera, Bostrichidae

Common bostrichid species are the lead cable borer (*Scobicia declivis* Lec.), the red shouldered hickory borer [*Xylobiops basilare* (Say)], and *Polycaon stouti* Lec.

INJURY. These powder post beetles cause typical powder posting damage, but their galleries are larger than those bored by *Lyctus*. Usually the larger ones are 1/16–1/8 inch in diameter. Some attack freshly cut and partially seasoned woods (*X. basilare*), whereas others infest only wood that has been in use for some time. The lead cable borer is so called because the beetles often bore into and damage the lead sheathing of telephone cables, apparently because the hardness of the lead is similar to that of some woods. Of course, they can not breed in lead, for they obtain no food from it.

HOSTS. Chiefly the sapwood of broad-leaf trees, but a few species infest conifers.

RANGE. Various species are found wherever the host woods occur. *X. basilare* is a native of eastern United States, whereas the other two species listed here occur in the West.

INSECT APPEARANCE. *Larvae:* small; 1/4 to 3/4 inch long when full grown; body whitish, curled C-shaped; head globular and greatly retracted into prothorax; mouthparts extend forward; small 4 segmented

FIG. 11.19. Red-shouldered hickory borer (*Xylobiops basilare*) (X7) and damage (X1½).

antennae; thorax enlarged; 3 pairs thoracic legs each with 4 segments; all spiracles of about same size. **Beetles:** 1/8 to 1/4 inch long, reddish brown to black; body cylindrical; head usually retracted beneath prothorax but is fully exposed in *Polycaon* spp.; antennae attached in front of eyes, with 11 segments, the last 3 enlarged to form a club; legs with anterior coxae large; tarsi with 5 segments with 1st segment smaller than 2nd; abdomen with 5 segments visible underneath, all are of about equal length. *Lead cable borer beetle*—about 1/4 inch long; dark brown to black; prothorax with many tubercles on front half. *Red shouldered hickory borer beetles*—about 3/16 inch long; black with basal third of elytra dull reddish or yellowish; posterior of elytra concave with 3 spines or tubercles on each side. *Polycaon stouti beetle*—about 3/4 inch long; black with body densely punctured; head fully exposed and not covered by prothorax; mandibles large.

LIFE CYCLE. A generation can develop from egg to adult in a minimum of one year under optimum conditions, but usually longer periods of time are required. The beetles emerge during the summer. Most species construct egg galleries in which the females lay their eggs, but a few deposit their eggs in wood pores. The resulting larvae bore long tunnels of 2 or 3 feet. Winter is passed in the larval stage and pupation occurs just prior to beetle emergence.

CONTROL. Infested material can be effectively treated by the methods described for killing *Lyctus* larvae.

ANOBIID POWDER POST BEETLES

Coleoptera, Anobiidae

Two common anobiid species are the deathwatch beetle (*Xestobium rufovillosum* Deg.) and the furniture beetle (*Anobium punctatum* Deg.).

INJURY. These insects infest only well-seasoned wood and cause damage that is very similar to that caused by *Lyctus*. After wood has been infested for many years, the internal parts may be completely reduced to fine borings with only a partially intact surrounding shell. Most damage is caused to old timbers containing some decay or to other old wood that has been in use for a long period of time. The borings consist of very small, oval, tapered or cylindrical-shaped pellets with the finer, powdery particles. These pellets make the borings feel gritty. Exit holes are circular and usually 1/32 to 1/16 inch in diameter.

The deathwatch beetles are so called because the beetles produce a tapping noise that is most often heard during the stillness of night by persons nursing seriously sick patients; consequently, the superstition developed that this tapping foretells that death is near.

HOSTS. These borers infest both the sapwood and heartwood of both conifers and hardwoods. Well-seasoned, moistened, and often partially decayed wood appears to be most susceptible.

RANGE. Various native species occur wherever the host woods are present. The two injurious species listed have been introduced from Europe into various places in eastern North America.

INSECT APPEARANCE. *Larvae:* small, up to ⅓ inch long when full grown; body grub-like; curved C-shaped; whitish; back of some or all segments roughened with minute points or hooks; body covered with fine hairs; head globular with mouthparts directed downward; thoracic segments enlarged making grubs appear humped; 3 pairs of thoracic legs each with 5 segments; posterior abdominal segments not enlarged. *Beetles:* small, up to ⅓ inch long; body cylindrical or subcylindrical; tan, reddish brown to almost black; head hidden beneath pronotum; antennae attached on sides of head in front of the eyes, with 11 segments, the 3 terminal ones somewhat enlarged and elongate; legs with front coxae small; tarsi 5 segmented with 1st segment longer than second; underside of abdomen with 5 visible segments all of which are about equal length. *Furniture beetle*—1/10 to ⅕ inch long; reddish to dark brown; clothed with fine short yellow hairs; pronotum is raised along middle and the wing covers (elytra) have many longitudinal lines formed by rows of pits. *Deathwatch beetle*—⅓ to ¼ inch long; dark brown; patches of short yellowish hairs makes the insects appear mottled or variegated.

LIFE CYCLE. Usually 2 years, but sometimes as many as 10 or more, are required for 1 generation to develop. The beetles emerge during the spring and early summer, and, after mating, the females deposit their eggs in checks, cracks, or crevices in the wood. The resulting larvae may crawl about for a time before entering and boring into the wood. Pupation occurs during the late summer or fall, and the insect lives through the last winter in this stage.

CONTROL. Infested wood that is still serviceable can be treated with the insecticides listed for the control of *Lyctus.* Heavily damaged timbers should be replaced with treated wood.

1943. *British Forest Products Research Leaflets* 4 and 8.

PIN-HOLE BORERS

AMBROSIA BEETLES

Coleoptera, Scolytidae and Platypodidae

INJURY. Ambrosia beetles cause defects named variously as pin holes, shot holes, black holes, and grease spots. The damage is caused mostly

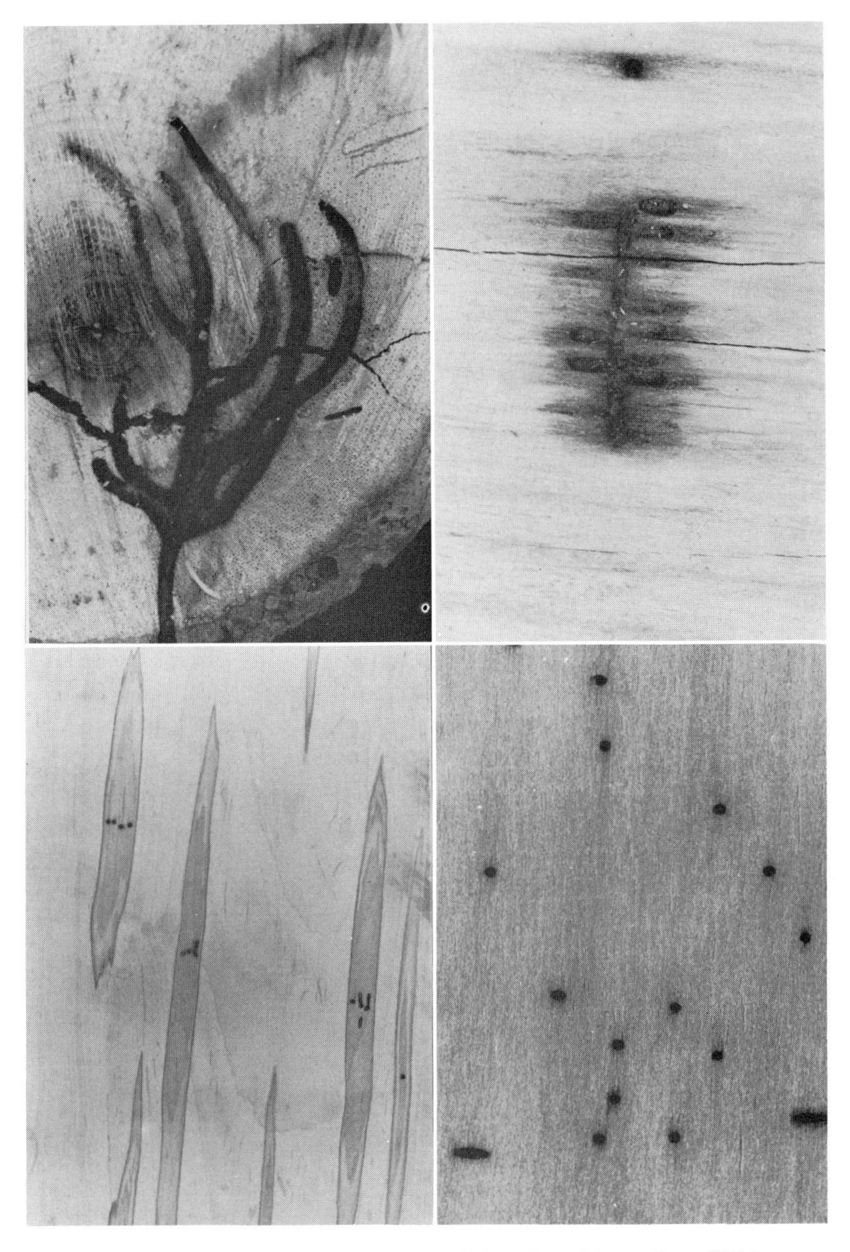

FIG. 11.20. Ambrosia beetle damage. Upper left: a branching gallery (X1¼); upper right: larval cradles (X1¼); lower left: extended stained areas around the galleries made by the Columbian timber beetle (X⅓); lower right: usual appearance of damaged wood (X1).

by the adult beetles that bore small (1/32 to 1/8 inch diameter) round holes in the wood of green logs, dying trees, stumps, and sometimes in moist freshly sawn lumber. Most damage occurs to logs. Only one North American species, the Columbian timber beetle (*Corthylus columbianus* Hopk.), infests, but does not kill, healthy living trees. The holes made by each species are rather uniform in size because they are bored by the beetles and not the larvae. The galleries are kept free of borings, and the walls generally become stained by a layer of fungus. These molds penetrate the wood cells for a depth of about 1/8 inch, but they are not wood destroyers. The insects of each genus bore a characteristic type of gallery as indicated in Table 11.8. The larvae of some bore short chambers (cradles) that are about 1/4 inch long and are at right angles to the main galleries. In these the larvae live and grow to maturity. The fibrous (*Platypus* spp.) or granular boring dust is always extruded from the galleries. It either falls in loose piles or, during humid weather, sticks together to form string-like masses. Inasmuch as these borings are derived from the xylem tissues, they are always the color of the wood. Inner bark borers produce dark reddish brown boring dust.

TABLE 11.8 GALLERY CHARACTERISTICS OF SOME COMMON GENERA OF AMBROSIA BEETLES*

1. Bore in the wood of living hardwoods; a thin, lenticular streak of stain extends 3 to 6 or more inches above and below each gallery . . . . . . *Corthylus columbianus* Hopk.
1. Infest the wood of dying or dead trees and fresh logs; wood staining not in elongated streaks, as indicated in 1 above, but is restricted to the area immediately surrounding tunnels 2
2. Egg galleries extend longitudinally with the grain of the wood; beetles stout; pronotum almost square . . . . *Ambrosiodmus*
2. Egg galleries extend across the grain of the wood 3
3. Rows of oval side pockets (cradles) present along top and bottom of egg galleries 4
3. Cradles absent 7
4. One row of cradles at the top and another at the bottom of each egg gallery . . 5
4. Two rows of cradles at both the top and bottom of each egg gallery; beetles with 2 pairs eyes . . . . *Xyloterinus*
5. Egg galleries each with several lateral side branches; beetles with a raised line across posterior margin . . . . *Gnathotrichus*
5. Egg galleries fork, and each branch follows an annual ring of the wood 6
6. Mostly infest hardwoods; egg galleries small, about 1/16 inch in diameter; beetles with tip of elytra having a thin ridge; antennal funicle 2 segmented . . . *Pterocyclon*
6. Mostly infest conifers; egg galleries larger, about 3/32 inch in diameter; beetles with 2 pairs of eyes . . . . *Trypodendron*

7. Egg galleries branch repeatedly and indiscriminately throughout length; borings fibrous; beetles elongate with elytral tips often prolonged into small spines; head visible from above and as wide as thorax; basal tarsal segment as long as the others combined . . . . . *Platypus*
7. Egg galleries divide into several to many simple branches at a place near the main entrance 8
8. Egg galleries follow the annual rings; pronotum of beetles with small teeth on front margin . . . . . *Anisandrus*
8. Egg galleries do not follow the annual rings . . . . . *Xyleborus*

* Adapted from Beal, J. A., and C. L. Massey. 1945. *Duke Univ. School of Forestry Bull.* 10.

IMPORTANCE. Ambrosia beetle problems are most troublesome in the South and on the Pacific Coast. In these regions the losses frequently are serious, for often they amount to as much as 5 per cent of the value of the lumber cut. The greatest loss occurs to the clear grades because only appearance and not structural value is affected. On the West Coast another form of loss occurs because the scaling rules provide for a deduction of 1 or 2 inches from the diameter of infested logs.

HOSTS AND RANGE. Moist fresh wood of both conifers and hardwoods, wherever the host trees grow. Different ambrosia species attack different hosts. The wood moisture content must be above fiber saturation.

INSECT APPEARANCE. All stages of ambrosia beetles resemble the corresponding stages of bark beetles; therefore, the reader is referred to the general section dealing with bark beetles. Other characteristics of the various genera are presented in Table 11.8.

LIFE CYCLE. Both larvae and adults feed on mold-like fungi. These are species of *Monila, Leptographium* and possible others, which grow in the galleries. The insects carry the inoculum from one tree to another and grow the fungi in pure cultures on specially prepared beds and on the walls of the tunnels—a good example of mutualistic symbiosis. The larvae of some species live in the main beetles galleries; whereas for others each larva bores and lives in its own short gallery (cradle). In the South the beetles are active most of the year, but in colder climates they usually hibernate in the host trees during winter. Species of *Trypodendron,* however, winter in the ground litter. The males usually initiate the attacks and then are joined by the females. The latter do most of the boring. Some cut niches in which they deposit their eggs, whereas others simply lay them in the main tunnels. During summer the time required to complete 1 generation varies from 6 weeks to several months; therefore, there may be one to several generations per year depending on species and climate.

FIG. 11.21. Ambrosia beetles. Left: *Platypus* spp. (X8); right: the Columbian timber beetle (X14).

CONTROL. Rapid utilization of cut timber and fast drying of the lumber will prevent damage. Sometimes this is not practical. Winter harvesting and water storage are also effective. With water storage, however, only the submerged portions of the logs are immune.

Effective insecticidal methods consist of thoroughly spraying the cut logs within 48 hours with benzene hexachloride (BHC) (½ per cent gamma isomer) dissolved in diesel oil. When power sprayers are used the logs can be decked before spraying. For protecting green lumber a dip containing ¼ per cent gamma isomer of BHC formulated as an emulsion can be used. In tropical countries, however, several times the concentrations indicated above may be required to obtain adequate protection.

Fisher, R. C., G. H. Thompson, and W. E. Webb. 1953, 1954. *Forestry Abstracts* 14:381–389 and 15:3–15.

Johnston, H. R., and R. J. Kowal. 1949. *Southern Lumberman.* December 15.

Leach, J. G., A. C. Hodson, S. P. Chilton, and C. M. Christensen. 1940. *Phytopathology* 30:227–236.

Prebble, M. L., and K. Graham. 1957. *Forest Science* 3:90–112.

TIMBER WORMS

Coleoptera, Lymexylidae, Brentidae, and Tenebrionidae

Common species are *Melittomma sericeum* (Harr.) and *Hylecoetus lugubris* Say, Lymexylidae; *Arrhenodes minuta* (Drury) Brenthidae; and *Strongylium* spp., Tenebrionidae.

INJURY. Timber worms make pinholes that differ from those bored by ambrosia beetles, since the holes made by any one species vary greatly in size. This occurs because only the larvae are wood borers. The larger holes are 1/32 to 1/8 inch in diameter. The galleries are kept free of borings, and the walls usually are not stained. These insects infest wounded living trees, fresh logs, and dying trees. *M. sericeum* attacks have occurred in most of the blight-killed chestnut and the damaged wood is known and valued as wormy chestnut. Of course, this tree species is no longer of commercial importance. These insects are not as injurious as ambrosia beetles, but at times they have caused much damage to the wood of injured living trees. Damage to logs seldom is severe because the insects develop too slowly.

HOSTS AND RANGE. Hardwood species such as oaks, beech, poplar, birch, and formerly chestnut, throughout much of eastern United States wherever the hosts grow.

INSECT APPEARANCE. *Larvae: Lymexylidae*—body elongate and cylindrical; skin armed with small hardened points; head globular with mouthparts directed downward; 3 pairs of thoracic legs, each 5 segmented; 9th abdominal segment of *Melittoma* has a hard median toothed ridge; *Hylecoetus* terminal segment forms a slender tail. *Brentidae*—

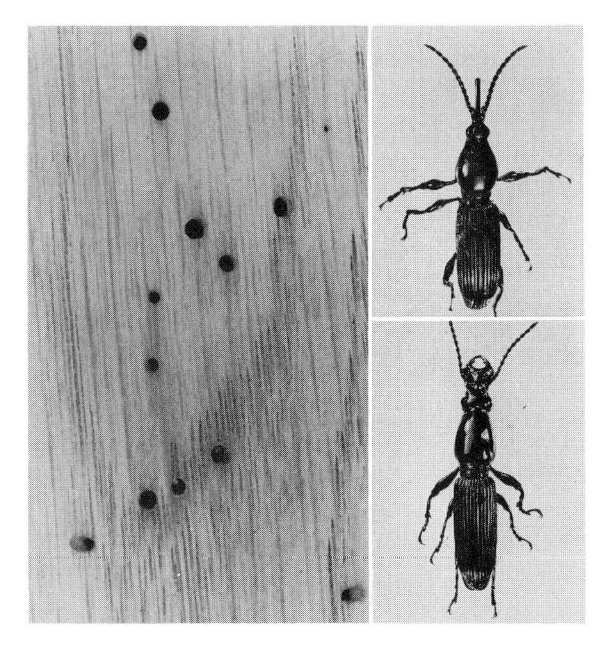

FIG. 11.22. Timber worm (*Arrhenodes minuta*) damage, (X1) and beetles (X2). Note the variation in size of larval holes. The female beetle has the beak.

body somewhat similar to *Lymexylidae,* but lacks posterior abdominal characteristics noted above; thoracic legs are very small, each with only 2 segments; last abdominal segment somewhat wider than others. Beetles: *Lymexylidae*—body slender and elongate; about ⅜ inch long; brown; antennae short, serrate, and with 11 segments; maxillary palps long, has 4 segments; pronotum with raised edge along sides; legs long slender; tarsi 5 segments. *Brentidae*—¼ to ⅝ inch long; color reddish brown; front of head of female prolonged into a snout that extends straight ahead; snout on male is broad and flat; antennae straight and not clubbed; prothorax longer than wide; elytra not wider than thorax and twice as long as wide. See Table 11.2.

LIFE CYCLE. The adults appear in the spring. The *Arrenodes* females gnaw small holes in which they deposit their eggs. The other species lay their eggs in cracks. The resulting larvae bore deeply through both the sapwood and heartwood. Other details regarding the life cycle of these insects are not available. Probably they require several years to complete 1 generation.

CONTROL. There are no practical methods for treating infested forest trees. If injury to standing trees can be avoided, the insect attacks can be prevented.

TERMITES OR WHITE ANTS

Isoptera

Termites are most abundant and troublesome in the tropics, but they also are common and injurious throughout most of the United States. In the Southeast these wood borers probably infest or have infested a large proportion of all wooden buildings. Sometimes these infestations are light and may go unnoticed by the human inhabitants. Often, however, the damage is severe; therefore termites are the most feared wood-boring pests that infest buildings. Termites also attack other cellulosic materials such as books, paper, and even herbaceous plants, especially those growing on recently cleared fields.

The taxonomic and biological characteristics are presented in Table 11.1 and in the discussions that follow.

SUBTERRANEAN TERMITES

Reticulitermes spp., Isoptera, Rhinotermitidae

INJURY. Subterranean termites cause an easily recognizable type of boring damage. Often, however, it is not discovered until in an advanced state, because termites are insidious workers and often produce little external evidence of their presence. Their extensive, irregular galleries more or less honeycomb the wood. Most of the tunnels follow the softer spring wood but interconnect with one another by means of holes through the harder layers. The walls of the galleries are coated with a

FIG. 11.23. Termites (Isoptera) and damage. Upper left: dry wood termite damage (X½). Containing fine pellets in the galleries; upper right: subterranian termite galleries (X⅔); lower left: a winged reproductive (X4); lower left: soldiers and workers (X5).

hardened, whitish, originally paste-like excrement. Frequently the cavities contain various amounts of soil, which is carried in by the workers and used for the water it contains. The most easily detected external symptoms of termite infestation are the short-to-long circular, half circular, or flattened earthen tubes (⅛ to ¾ inch or more wide), which are constructed by the termites for the purpose of forming passageways over masonry or other obstructions between the soil and the wood. Inside these earthen tubes the termites can travel without exposing themselves. Sometimes, these tubes are even built on wood surfaces. They are usually located in dark, moist locations such as inside foundation walls. Another characteristic indicating termite infestation is the emergence and swarming of the black-winged reproductives. Swarming commonly occurs during the spring season at a time when both temperature and humidity are high. After flying a short distance, the insects quickly shed their wings and disappear, leaving only an abundance of shed wings.

HOSTS. The wood of most woody plant species as well as other cel-

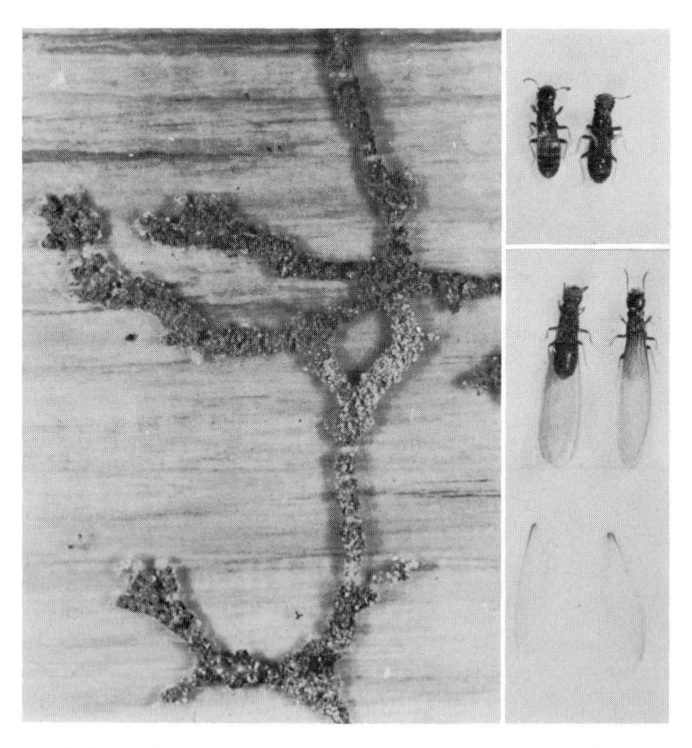

FIG. 11.24. Left: earthen termite tubes made by subterranean termites (X½); right: winged and dewinged termite reproductives (X2½).

lulosic materials is susceptible to attack. The softer springwood of the sapwood is preferred, but heartwood sometimes also is consumed. The few native woods that are somewhat resistant include the heartwood of redwood, cypress, cedars, and pitch-soaked pine.

RANGE. These insects occur throughout most of the United States, but they cause little trouble in the northern tier of states. The most severe damage occurs in California and along the coastal plain of the Southeast.

INSECT APPEARANCE. Termites are social insects. They live in large colonies consisting of thousands of individuals that occupy extensive galleries in both soil and wood. There are several types of individuals (castes) in each colony. These are named according to the functions performed, which are namely: the reproductive castes, the worker caste, and the soldier caste. *Workers:* comprise most individuals of the colonies; size mostly small, about $\frac{1}{8}$ to $\frac{5}{32}$ inch long when mature; body soft, color a dirty white; jaws with saw-tooth edges; blind; antennae prominent, with 12 to 25 bead-like segments (moniliform); thorax with 3 pairs of well-developed legs; wingless; thorax broadly joined to abdomen; both sexes represented, but all are sterile. *Worker nymphs:* similar to adults but are smaller in size. *Soldiers:* similar to workers in most respects, except that they occur in fewer numbers and have enlarged heads and large saber-tooth jaws that lack prominent marginal teeth; head may be almost as long as and as wide as the remainder of the body. *Primary reproductives:* (kings and queens). *Nymphs:* similar to worker nymphs except that they have smaller heads, are cream colored and as they grow older the abdomens become enlarged and wing pads develop. *Adults:* virgin individuals slightly larger than workers, about $\frac{3}{16}$ inch long; black to dark brown in color; after they mature and reach their full reproductive powers the abdomen of females becomes greatly enlarged so that the total length may be $\frac{3}{8}$ to $\frac{3}{4}$ inch long; these functional individuals are seldom seen; compound eyes present; 2 pairs of membranous wings, both pairs similar in size, shape, and veination; when at rest the wings project beyond posterior of abdomen by half or more of the total length; wings lost immediately following swarming flight; 3 pairs of well-developed thoracic legs; abdomen broadly joined to thorax. This is the only caste that voluntarily leaves the nests in swarms at certain times of the year for the purpose of starting new colonies. *Secondary reproductives:* sexually mature individuals similar in appearance to nymphs of the primary reproductives; whitish colored; wing-pads present, but they never develop functional wings; compound eyes are only slightly pigmented; these individuals do not leave the galleries of the parent nest.

LIFE CYCLE. All stages except possibly eggs are present during winter.

The black, winged, primary reproductives mature during the spring and soon thereafter leave the parent nests in swarms. They are weak fliers and travel only a short distance before shedding their wings and pairing. Each successful pair burrows under a piece of wood or other debris where they form a small cell and immediately start to breed. Only a few to a dozen offspring are produced in the first brood. These are cared for by the reproductives, but as soon as the nymphs mature sufficiently they take over all the work of the colony. The reproductives then function only to reproduce all the other individuals in the colony. Mating occurs repeatedly. Each queen of the temperate species produces only a few to a couple of dozen eggs per day. The eggs hatch in 1 to 3 months, and the resulting nymphs mature to the 4th or 5th instar in about 4 months. In young colonies the worker nymphs stop growing at this stage, but in older colonies some individuals continue growing for 8 or more months to become 6th or even 7th instar workers. The mature reproductives are also 7th instar insects. Termite reproductives may have a rather long life span of 3 to many years. During the colder months the insects usually descend in the soil to where the temperatures are warmer.

Forty to 100 secondary reproductives develop when the primary king or queen dies or parts of the colony become separated; consequently, the rate of growth in secondary colonies may be much greater than in primary colonies. Unlike primary reproductives, the secondary reproductives are polygamous.

A most interesting aspect regarding the utilization of wood by termites is the role played by the protozoan symbionts which inhabit the intestinal tract of termites. These protozoa apparently secrete enzymes needed for the reduction of cellulose and hemicelluloses to smaller carbohydrate units that can be utilized by the insects. The termites, in turn, supply the protozoa with food and a suitable environment. The protozoa species found in termites have not been found elsewhere.

CONTROL. The use of good sanitation and correct structural methods are the best methods for preventing attacks by subterranean termites. No wood should be allowed to accumulate beneath or near buildings. The wood underframing of buildings should be placed 2 feet or more above the ground and isolated from the supporting masonry by correctly installed metal termite shields. A most vulnerable point of termite entry, present in most dwellings built during the past 30 years is associated with the concrete slab type of porch floor. These slabs are commonly laid on earth fills, which often contain some to much wood debris. Large termite populations develop in these places within a few years and then infest the house proper via the separation cracks which invariably develop between the porch slab and the building. These vulnerable places can be adequately protected by using metal shielding, but seldom is the work done properly. A satisfactory alternative is to use a reinforced concrete slab supported on piers and walls so that there is an accessible crawl space beneath. Concrete slab floor construction for the whole dwelling is becoming popular in many places but is not recommended. Cracks form through the concrete and poor sealing of the holes cut for utilities usually provide numerous entrances for the insects; therefore, if this type of construction must be used, good sanitation as described previously must be practiced, and soil poisons should be applied to the soil before the slab is laid. All of the soil poisons listed below are satisfactory. One gallon of the insecticide mixture should be uniformly applied to every 5 square feet of soil before the slab is poured.

The use of pressure-treated wood, impregnated with any of the nonodorous preservatives, for the underframing and subfloors in newly constructed buildings is another method whereby the most vulnerable portion of wood buildings can be protected. The cost of this method should be only about 1 to 2 per cent of the total cost of the structure.

Preventing infestations also requires continuous vigilance on the part of the owners. Annual inspections should be made of the underframing and walls and proper insecticidal controls initiated as soon as any termite activity is detected.

When undertaking control operations the first thing that must be done is to make structural changes needed because of wood failure or for the purpose of reducing the possibility of future attacks. Next, the termites should be deprived of water by isolating the parts of the colonies in the buildings from the soil and preventing other sources of moisture (leaking roofs or pipes) from reaching the wood where the insects are located. Isolation of the part of the colony in the building from the part in the soil often can be done most readily by using insecticidal barriers. Suitable insecticides for establishing these barriers include oil solutions or emulsions of one of the following: chlordane (1 per cent), DDT (8 per cent), benzene hexachloride (1 per cent gamma isomer), dieldrin (½ per cent), or pentachlorophenol (5 per cent). About ½ gallon of solution should be supplied per lineal foot of foundation wall wherever the termites are active or are entering the structure. The poisons should be intermixed with the soil removed from along the wall in the form of a trench (½ foot wide by ½ foot deep). When areas beneath concrete slabs are the sources of infestations, holes about a foot apart must be drilled through the concrete and the chemicals added at the above rate. In addition to using soil poison barriers the direct injection of any of the previously listed insecticides into the termite galleries is sometimes possible and will cause a more rapid decimation of that part of the colony reached. On the other hand, the practice of drilling holes in sound wood at regular intervals and applying poison under pressure does little good except in those places where the galleries are actually penetrated.

Anon. 1949. *U.S.D.A. Farmers' Bull.* 1911 (revised).

Kofoid, C. A., S. F. Light, et al. 1934. *Termites and termite control.* Univ. Cal. Press., Berkeley, Cal. 734 pp.

St. George, R. A., H. R. Johnston and R. J. Kowal. 1960. U.S.D.A. *Home and Garden Bull.* 64.

Snyder, T. E. 1916. *U.S.D.A. Bull.* 333.

Snyder, T. E. 1948. *Our enemy the termite.* Comstock Publishing Co., Ithaca, N. Y.

DRY WOOD AND DAMP WOOD TERMITES

Kalotermes spp. and *Cryptotermes* spp., Isoptera, Kalotermitidae

INJURY. Nonsubterranean termites do not use soil in forming part of their gallery systems. They enter the wood directly and consume both the summerwood and springwood. Their excrement consists of characteristic small, smooth, oval, indented excrement pellets. (See Figure 11.23.) Some pellets are left inside the galleries, some are ejected to the outside, and some are used as building material for the small tubes they sometimes build on the outside of infested wood. Often the galleries are bored so close to the surface that the thin remaining wood layers blister. Small entrances or other openings are sealed with thin blackish or brown plugs. Generally these termites are not as destructive as the subterranean species, but in the San Francisco Bay area the damp wood termites infest more buildings than subterranean termites.

HOSTS. Most species of wood are susceptible. The resistant native species listed for subterranean termites are also somewhat resistant to nonsubterranean termites. Certain tropical woods used in furniture are also resistant.

RANGE. In North America dry wood and damp wood termites occur only in the area along the southern edge of the United States from North Carolina along the coastal regions west to Texas and across the Southwest to California. Most damage by these species occurs in southern California and southern Florida.

INSECT APPEARANCE. These termites are similar in many respects to the subterranean species, but there are only two castes—the reproductives (⅓ to 1 inch long) and the sterile wingless soldiers. The older nymphs of the reproductives do all the work of the colony. The soldiers are comparatively large and have prominent teeth along the inner edges of the jaws.

LIFE CYCLE. These termites have habits and life cycles similar to those of the subterranean species. They differ in that they never make galleries in the soil, and they produce characteristic excrement pellets. The dry wood species attack dry sound wood, whereas the damp wood species require additional moisture.

CONTROL. Preventing attacks to wood inside buildings can be accomplished by screening all openings to exclude the insects. Exposed wood can be protected by good coatings of paint or by using wood adequately treated with any of the usual wood preservatives.

Destruction of the colonies established in wood requires the injection of poisons directly into the galleries. All of the insecticides listed for use against the subterranean termites are effective. In addition the insecticide trichlorobenzene is also recommended because it is a fumigant whose vapors penetrate throughout the chambers and kill more quickly. Dusts such as 50 per cent DDT can also be used.

Snyder, T. E. 1950. *U.S.D.A. Farmers' Bull.* 2018.

WOOD-BORING CATERPILLARS

Lepidoptera, Cossidae and Aegeriidae

Only a few wood-boring caterpillars are serious pests, and these are especially troublesome as borers in living hardwood trees. Only the larvae are borers. The general order characteristics are presented in Table 11.1, and the species characteristics are outlined in Table 11.9. Other biological characteristics are presented in the discussions that follow.

TABLE 11.9 THE COMMON WOOD-BORING LEPIDOPTERA CATERPILLARS

1.	Minute hooks (crochets) arranged so as to form a circle on all except the last pair of fleshy abdominal legs (prolegs); crochets of 2 or 3 intermixed lengths	2
1.	Crochets on each of the first 4 pairs of prolegs in 2 transverse rows; crochets not of intermixed lengths	3
2.	Pronotum smooth (page 325)	LARGE CARPENTER WORM
2.	Posterior of pronotum roughened (page 329)	LEOPARD MOTH WORM
3.	Posterior of pronotum roughened (page 326)	PECAN CARPENTER WORM
3.	Posterior of pronotum smooth (page 330)	CLEAR-WING MOTH WORM

LARGE CARPENTER WORM

Prionoxystus robiniae (Peck), Lepidoptera, Cossidae

This caterpillar often is found boring in living trees, but its damage can not always be differentiated readily from that caused by certain cerambycid borers. Recent studies suggest that this insect is not as important as the cerambycid trunk borers in the black oaks of the central hardwood region.

INJURY. The tunnels are large (½ to ⅞ inch in diameter), oval, irregular, and wind throughout the wood. They occur in both the sapwood and heartwood of living trees. Borings and globular excrement pellets are mostly ejected to the outside; therefore, in lumber the damage can not always be readily distinguished from that produced by certain other lepidopterous and cerambycid borers. The tunnels bored by older larvae are always larger than those bored by beetle larvae.

HOSTS AND RANGE. Many species of hardwoods throughout the United States and southern Canada.

INSECT APPEARANCE. *Caterpillars:* large, 2 to 3 inches long when full grown; body segmented; naked, soft, whitish, and wormlike; pronotum smooth; 5 pairs of fleshy abdominal legs (prolegs) with the first 4 pairs each with a circle of minute hooks (crochets) of 2 or 3 intermixed lengths. *Moths:* large, wingspread 2 to 3 inches; body stout; color grayish with forewings slightly translucent and mottled with a few irregular gray spots; hindwings grayish with outer portions yellowish on males.

LIFE CYCLE. The moths emerge during June and July. After mating, the females lay several hundred greenish to brown oval eggs on the bark of the host trees. These hatch after 1½ to 2 weeks, and the resulting larvae mostly enter the trees via wounds made by other insects. Apparently very few enter directly by boring through the outer bark. They bore in the inner bark for a short time and then enter the sapwood where they spend the first 2 years. The older larvae bore in the heartwood. The time required to complete 1 generation is 3 or 4 years.

CONTROL. There is no practical way to protect forest trees from the attacks of this insect. Infested shade trees, however, can be treated by enlarging the openings to the galleries at the entrances and injecting a fumigant, such as carbon disulphide or calcium cyanide. After treating, the entrances should be plugged with putty, grafting wax, or some other similar plastic materials.

PECAN CARPENTER WORM

Cossula magnifica (Stkr.), Lepidoptera, Cossidae

INJURY. Caterpillars of the pecan carpenter worm bore small circular holes about ¼ inch in diameter in the twigs, branches, and trunks of the host trees. These long, rather straight holes are bored mostly with the grain of the wood. The wood surrounding the holes is stained brown. The larvae keep the galleries free of borings and excrement pellets by extruding these to the outside. Sometimes this insect causes much loss

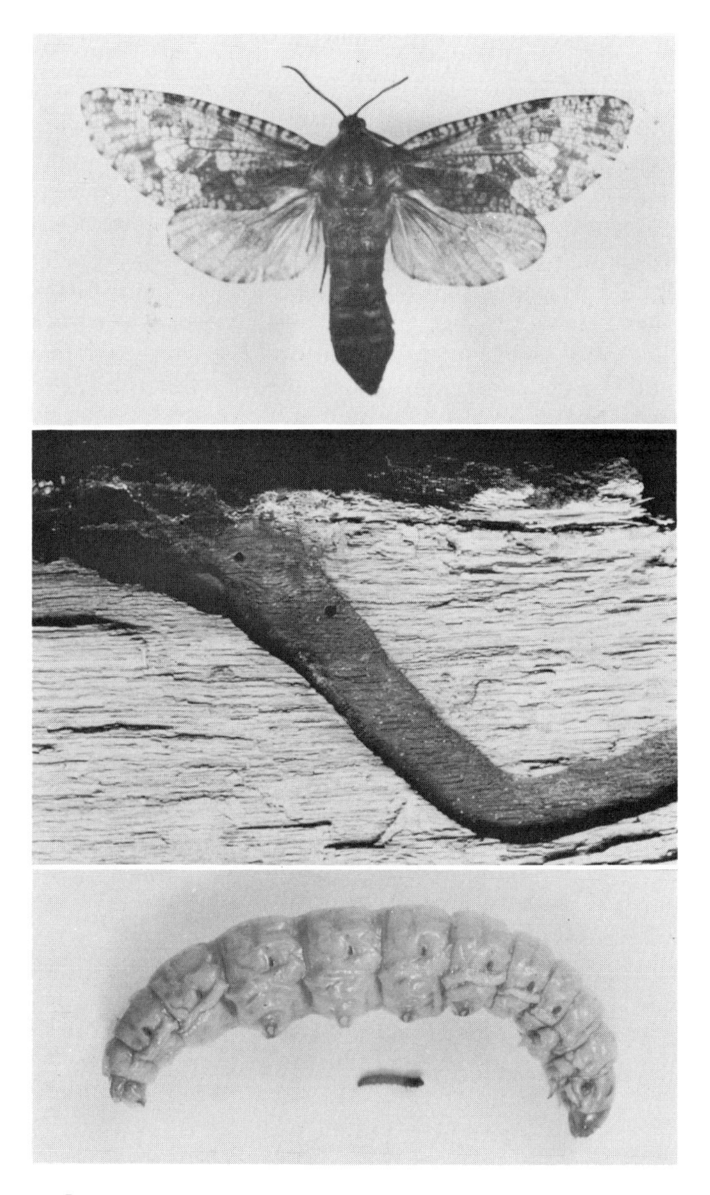

FIG. 11.25. Large carpenter worm (*Prionoxystus robiniae*) moth (X1), damage (X½), and caterpillar (X1).

to buyers of oak stumpage because the old damage can not be detected from outside.

HOSTS AND RANGE. Hickories and oaks throughout the Southern States wherever the hosts grow.

INSECT APPEARANCE. *Caterpillars:* typical; with 5 pairs prolegs; body naked, soft, worm-like, about 1½ inches long when full grown; whitish colored with head, posterior abdominal tergite, and pronotum brown; other characteristics are presented in Table 11.9. *Moths:* wingspread about 1½ inches; head brown, wings ashy gray with outer fifth of forewings brown.

LIFE CYCLE. The moths emerge during May and June. After mating, the females lay their eggs on the bark of small twigs. The newly hatched larvae first enter and bore in the twigs, but as they grow older they emerge and crawl to larger branches and the main trunks. They re-enter

FIG. 11.26. Pecan carpenter worm (*Cossula magnifica*) insects (X1½) and damage (X2).

FIG. 11.27. Leopard moth (*Zeuzera pyrina*) (X1).

these parts and here cause the most severe damage. It has been stated that the length of time required to develop from eggs to adults is 1 year, but probably a longer period of time is required under some conditions.

CONTROL. Applied control methods are the same as those presented for the large carpenter worm.

LEOPARD MOTH

Zeuzera pyrina (L.), Lepidoptera, Cossidae

INJURY. Leopard moth caterpillars cause damage somewhat similar to that caused by the large carpenter worm but differ in that the galleries also occur in the twigs and small branches as well as in the main trunks. The weakened, tunneled branches often break and drop from the trees in large numbers. Wood chips and excrement are ejected to the outside; consequently, the galleries are kept clean. It is most injurious as a shade tree pest. Silk webbing commonly is present.

HOSTS. Many deciduous tree species are attacked, but elm and maples are preferred.

RANGE. This European species was introduced into the United States sometime during the last century. It now occurs in the Northeast from Massachusetts to Pennsylvania.

INSECT APPEARANCE. *Caterpillars:* typical; with 5 pairs prolegs; body naked, soft, and worm-like; about 2 inches long when full grown; color pale yellow to pink with head, pronotum, and last abdominal tergite brown; other characteristics are presented in Table 11.9. *Moths:* large, 2½-inch wingspread; heavy bodied; thorax white with 7 black spots on back; abdomen white with black crossbands; wings white, semitransparent and thickly dotted with dark blue or greenish black spots.

LIFE CYCLE. The moths emerge throughout the summer from May through September. After mating, each female lays many hundreds of eggs in small groups of 3 or more in the bark crevices. The eggs hatch in about 1½ weeks and the resulting larvae immediately bore into the tree. Sometimes the caterpillars emerge and crawl to other places where they re-enter. Pupation occurs within the burrows just prior to adult emergence. Two years are required for the insects to complete 1 generation.

CONTROL. Heavily infested branches of shade trees should be cut and burned. When infestations are light the trees can be treated in the same way as described for the carpenter worm.

Howard, L. O., and F. H. Chittenden. 1916. *U.S.D.A. Farmers' Bull.* 708.

CLEAR-WINGED MOTH BORERS

Lepidoptera, Aegeriidae

Caterpillars of the common species of clear-winged moth can be tentatively identified on the basis of host infested. These are as follows: ALDER BORER [*Thamnosphecia americana* (Beut.)], alder; ASH BORER [*Podosesia fraxini* (Lugger)]; LILAC BORER [*P. syringae* (Harr.)], species of ash, mountain ash, and lilac; HORNET MOTHS [*Aegeria apiformis* (Clerck)] and (*A. tibialis* Harr.), willows and poplars; LOCUST CLEAR-WING [*Paranthrene robiniae* (Hy. Edw.)], locust and poplar; MAPLE CALLUS BORER [*Conopia acerni* (Clem.)], maples; PERSIMMON BORER (*Sannina uroceriformis* Wlkr.), persimmon; PITCH MASS BORER [*Parharmonia pini* (Kellicott)], pines and spruces; and SYCAMORE CLEAR-WING [*Ramosia mellinipennis* (Blv.)], sycamore and oaks. Also see page 260.

INJURY. The boring damage caused by clear-winged moth caterpillars often can be identified because of the rough, callused, often gall-like appearance of the outer bark. Other evidence consists of the presence of cocoons, excrement pellets, and sometimes a small amount of silk. For a positive diagnosis, however, the caterpillars must be examined. These insects are only of minor importance as forest pests, but their damage may be encountered frequently.

HOSTS AND RANGE. The hosts of the various common species are listed above. They occur wherever the hosts grow.

INSECT APPEARANCE. *Caterpillars:* typical with 5 pairs prolegs; medium size, ¾ to 1½ inches long; body naked, soft, whitish, worm-like; other characteristics are presented in Table 11.9. *Moths:* superficially they resemble wasps; wings narrow with forewings narrower than hind-

FIG. 11.28. Clear-wing (*Aegeriidae*), moth caterpillar (X3), and damage (X¾).

wings; wings mostly devoid of scales and transparent, scales mostly present around edges and along wing veins; bodies often brightly colored; antennae taper toward both tip and base, tip with a tuft of fine hairs; body colored tan to blue-black with a metallic iridescence and frequently marked with yellow, orange, or red spots or bands.

The general characteristics of the moths of the common species are as follows: *Alder borer*—wingspread ¾ to 1 inch; color blue-black with an orange-red spot on each side beneath thorax, 4th abdominal segment red. *Ash borer*—wingspread ⅞ to 1¼ inches; color black; forewings opaquish, blackish brown with violet reflections, a red cross vein near center of wing; hindwings transparent with a narrow black border. *Hornet moths*—wingspread 1¼ to 1⅔ inches; color brown, yellow spot on each side of thorax; yellow bands on front part of each abdominal segment except the last 2, which are completely yellow; legs yellow; wings transparent with brown borders. Many other species are similar to the hornet moth except that the yellow markings on the abdominal segments are different. *Locust clear-wing*—wingspread 1 to 1⅓ inches; color of abdomen mostly yellow with the first 3 segments black; forewings opaque and yellowish. *Maple callus borer*—wingspread ¾ to 1 inch; color brownish; wings hyaline tinged with yellow; anal tuft light red. *Persimmon borer*—wingspread 1 to 1¼ inches; color bluish-black; 4th abdominal segment reddish on top with a narrow dark center line; wings opaque except for a transparent area near base of hindwing. *Pitch mass borer*—wingspread 1 to 1¼ inches; color bluish-black with the following yellow or orange markings; cross bands on front of prothorax and the 4th abdominal segment; underside of abdomen and anal tuft also yellow or orange; forewings opaque, black, with iridescent blue or green reflections. *Sycamore borer*—black and yellow with 4 or 5 broad yellow bands on abdomen; forewings transparent with margins reddish, yellowish, or bronze.

LIFE CYCLE. These swift flying, diurnal moths emerge mostly during the early summer. They feed on nectar. After mating, the females deposit their eggs on the bark of the host trees. The length of life cycle varies from 1 to several years depending on the species.

CONTROL. Applied methods have never been needed to kill these borers in forest trees. Infested shade or ornamental trees can be successfully treated by injecting carbon disulphide or calcium cyanide into the burrows at the places from which borings are being ejected. The holes should then be plugged with putty or wet clay. To prevent reinfestation the trunks and branches of the trees should be thoroughly sprayed with a spray consisting of ½ per cent DDT either as a wettable powder or emulsion. The spray should be applied at the time the first moths appear and thereafter every 2 weeks throughout the flight period.

Blakeslee, E. B. 1915. *U.S.D.A. Bull.* 261.
Brooks, E. F. 1920. *U.S.D.A. Bull.* 887.
Peterson, L. O. T. 1951. Development and control of the ash borer. *Canadian Dept. Agri. Publ.* (Unnumbered).
Schread, J. C. 1956. *Conn. Agri. Exp. Sta. Circ.* 199.
Snapp, O. I. 1943. *U.S.D.A. Tech. Bull.* 854.
————. 1954. *U.S.D.A. Farmers' Bull.* 1861.

HYMENOPTEROUS WOOD BORERS

Hymenoptera

The wood-boring Hymenoptera include the horntails, the carpenter bees, and the carpenter ants. Members of the latter two do not eat wood but merely use it as a place in which to live. The identifying characteristics and habits of these borers are presented in Table 11.1 and in the discussions that follow.

HORNTAILS

Tremex spp., *Sirex* spp., *Urocerus* spp., and others, Hymenoptera, Siricidae

INJURY. The larvae of horntails bore through the wood of dying and recently felled trees making circular tunnels up to about 3/16 inch in diameter. These are tightly packed with fine borings. The adult wasps frequently are attracted to fires and often are observed ovipositing on still smoldering trees. Most damage occurs to fire killed timber that cannot be salvaged immediately.

HOSTS AND RANGE. The species of greatest importance infest the Pinaceae conifers, but some, including the common pigeon tremex, infest hardwoods. They occur wherever the hosts grow.

INSECT APPEARANCE. *Larvae:* body elongate, up to 2 inches long; soft, wrinkled, cylindrical, and whitish; head globular somewhat retracted into prothorax with mouthparts directed downward; 3 pairs of fleshy, unjointed, thoracic legs; posterior of abdomen with a small backward projecting spine. *Adults:* wasp-like; body elongate, 7/8 to 1 1/3 inches long, slender, cylindrical, and thick waisted (thorax and abdomen broadly joined), color black or metallic blue-black; some species marked with brownish, reddish, yellow, or whitish cross bands or spots; wings sometimes dusky; back of posterior abdominal segment with a short backward projecting spine; long ovipositer also projects backward from the under parts of the posterior abdominal segments. *Pupae:* resemble the adults but have wing pads instead of wings and are of a whitish color; appendages are free of body but held rather close to the body.

LIFE CYCLE. The adult wasps of the various species appear at different times during the summer. The females often are much more numerous than the males, which suggests that parthenogenetic reproduction may occur. They prefer to attack freshly felled or dying trees. When laying eggs the females insert their ovipositers into the wood for a distance of 1/4 to 3/4 inch and deposit 3 or 4 eggs in each hole. The resulting larvae bore through the wood and make the characteristic galleries. They spend about half their larval life in the sapwood and then make a loop

FIG. 11.29. Horn tail (*Siricidae*) adults (X1) and damage (X1½). Note the short sharp projection on the posterior abdominal segment and also the longer ovipositor originating from lower parts of the abdomen.

into the heartwood. Their pupal chambers are formed near the surface. The period of time required to complete a life cycle is 1 or 2 years depending on the species.

CONTROL. Reasonably rapid utilization of all dead timber is the only way to avoid horntails damage.

CARPENTER ANTS

Camponotus herculeanus L. (many varieties), Hymenoptera, Formicidae

INJURY. Carpenter ants bore large irregular cavities which "honeycomb" the wood of posts, timbers, and even living trees. The galleries are kept clean of borings and other debris. The walls are rather smooth and never covered with a coating—a characteristic present in all subterranean termite galleries. Coarse fibrous borings are extruded from the galleries and sometimes these may accumulate in quantity around infested wood, but excrement pellets are never produced.

At one time about a fifth of the standing white cedar in parts of Minnesota was reported to be infested, and about 10 per cent replacement of chestnut telephone poles in the Northeast was necessitated because of carpenter ant damage. Nevertheless, they are much less injurious than termites.

HOSTS. All types of wood are subject to attack. These include posts, poles, logs, timbers, and living trees. In living trees, however, the entrances are always made through wounds. Only the adult ants do the boring. This is done only for the purpose of making a cavity in which to live. The wood removed is never eaten.

RANGE. Different species and varieties occur in various regions throughout North America.

INSECT APPEARANCE. Ants are well known to most people. They are social insects which live in colonies consisting of hundreds to thousands of individuals. Each colony contains several types or castes. These are the reproductives, which are commonly called "kings and

FIG. 11.30. Carpenter ants (*Camponotus herculeanus*) adults (X1⅓) and damage (X⅔). Insects: winged male and larger winged female (queen) are shown above; below are two workers.

queens," and various sized workers. The larger workers are commonly called the major workers and the smaller ones the minor workers, but usually there are many intergrading sizes. *Reproductives:* females are large, ½ to ¾ inch long; males are smaller, ⅜ to ½ inch long; color mostly black but some varieties are brownish and others have reddish legs; wings are present until after the queens have mated; body greatly constricted between thorax and abdomen with a small segment (node) joining the two parts; antennae with elongate basal segment (scape), the remaining series of small segments connected to the scape at a pronounced angle (elbowed); tip of abdomen with a circle of hairs around anus. *Workers:* sterile workers similar in appearance to the reproductives but are always wingless; ¼ to ½ inch long; crushed bodies exude a distinct formic acid order. The males are parthenogenetically produced (have only a haploid number of chromosomes). *Larvae:* bodies soft, whitish, and segmented; legless; thoracic segments smaller in girth than the large sac-like abdomen; head globular with small jaws; found inside the nests with adults. *Pupae:* whitish, immobile, soft-bodied insects that resemble adults; appendages not closely appressed to body; enclosed in a whitish, silk cocoon; occur inside the nests with the adults.

LIFE CYCLE. Several hundred winged reproductives are produced annually in each of the older colonies. These develop during the summer but do not leave the nest until the following spring or summer. Mating occurs in flight, and the successful males soon die. Each queen mates only once; therefore, she receives sufficient sperm to produce offspring during her 2 to 15 year life span. Shortly after mating each queen sheds her wings and bores into a suitable large piece of wood as described previously. Here she excavates a small cell, lays about 20 eggs, and cares for the resulting larvae. The larval food apparently is secreted by the queens for they do not feed while caring for this first brood. The workers produced during the first 2 years are all small (minor workers). About 2 months are required for the complete development from egg to adult.

As their name suggests, the workers do all the work of the colony such as enlarging and cleaning the nest cavity, foraging for food, and feeding and caring for the eggs, larvae, and queen. Carpenter ants do not eat wood. Their food consists mostly of the sweet, liquid, secretion (honeydew) produced by plant lice (aphids), but dead insects and succulent plant foods are also used. The smaller workers collect the honeydew and the larger ones transport it to the nests. Those that are extremely large or very small probably do the work inside the nest. The animal and plant foods used also commonly are extracted as liquids before being carried to the nests. Sometimes, however, solid food is carried home.

CONTROL. Any of the insecticides suggested for use against subterranean termites also are effective for killing carpenter ants. These poisons must be injected directly into the cavities housing the colonies. The tops of the nests can be located by percussing the poles or timbers with a hammer and/or boring trial holes. The insecticides then can be injected into the nests by means of hand-operated oil guns similar to the type used to fill automobile transmissions. Dusts of these insecticides may also be used, but the concentrations of the active ingredients should be about twice that for the liquids.

Friend, R. B., and A. B. Carlson. 1937. *Conn. Agri. Exp. Sta. Bull.* 403.
Furniss, R. L. 1944. *Oregon State Coll. Stat. Circ.* 158.
Pricer, J. L. 1908. *Biol. Bull.* 14:177–218.

LARGE CARPENTER BEES

Xylocopa spp., Hymenoptera, Xylocopidae

X. orpifex Smith and *X. virginica* (Drury) probably are the most common.

INJURY. The adult bees bore large circular holes which are 3/8 to 9/16 inch in diameter and 5 to 18 inches long. These tunnels penetrate inwards for an inch or so and then extend with the grain of the wood. Occasionally they may branch. Galleries are slightly expanded every 1/2 to 3/4 inch in a manner which suggests that walls have been cut with a gouge. They are always connected with a circular entrance hole of about the same diameter as the galleries. The burrows are divided into individual cells 1/2 to 3/4 inch long by means of cross walls, which consist of cemented, compacted, macerated wood. Each cell houses one larva. Even after the brood has matured and left, parts of the cross walls sometimes remain. Usually the damage these bees caused is not very serious, but at times the galleries may be so concentrated that structural weakening results.

HOSTS. All species of dry, seasoned wood may be attacked, but the softer woods such as cedars, redwood, pines, and firs are preferred.

RANGE. *X. virginica* occurs throughout the East, whereas *X. orpifex* and others are western species.

INSECT APPEARANCE. *Bees:* large, 1/2 to 1 inch long, and robust; somewhat resemble bumble bees; color of females greenish black (*X. orpifex*) or black (*X. virginica*); males often with yellow or pale hairs; other species metallic purplish or bluish black; females with dense brush or hairs on hind legs instead of pollen baskets; back of abdominal segments are sparsely covered with hairs and appear shiny, whereas the abdominal terga of bumble bees are densely covered with hair. *Larvae:* typical

FIG. 11.31. Large carpenter bee (*Xylocopa virginica*) (X1) and damage (X⅔). Note the gouged appearance of the walls. The cross walls are destroyed when the adult bees emerge.

Hymenoptera grubs; body sac-like; legless; head globular with small mouthparts; thoracic segments smaller than those of abdomen. **Pupae:** also typical hymenopterous type; robust, always inside silk cocoons. Both larvae and pupae always found in the wood as described above.

LIFE CYCLE. The adult bees emerge in the spring. After mating, the females bore the burrows in the wood, which are used solely for the rearing of the brood. The wood removed is never eaten. After a female completes a gallery she collects a small quantity of pollen and nectar which is placed near the closed end of the gallery. An egg is deposited on this food mass and then sealed off with a wooden cross wall placed ½ to ¾ inch from the end. This process is repeated until each gallery is filled with these provisioned cells. The developing larvae eat this pollen and nectar and grow to maturity in 7 or 8 weeks. Pupation follows immediately, and the young adults remain in the chambers until the following spring. Sometimes, however, a few close to the entrance emerge during the fall. All escape through the entrance holes made by their parents; therefore, the youngest bee emerges first and the oldest last. There is only 1 generation per year.

CONTROL. Applied control is seldom needed. Coating the wood with paint is effective for preventing attacks. The insects within the galleries can be killed by injecting a fumigant such as calcium cyanide or carbon disulphide in the tunnels and then plugging the entrances. Of course, this treatment does not reduce the damage already done. Five per cent DDT dust or spray injected into partially constructed galleries will kill the adult bees doing the boring.

Ackerman, A. J. 1916. *J. N. Y. Entomol. Soc.* 24:196–232.
Davidson, A. 1893. *Entomol. News* 4:151–153.

MARINE BORERS

Wood placed in sea or brackish water is subject to rapid and extensive damage by wood-boring animals that are not insects. These attacks are caused chiefly by shipworms (Phylum Mollusca) and by wood lice (Phylum Arthropoda, Class Crustacea). In the coastal waters of North America they cause an annual estimated loss of about 50 million dollars.

SHIPWORMS

Teredo spp. and *Bankia* spp., Phylum Mollusca, Class Pelecypoda, Order Eulamellibranchia, Family Teredinidae

INJURY. Shipworms cause serious damage to wood whenever it is placed in sea water. They bore smooth, circular holes that are ⅛ to ⅞ inch in diameter, up to several feet long, and are always lined with a whitish calcareous deposit. These holes enter the wood at right angles but soon turn so as to extend lengthwise with the grain of the wood. Usually, however, they twist and turn so as to avoid adjacent tunnels. The entrance holes are very small (less than ⅛ inch in diameter); therefore, the extent of damage is not very evident from the outside appearance of the wood. Inside, however, the holes increase in size rapidly and usually are so numerous that the wood is honeycombed. Untreated piling may be destroyed within a period of time of only few months to a few years. Only wood exposed in sea or brackish waters is attacked.

HOSTS. Most species of wood are susceptible although certain tropical woods are somewhat resistant.

RANGE. Throughout the seas of the world, with most damage occurring between the north and south latitudes of 45° to 60°.

ANIMAL APPEARANCE. Shipworms live most of their life with almost all of their bodies concealed in the galleries. They are related to the clams and oysters but differ considerably in general appearance. Their bodies are worm-like, elongate, about as long and large as the burrows each inhabits, soft, whitish, and unsegmented; a pair of small 3-lobed calcareous shells at the anterior; a fleshy locomotive and holding organ

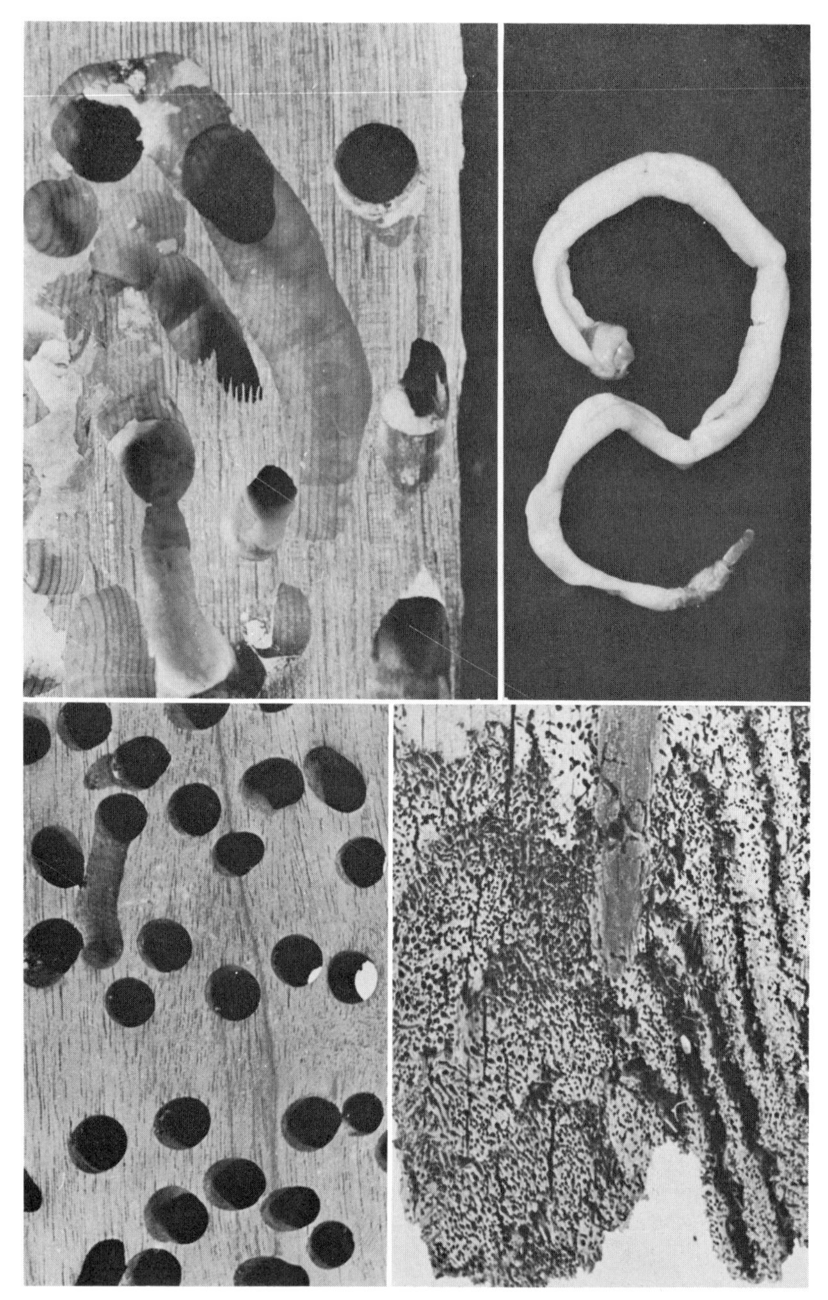

FIG. 11.32. Marine borers. Left: shipworm damage (X1); upper right: a shipworm (*Bankia* spp.) (X½); lower right: wood lice damage (X⅔).

(foot) is located between the 2 shells; a pair of short tubes (siphons) and a pair of feathery (*Bankia*) or spade-like (*Teredo*) calcareous structures (pallets) project from the posterior end. The pallets are used for closing the minute entrance holes whenever the animals are disturbed or for retaining water in the galleries when the tide drops and the water level is below the entrances. The siphons transport water in and out of the animals.

LIFE CYCLE. *Bankia* species discharge both eggs and sperm into the water where fertilization and subsequent development starts. Reproduction of *Teredo* differs in that the females retain their eggs by holding them between their gills. Here fertilization and ultimate hatching occur with the spermatazoa being carried to the females through the water. The resulting offspring escapes into the sea via the discharge siphons. Each female produces a tremendous number (millions) of eggs. Reproduction occurs throughout the warmest 5 or 6 months of the year.

The minute young, enclosed by 2 shells, first swim about searching for wood, but as soon as they locate a suitable piece they start boring. From then on for the remainder of their lives they are imprisoned in their burrows. They grow rapidly under favorable conditions so that some are as long as 1½ feet after 8 months. Others in less favorable locations require 2 years to reach this size. Most individuals that become established in wood probably die within 6 months, and probably less than 10 per cent ever live until the following year. Usually the life cycle is completed within 1 year.

Food probably consists mostly of the minute floating or weakly swimming organisms (plankton) that are drawn in via the intake siphons and partly by digesting the wood removed in boring. The ingested sea water passes first into a chamber and then past the filamentous gills, which are arranged inside along the sides of the body. These organs absorb oxygen from the water. The water then passes into a second chamber and finally out through the discharge siphon.

CONTROL. Prevention of shipworm damage consists of either using wood preservatives or covering the surface of the wood with some inert material. For waterfront timbers the best method appears to be to use timber heavily impregnated with coal tar creosote, applied by a full cell pressure treating method. The minimum retentions should be 14 and 20 pounds per cubic foot for Douglas-fir and southern pine respectively. Cooper naphthenate is usually used for treating the wood of boats because it doesn't have a strong odor nor does it blacken the treated wood. Coatings of paint or durable metal such as copper sheathing are also effective but are difficult to maintain intact. Breaks in these coatings occur readily owing to rough usage or theft of the copper. Encasing piling timbers with reinforced concrete has had some success but, here again, maintenance of the coating is often difficult.

Am. Soc. Testing Materials. 1957. *Tech. Publ.* 200. Philadelphia, Pa.

Clapp, W. F., and R. Kenk. 1956 and 1957. *Marine borers—a preliminary bibliography.* Library of Congress. Tech. Inform. Div. Vol. I: 346 pp., Vol. II: 350 pp.

Hill, C. L., and C. A. Kofoid (editors). 1927. *Marine borers and their relation to marine construction on the Pacific coast.* University of California Press, Berkeley, Cal. 734 pp.

1950. *Great Britain Forest Products Research Leaflet* 46.

WOOD LICE (GRIBBLES)

Limnoria spp. and *Sphaeroma* spp., Phylum Arthropoda, Class Crustacea, Order Isopoda

INJURY. *Limnoria* bore holes 1/16 to 1/8 inch in diameter and about 1/2 inch deep, whereas the holes of *Sphaeroma* are 1/2 inch in diameter and 3 to 4 inches deep. (See Figure 11.32.) Only wood placed in sea or brackish waters are attacked but these commonly are infested so heavily that the exposed surfaces are completely honeycombed and resemble sponges. Subsequent wave erosion wears away the remaining thin walls so as to expose the deeper layers of wood. Continuous repetition of this process results in the erosion of 1/2 to 1 inch of surface wood per year. Wood lice in conjunction with the shipworms can destroy heavy unprotected timbers within a period of from 6 months to a few years. The heaviest attacks occur between the mean tide level and low tide, but heavy damage may occur at water depths of 40 to 70 feet or more.

HOSTS. Most species of wood placed in the seas.

RANGE. In sea waters throughout most of the world, except possibly the coldest regions. Even in the colder waters, however, they cause more damage than do the shipworms.

ANIMAL APPEARANCE. *Limnoria* are only 1/16 to 1/8 inch long, whereas *Sphaeroma* are larger (about 1/2 inch long). They resemble terrestrial pill bugs; body light colored, subcylindrical with 2 pairs short antennae; 7 pairs of equal-sized thoracic legs and 5 pairs of thin plate-like abdominal appendages; the latter serve for both swimming and respiration; mouthparts somewhat resemble those of insects; with a pair of palps on the main wood-chewing jaws (mandibles) but not on the accessory jaws (maxillae); the lower lips consist of a pair of structures; posterior abdominal segment is a plate-like (telson).

LIFE CYCLE. Not very much is known about the life history of wood lice. They differ from shipworms in their habits, for they can vacate one piece of wood and attack another. In some places they breed throughout the year, but in other waters there appears to be a definite breeding season that occurs during the warmer months. Ten to twelve eggs are produced at one time per female. These are placed in a brood pouch located on the underside of the thorax where they remain until they hatch. Gribbles bore in the wood only for the purpose of obtaining a place to live. They do not obtain nourishment from the wood.

CONTROL. The applied methods for preventing damage by wood lice are the same as those used against the shipworms.

See the references listed following the discussion of shipworms.

WOOD DISCOLORATIONS

Many species of insects and a species of woodpecker cause or are associated with discolorations of wood. Some of these insects, such as bark beetles (page 203) and ambrosia beetles (page 312), act as vectors of wood staining fungi and inoculate these organisms into living, dying, or dead trees. Other insect borers, including many of those that infest living trees (pages 264, 282, 292, and 328), simply allow air to enter the wood where the oxygen combines with certain chemicals present to form oxides that discolor the wood. As indicated by the page references given previously, the various insects that are associated with wood discolorations are discussed elsewhere in the book. Only birdpeck damage is discussed here.

BIRD PECK

Yellow-bellied sapsucker [*Sphyrapicus varius* (L.)] Phylum Chordata, Class Aves, Order Pici, Family Picidae. Several varieties include the eastern, the red-breasted, and the red-napped.

INJURY. These woodpeckers bore rows of small (3/8 to 1/2-inch diameter), conical to oval-shaped holes through the outer bark and into the inner bark of the boles of trees. Sometimes this injury girdles and kills parts or all of the tree crowns, but usually it causes only dark discolorations in the wood. Subsequent tree growth embeds these stained spots in the wood. In the Piedmont region of the Southeast bird peck causes a serious defect in many hardwoods. Almost half the trees of hickories and red maple have been attacked in some areas.

HOSTS AND RANGES. The western varieties of the yellow-bellied sapsucker are the Rocky Mountains inhabiting red-naped sapsucker and the red-breasted sapsucker, which ranges through the far western states and Canadian Provinces. The eastern variety occurs throughout the East.

BIRD APPEARANCE. Back and wings are black varied with white;

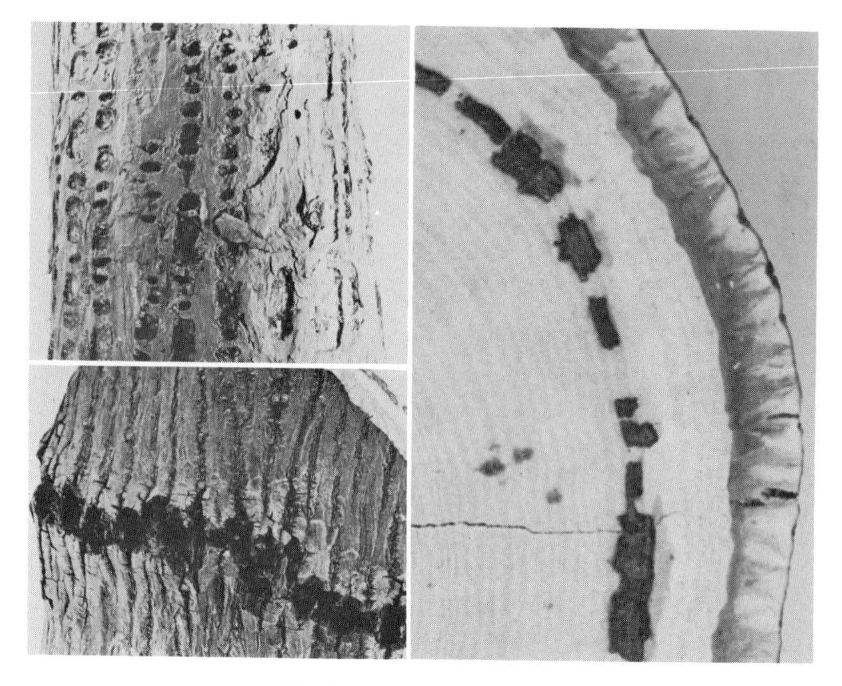

FIG. 11.33. Birdpeck damage.

crown of head red, throat and chin red on males but white on females; belly yellowish; white elongated patch on each wing; 2 white stripes on each side of head; black patch on breast of both sexes. The red-naped variety is similar to the eastern except that there is another red band on the back of the neck and the belly is very pale yellow (almost white). The red-breasted sapsucker has all of the head, neck, and breast red.

LIFE CYCLE. These sapsuckers breed in Canada and northern United States. The adult birds excavate cavities in dead trees in which they make their nests. Five to seven eggs are laid per clutch. The young are featherless and helpless when first hatched and do not leave the nest until full grown. Food consists chiefly of sap and the insects which are attracted to the sap. The diet of the young consists chiefly of insects. In the fall the birds migrate to the South where they spend the winter.

CONTROL. Ornamental or shade trees can be protected by shooting the offending birds or by scaring them away by means of some mechanical device. Shooting the woodpeckers, of course, is illegal, because woodpeckers are considered songbirds and, in some places (cities and villages), the use of firearms is prohibited.

CHAPTER 12

Sapsucking Insects

Many insects and some other Arthropods feed only on plant sap. The injury consists of enlarged growth (galls), foliage disturbances such as bleaching or yellowing, and/or deformations such as curling. All parts of all plant species are subject to attack by these pests, but usually a given species infests only a particular plant part. In forests conifers are more severely injured than hardwoods.

Most sapsucking insects belong to the order Hemiptera (Homoptera), but some are mites (Class Arachnida) and many of the gall formers are wasps (Hymenoptera) and flies (Diptera). The characteristics of the various types of these insects are presented in Table 12.1. A few sapsucking insects are not considered here because the main injury they inflict consists merely of slits made with their ovipositers in twigs. See page 392.

TABLE 12.1 COMMON GROUPS OF TREE-INFESTING SAPSUCKING INSECTS

1\. Leaves, twigs, and/or roots become enlarged vegetative growths (galls) inside of which the insects live (page 369) . . . . GALL INSECTS

1\. Enlarged growths not formed but curling and deformation of the leaves or young shoots may occur 2

2\. Adults scale-like; flattened to globular, usually immobile and rather firmly attached to the leaves, twigs, or branches; frequently appear like a bark scale; some secrete a white woolly coating; tarsi 1-segmented (page 354) . . . . SCALE INSECTS

2\. Insects always move about readily, frequently are very agile 3

3\. Antennae absent; minute (1/60 to 1/80 inch) oval globular animals; adults with 4 pairs of legs, nymphs with 3 pairs; head indistinct, combined with thorax (caphalothorax); fine silken webbing commonly present on the infested plants (page 372) . . . . SPIDER MITES

3. Antennae always present but may be small and bristle-like; adults with only 3 pairs of legs; head always distinct; silk webbing never present 4
4. Small insects; forewings net veined, held horizontal, and appear rather square when viewed from above; prothorax with a lateral wing-like, net-veined outgrowth on each side; feeding causes small, whitish spots on upper leaf surfaces; frequently leaves appear bleached; numerous small blackish excrement spots on lower leaf surfaces (page 368) . . . . LACE BUGS
4. Wings, when present, not as in 4 above; no black excrement spots on leaves . . 5
5. Tarsi 2-segmented, antennae thread-like; pair of gland-like or tube-like structures (cornicles) often present on back of 6th abdominal segment; hind legs (femora) usually not enlarged and adapted for jumping (page 347) . . PLANT LICE OR APHIDS
5. Tarsi 3-segmented; antennae bristle-like; hind legs (femora) always enlarged and adapted for jumping 6
6. Edge of hind tibia each with 1 or 2 stout teeth; tip also crowned with short, stout spines; nymphs live enveloped in a sputum-like mass on the host plants (page 364) . . . . SPITTLE BUGS
6. Each hind tibia with a row of spines along edge; nymphs do not live in a spittle mass; upper leaf surface with small whitish spots often making the leaves appear bleached; dark excrement spots never present on lower leaf surfaces (page 367) . . . . LEAFHOPPERS

HEMIPTEROUS SAPSUCKERS

Hemiptera

The order Hemiptera includes most of the sapsucking insects and some of the gall formers. Mites are not insects. They belong to the class Arachnida.

INJURY. When feeding, the mouthpart stylets are inserted into the host tissues, and the sap is extracted by aspiration. The madibular stylets cut a minute puncture through which the maxillary stylets follow. The latter pair are joined lengthwise to form a tube through which the plant sap is ingested. The saliva pumped into the host when feeding usually is toxic to the plant cells and, thereby, facilitates extraction of the cell contents. When the insects are not feeding, the stylets are all housed in the tube-shaped labium.

Many sapsucking insects excrete a large quantity of sweet liquid known as honeydew. This substance attracts and is utilized by many insects such as bees, wasps, flies, and ants. It also falls on the foliage below, where it becomes a medium on which black sooty molds grow. Frequently this mold completely covers the upper surfaces of the leaves and twigs so that these parts appear black.

INSECT APPEARANCE. Hemipterous insects are of various sizes and shapes, but all have elongate, slender, tube-like mouthparts that consist of several needle-like stylets enclosed within the tube formed by the lower lip (labium). The adults may be either winged or wingless. Some have clear membranous wings, whereas others have the basal two third thick and leathery. Other characteristics are that the terminal segment of each tarsus is never bladder-like, the compound eyes are large, each antenna has from 4 to 10 segments, there are no cerci, and the metamorphosis generally is of the gradual type.

PLANT LICE OR APHIDS

Hemiptera (Homoptera), Aphidoidea

INJURY. Aphids are sapsucking insects that cause a general devitalization of the parts fed upon. They cause the foliage to yellow, and if infested while young and growing the leaves curl and become deformed. Some are true gall formers. When feeding is sufficiently intense and sustained, branch killing and even tree killing results. Various species feed on different parts of the hosts, but the foliage, twigs, and roots are most commonly infested. Since plant lice commonly produce large quantities of honeydew, sometimes during heavy infestations the excreted liquid may appear as mist falling from the trees. This honeydew often spots or forms a glistening coat on the leaves, cars, and other objects located below. Various insects utilize honeydew as food and one group of these, the ants, also cares for and protects the aphids, thereby forming a mutually beneficial symbiotic relationship.

HOSTS AND RANGE. All species of plants are attacked by aphids. They occur wherever the hosts grow. The more injurious infest conifers; therefore, they are considered here in some detail. Identification of the commoner species can be facilitated by means of Table 12.2. None of the numerous species that infests broad-leaved trees causes serious forest damage.

TABLE 12.2 COMMON SPECIES OF APHIDS THAT INFEST CONIFERS*

1.	Infest spruce	2
1.	Infest firs, pines, larches, or hemlocks	12
2.	Cause formation of enlarged plant growths (galls)	6
2.	Do not cause the formation of galls	3
3.	Aphids feed on both needles and twigs; causes roughening of twigs; aphids pale green	BALSAM TWIG APHID (*Mindarus abietinus* Koch)
3.	Aphids feed on either needles or twigs but not on both	4
4.	Aphids feed only on needles; aphids dark dull green (page 351)	SPRUCE APHID
4.	Aphids feed only on twigs	5
5.	Aphid colored brown with 4 rows of black spots on abdomen; many covered with white woolly wax	*Cinara Abietis* (Fitch)
5.	Aphids colored pale green	BALSAM TWIG APHID (*Mindarus abietinus* Koch)

6. Entire new growth galled 7
6. Only part of new growth galled 9
7. Gall consisting of a loose growth (page 352) . . . . *Pineus similis* (Gill.)
7. Cone-like galls 8
8. Galls about 1 inch long; reddish or purplish; needles bract-like (page 352) . . . . *Pineus floccus* (Patch)
8. Galls about 2½ inches long (page 352) . . . . COOLEY'S GALL APHID
9. Galls terminal 10
9. Galls not terminal 11
10. Galls cone-like; reddish or purplish; needle bract-like; about 1 inch long; chambers poorly formed and interconnecting; only 1 or 2 aphids per chamber (page 352) . . . . *Pineus pinifoliae* (Fitch)
10. Galls small, not cone-like (page 352) . . . . *Chermes strobilobius* Kalt.
11. Galls with very short needles (page 352) . . . . *Chermes lariciatus* Patch.
11. Galls with long needles (page 352) . . . . *Chermes abietis* L.
12. Infest pines 13
12. Infest firs larch or hemlock 17
13. Infest only needles or both needles and bark of twigs 14
13. Infest only bark of twigs, branches, or main bole 16
14. Aphids covered with woolly wax; infest both twigs and needles (page 352) . . . . *Pineus pinifoliae* (Fitch)
14. Aphids naked 15
15. Aphids brown; infest only needles . . . . *Eulachnus rileyi* Wms.
15. Aphids pale green; infest both needles and twigs . . . . . . *Mindarus abietinus* Koch
16. Aphids covered with woolly wax; host chiefly white pines . . . *Pineus strobi* (Hartig)
16. Black aphids with powdery spots on sides and white mid-line along back . . . . *Cinara strobi* (Fitch)
17. Infest firs 18
17. Infest larch and hemlock 21
18. Body color aphid brownish or purplish black 19
18. Body color green 20
19. Body purplish black, generally covered with a whitish, floculent, woolly wax (page 353) . . . . BALSAM WOOLLY APHID
19. Body brownish black . . . . *Cinara curvipes* (Patch)
20. Body color pale green . . . . *Mindarus abietinus* Koch
20. Body color dark green; host Douglas-fir (page 352) . . . . COOLEY'S APHID
21. Infest bark of twigs and needles of western hemlock; aphids covered with white cottonly encrustation . . . . *Chermes tsugae* (Annand)
21. Infest larch 22
22. Aphids dark brown to black with brown spots on abdomen; infest twigs and bases of needles . . . . *Cinara laricis* (Wlkr.)
22. Aphids dark, sometimes covered woolly wax; back brown spots abdomen (page 352) . . . . *Chermes strobilobius* Kalt.

* Adapted from Baker, W. L., P. W. Oman, and T. J. Parr. *U.S.D.A. Misc. Publ. 657.*

INSECT APPEARANCE. *Adults:* small, 1/32 to 1/4 inch long; delicate, soft bodied, globular to pear-shaped insects; colored variously, yellow, red, green, gray, blue, or black; adults occur as both winged and wingless forms; winged forms with 2 pairs of delicate membranous wings usually held roof-like over body; 3 pairs of long and slender thoracic legs; tarsi 2-segmented; antennae prominent thread-like, each with 3 to 6 segments;

FIG. 12.1. Plant lice or aphids (Aphididae). Upper left: wooly aphid; upper right: leaf curled and distorted by aphid feeding; lower: winged and wingless giant hickory aphid (*Longistigma caryae*) (X4).

beak apparently arising from between the front legs, 4-segmented; frequently a pair of gland-like or tube-like structures (cornicles) located on back of 6th abdominal segment; some species secrete a woolly appearing wax that covers the body. **Nymphs:** resemble adults in most characteristics except size, always lack wings and are sexually immature. **Eggs:** commonly rather large and often black.

Many species are naked and live exposed on various parts of the plants; others secrete a protective, floculent, woolly wax that covers their bodies; some are gall formers. The latter usually can be identified most readily by means of the shape, form, and other characteristics of the galls. For this purpose the keys presented by Felt (1940) are most useful. Identification of other species is more difficult and usually requires the services of specialists.

Aphids are commonly known as plant lice because they resemble the lice that infest mammals and birds. They are small, globular appearing, occur in groups, and are rather inactive. Even when touched they move reluctantly and withdraw their mouthpart stylets only when greatly disturbed.

LIFE CYCLE. Some aphids (*Cinara* spp.) have a simple life cycle, whereas others have rather complex type of development. The latter can be outlined in a general way as follows: winter is passed as eggs on the primary host. The eggs hatch in the spring to produce nymphs, which all mature into wingless females. These reproduce without mating (parthenogensis) and some (Aphididae) give birth to living young (viviparous), whereas other (Adelgidae) always lay eggs (oviparous). Several additional similarly produced asexual, wingless generations may be produced on the primary host before a winged asexual generation develops. These migrate to the secondary host, where wingless females are produced again for a number of generations. A second winged, migrating, generation consisting solely of females develops later in the year and these return to the primary host. A bisexual generation then develops. These mate and each female lays one overwintering egg.

CONTROL. Applied methods have not been used against aphids infesting forests and seldom are used even for protecting shade trees. In the future, however, it is very likely that insecticidal sprays will be used more. The organic phosphate insecticides are very effective against these insects. Of these, malathion is preferred because it is less dangerous to humans. Effective dosages are ½ to ¾ pound of actual malathion per 100 gallons of spray formulated either as an emulsion or a wettable powder. Other suitable insecticides are benzene hexachloride (2/10 to 4/10 pound of the gamma isomer per 100 gallons spray) and nicotine sulfate (1 pint of the 40 per cent solution per 100 gallons spray plus 1

FIG. 12.2. Aphid predators. Above: lacewing adult, (*Chrysopa* spp. Neuroptera) (X2½); below: ladybird beetle larvae (X3).

pound of hydrated lime). Soap can be used instead of the lime for the purpose of alkalizing the nicotine spray, especially when only a small amount is to be mixed.

Felt, E. P. 1940. *Plant galls and gall makers.* Comstock Publishing Co., Ithaca, N. Y. 364 pp.

SPRUCE APHID

Aphis abietina Wlk., Hemiptera (Homoptera), Aphidae

INJURY. This injurious species feeds on the needles causing them to turn brown and drop. There have been several serious spruce aphid epidemics, and much timber has been killed in the tideland sitka spruce region of the Pacific Northwest.

HOSTS AND RANGE. All species of spruce wherever the hosts grow.

INSECT APPEARANCE. Typical; about 3⁄16 inch long when full grown; wingless; dull dark green; cornicles present.

LIFE CYCLE. The seasonal history of the spruce aphid is not known. The insects appear on the spruce in the spring and disappear about midsummer. This suggests that there is an alternate host, but the identity of the other host is unknown.

CONTROL. On shade and ornamental trees the insects can be killed by any of the contact sprays listed in the general discussion on aphid control. Protection of forests has not been attempted, but it is possible that aerial application of one of the organic phosphate insecticides would be practical.

SPRUCE GALL APHIDS

Chermes (*Adelges*) spp. and *Pineus* spp., Hemiptera (Homoptera), Chermidae

INJURY. Feeding of the aphid nymphs stimulates the host trees to produce characteristic enlarged growths called galls. These cone-like structures develop very rapidly requiring only a few days to become full grown. They vary in length from a third of an inch up to several inches. First they are green or purple but later become reddish brown. Inside the galls are numerous closed chambers in each of which many nymphs develop. The characteristics for the various species are presented in Table 12.2.

Seldom are these galls sufficiently abundant to cause serious damage to forest-grown trees, but often they are severe pests of ornamentals. They disfigure the trees and also cause bole deformation by destroying the terminal buds.

HOSTS AND RANGE. These gall-forming aphids occur wherever spruces and the alternate hosts grow. The alternate hosts for various species are as follows: Cooley gall aphid, *Chermes cooleyi* Gill., Douglas-fir; eastern spruce gall aphid, *C. abietis* L., no alternate host; *C.*

FIG. 12.3. Eastern spruce gall aphid (*Chermes abietis*) (X1).

strobilobius (Kalt.) and *C. laricatus* (Patch), larch; *Pineus similis* (Gill.), no alternate host; *P. floccus* (Patch) and *P. pinifoliae* (Fitch), white pine.

INSECT APPEARANCE. Very small; winged or wingless; body soft, fragile, oval, and indistinctly segmented; non-gall producing stages on alternate hosts often secrete a woolly wax from glands arranged in transverse rows on all parts of body; beak arising from between front legs; antennae 5-segmented on winged forms, 3 on summer forms, and greatly reduced on overwintering forms; compound eyes plus 3 ocelli on winged forms; others have only 3 ocelli on each side; back of head and prothorax with hardened shield-like area; tarsi 2-segmented; *Chermes* has 5 abdominal spiracles, whereas *Pineus* has 4. The common species are presented in Table 12.2.

LIFE CYCLE. Gall-forming aphids have complex life histories. The general pattern is similar to that indicated in the general discussion but differs in one respect—the females of all forms are oviparous. A few species, as indicated previously, have only one host.

CONTROL. Practical insecticidal control methods for forest have not been developed, but shade and ornamental trees can be protected by spraying with any of the materials indicated in the general discussion of aphids. It is imperative to apply the spray at the correct time if gall formation is to be prevented. This time is just before the buds begin to swell in early spring.

Wilford, B. H. 1937. *Univ. Mich. School For. and Conserv. Circ.* 2.

BALSAM WOOLLY APHID

Chermes (Adelges) piceae (Ratz.), Hemiptera (Homoptera), Chermidae

INJURY. The injurious balsam fir aphid feeds by sucking the sap from the parenchyma cells beneath the bark of twigs, limbs, and trunks of all size trees, provided that the outer dry bark is not more than about 1/25 inch thick. The saliva injected during feeding stimulates plant tissue growth so that the twigs and stems become enlarged, contorted, gouty, and gall-like (hypertrophy and hyperplasia). The damage first appears as reduced needle growth on the galled twigs and is followed by dying of foliage, twigs, and patches of bark on the larger stems. Several to many years of heavy attack kills the trees. The wood produced in infested trees becomes reddish, brittle, and the cells thick walled. These conduct water poorly. The larger branches and main boles often appear whitish owing to the woolly wax secreted by the insects.

HOSTS. All species of firs are infested, but noble fir and shasta red fir appear to be somewhat resistant.

RANGE. This introduced European pest now occurs in eastern Canada

(Nova Scotia and New Brunswick), the New England States, North Carolina, and in western Washington and Oregon.

INSECT APPEARANCE. Small, less than 1/32 inch long; globular, soft bodied; usually wingless; color purplish black; usually completely covered with long, slender threads of wax; mouthpart stylets long, originate from between front legs.

LIFE CYCLE. This aphid does not have an alternate host. There are 2 generations per year and the females are always oviparous. Each female lays from 50 to 200 eggs. Winged forms are seldom numerous, but usually some can be found. These have 4 nymphal instars, whereas the wingless forms have only 3. The spread of the wingless forms from one tree to another is chiefly by means of winds which carry the young nymphs. Winter is passed in the nymphal stage.

CONTROL. Temperatures lower than −30° F kill all of the overwintering aphids. At the present time applied methods of control are impractical under forest conditions. For killing the aphids on shade trees various sprays listed in the general discussion of aphids have proven effective.

Balch, R. E. 1952. *Can. Dept. Agri. Publ.* 867.
Johnson, N. E. 1957. *U.S.D.A. Forest Service, Penna. Northwest Forest and Range Exp. Sta. Research Paper* 18.

SCALE INSECTS

Hemiptera (Homoptera), Coccoidea

INJURY. Scale insects are so called because many species secrete a scale-like wax coating over their back and others resemble bark scales. These insects cause general devitalization and death of the infested parts by extracting plant sap and by injecting toxic saliva into the host tissues. Different species attack different parts of the hosts, but most infest the twigs and smaller branches. The soft scales frequently produce large amounts of honeydew, which attracts nectar-feeding insects. That which is not used falls on the leaves, twigs, and branches located below, where it forms a medium on which black sooty molds grow. Only a few species are troublesome forest pests, but many are most injurious to shade and ornamental trees. A few species produce useful products such as shellac, and in past times some were the sole source of certain brilliant dyes.

TABLE 12.3 COMMON TYPES OF SCALES THAT INFEST TREES

1. Adult females retain appendages throughout life and can move about slowly; wax filaments project out from all sides forming a fringe around each insect and

on some species forming tail-like projections; bodies oval generally heavily covered with powdery wax (page 356) . . . . Mealy bugs

1. Adult females stationary; some are covered with wax 2
2. Adult female scales or egg cases more or less covered with a woolly wax (page 356) . . . . WOOLLY SCALES
2. Not as in 2 above, but bodies may be more or less covered with powdery wax, a wax scales, or the males may form minute wax cocoons 3
3. Small pits or ring-like galls formed in the host plant tissues in which part of the scale bodies are embedded . . . . PIT-MAKING SCALES (*Asterolecanium* spp.)
3. Not as in 3 above 4
4. Back of female scales appear soft or leathery (page 359) . . . . SOFT SCALES
4. Back of females scales covered with a shell-like wax scale (page 362) . . . . ARMORED SCALES

HOSTS AND RANGE. All species of plants wherever the hosts grow. Most damage occurs in warm, arid regions.

INSECT APPEARANCE. *Adults: Females*—body flattened, globular, hemispherical, sac-like, elongate, or circular; sometimes covered with wax in the form of powder, cottony masses or a continuous scale-like layer; always wingless; legs, antennae, and compound eyes reduced or absent; most species sedentary and cannot move from the place where settled. *Males*—seldom observed; usually with a pair of membranous wings with 2 simple veins and 1 pair of club-shaped appendages (halters) where second pair of wings, if present, would be attached; legs well developed; tarsi with 1 segment and 1 claw; abdomen with a posterior tubular process (stylus); antennae long, each with 6 to 13 segments; beak absent. *Nymphs:* first instar (crawlers) of both sexes have legs, antennae, functional mouthparts, and are mobile; other nymphal instars of both sexes scale-like with male scales being smaller and often more elongate.

LIFE CYCLE. Most species lay eggs, but a few reproduce viviparously. The resulting crawlers disperse to new places on the same host plant or get carried by winds, birds, or other animals to other hosts. During subsequent development the females have 1 to 3 additional instars, and for most species they become immobile, sedentary, and quiescent. The males also have several scale-like nymphal instars. There may be 1 to 6 generations per year depending on the species and the length of the summer season.

CONTROL. Insecticides seldom have been used for protecting forest trees but commonly are used on shade trees. Mineral oils (3 per cent emulsions) can be used as a dormant spray on most deciduous trees except walnuts, beech, and sugar maple. These tree species are injured by oils. The heavy dormant oil sprays also may injure evergreens;

therefore, only the lighter, milder, summer oil emulsions are suitable. Dormant oils are best applied in the spring before the buds begin to expand but during a time when the temperature does not drop below freezing for at least 24 hours following application. Other insecticides such as DDT or one of the organic phosphate insecticides (malathion) can also be used effectively. These materials must be applied at the time the crawlers are active. They are commonly formulated as emulsions. Dosages should be one pint of actual malathion (2 pints 50 per cent emulsion concentrate) or 2 pounds DDT per 100 gallons of water. Nicotine sulfate has also been used effectively against the crawler stage. A mixture consisting of 2 pints of nicotine sulfate, 2 pounds of hydrated lime, and 7 pounds of summer oil emulsion per 100 gallons is a suitable formulation.

MEALY BUGS

Pseudococcus spp. Hemiptera (Homoptera), Pseudoccidae

INJURY. The coccids are not very serious forest pests but do cause much injury to plants in warmer regions and those growing in greenhouses. Often they secrete much honeydew.

HOSTS. Many species of plants are hosts of these cosmopolitan insects but cypresses, redwoods, and oaks in the far West are most commonly infested.

RANGE. Wherever the hosts grow, but damage is more severe in warmer climates.

INSECT APPEARANCE. *Adult females:* see characteristics presented in Table 12.3. Antennae with 7 to 9 segments; each tarsus with a single claw not toothed on inner margin; size up to about ¼ inch long. *Adult males:* active 2-winged insects that do not feed as adults. *Eggs:* yellowish to purple, laid in cottony masses; some species viviparous.

LIFE HISTORY. Mealy bugs breed throughout the warm seasons, and some eggs and young are always present. One generation develops every month during the summer.

CONTROL. Ornamentals can be freed of these insects by spraying with malathion or nicotine sulfate as described in the general discussion of scale insects.

WOOLLY SCALES

Hemiptera (Homoptera), Coccidoidea

Many scales secrete a copious amount of woolly or cottony appearing wax, which more or less covers the insects or forms conspicuous egg cases. The females usually retain their legs and antennae even in the adult stage but become stationary after the first instar. The males transform to adult winged insects beneath small flat scales. Sometimes much honeydew is produced.

Some of the common species are listed in Table 12.4, and two of these are considered below in detail.

TABLE 12.4 SOME COMMON WOOLLY SCALES THAT INFEST TREES

	Characters	Go to / Species
1.	Infest conifers	2
1.	Infest hardwoods	4
2.	Infest pines	3
2.	Infest species of Cupressaceae (cedars, cypress etc.)	CYPRESS BARK SCALE [*Ehrhornia cupressi* (Ehrh.)]
3.	Scales locate themselves at the needle base on new growth of pines	WOOLLY PINE SCALE (*Pseudophilippia quaintancii* Ckll.)
3.	Scales locate themselves beneath bark scales or in cracks and crevices of the bark of twigs, branches, and the main bole (page 358)	*Matsucoccus* spp.
4.	Female scales appear as if settled on a cushion of white cottony wax, which protrudes around sides; infest elms	EUROPEAN ELM SCALE [*Gossyparia spuria* (Mod.)]
4.	Wooly wax covering part or all of the back of the scales or over the egg mass	5
5.	Scales feed on the bark of trunks and branches of beech (page 357)	BEECH SCALE
5.	Infest other species of broad-leaved trees	6
6.	Yellow bodied scale covered with a cottony mass; chief host is hard maple	MAPLE FALSE MEALY BUG (*Phenacoccus acericola* King)
6.	Body colored brown or dark purple; long cottony egg sac at posterior end	7
7.	Adult female dark purple with a brown stripe along mid-back; infests only maples	MAPLE LEAF SCALE (*Pulvinaria acericola* Walsh and Riley)
7.	Adult females brown, infests many species of trees	COTTONY MAPLE SCALE [*Pulvinaria vitis* (L.)]

BEECH SCALE

Cryptococcus fagi (Baer.) Hemiptera (Homoptera), Dactylopiidae

INJURY. This scale insect and a fungus (*Nectria* spp.) commonly occur together, infesting and infecting the same bark areas. The exact role played by each organism has not been determined, but either and/or both have severely injured and killed a large quantity of beech in the Northeast. The insects feed by sucking the sap from the bark. Heavy infestations kill large areas of bark and sometimes also kill the main cambium.

HOST AND RANGE. This European beech-infesting insect was introduced into the Northeast during the early part of the present century. It now occurs in the New England States and adjacent parts of Canada.

INSECT APPEARANCE. A small, white, woolly scale. The individuals often are massed to form vertical lines.

LIFE CYCLE. There is only 1 generation per year. The eggs are laid in early summer and winter is spent as nymphs.

CONTROL. No practical applied controls have been developed for use on forested trees. For protecting shade trees lime sulfur spray (1 part to 19 parts of water) used during the dormant season is effective. Oil sprays should not be used for they injure beech. Malathion probably would be effective against the crawlers. See the general discussion on scale controls.

Brown, R. C. 1936. *Mass. Forest and Park Asso. Tree Pest Leaflet* 4.

MATSUCOCCUS SCALES

Matusucoccus spp. Hemiptera (Homoptera), Margarodidae

INJURY. There are many species in this genus, but only one eastern and one western species have caused sufficient forest tree damage to be considered here. The injury consists of twig dying, flagging, and the killing of small trees. The bark on the twigs and branches becomes swollen and cracked with areas of dead tissues occurring beneath the feeding punctures.

HOSTS. Only species of pines.

FIG. 12.4. Beech scale (*Cryptococcus fogi*) (X2).

RANGES. The Prescott scale (*M. vexillorum* Morrison) occurs in Arizona and New Mexico where its chief host is ponderosa pine. The red pine scale (*M. resinosae* Bean and Godwin) occurs in the East around Long Island Sound, where it infests red pines. Most of the remaining 12 known species are found only in the West.

INSECT APPEARANCE. *Adult females:* 1/16 to 3/16 inch long; body pear shaped, soft, brownish, and wrinkled; wingless. *Males:* transform within loose wax cocoons; typical 2-winged, midge-like, and have a brush of wax thread at posterior end. *Eggs:* yellowish, oval in loosely woven wax egg sac.

LIFE CYCLE. These scales have either 1 or 2 generations per year. The winter is passed in the nymphal or preadult stages.

CONTROL. Practical applied controls have not been developed. Two per cent oil emulsion, demeton, and ethylene dibromide have showed promising results against *M. resinosae.* Malathion should be tested against the crawlers.

Bean, J. L. 1956. *U.S.D.A. Forest Serv. Forest Pest Leaflet* 10.
Bean, J. L., and P. A. Godwin. 1955. *Forest Sci.* 1:164–176.
McKenzie, H. L., L. S. Gill, and Dan E. Ellis. 1948. *J. Agri. Research* 76:33–51.

SOFT SCALES

Hemiptera (Homoptera), Coccidae

Soft scales frequently resemble a miniature tortoise shell; therefore, these are commonly called tortoise scales. Their general body appearance is somewhat similar to that of the woolly scales but, of course, the woolly wax is lacking.

Some common species are listed in Table 12.5. One important genus is discussed below for the purpose of indicating the general biological characteristics for this group.

TABLE 12.5 SOME COMMON SOFT SCALES THAT INFEST SHADE AND FOREST TREES*

1. Infest conifers	2
1. Infest broad-leaved trees	7
2. Infest pines	3
2. Infest other species of conifers	6
3. Surface irregular or pitted; body flattened, grub-like, or hemispherical	4
3. Surface smooth, shiny; color reddish brown or black; shape semiglobular	*Physokermes* spp.
4. Body flattened or sac-like; located beneath bark scales or in cracks and crevices; less than 1/8 inch long (page 358)	*Matsucoccus* spp.

4. Scales exposed on twigs; 1/8 to 1/4 inch in diameter 5
5. Body color yellowish; host Monterey pine in California . . . . IRREGULAR PINE SCALE (*Toumeyella pinicola* Ferris)
5. Body color reddish brown (page 361) . . PINE TORTOISE SCALES *Toumeyella* spp.
6. Infest arborvitae; hemispherical, brown smooth and shiny . . . . EUROPEAN FRUIT SCALE (*Lecanium corni* Bouché)
6. Infest spruces; closely resemble buds; body globular, gall-like, reddish brown with irregular flecks of yellow; coated thinly with wax . . . . SPRUCE BUD SCALE [*Physokermes piceae* (Schr.)]
7. Back of scales ridged so as to form an H; dark brown or black; occurs only in milder climates . . . . BLACK SCALE [*Saissetia oleae* (Bernard)]
7. Back of scales not ridged so as to form an H mark 8
8. Black wrinkled 9
8. Back smooth but may be lobed, ridged, or furrowed 10
9. Female scales less 1/4 inch long . . . . BROWN ELM SCALE (*Lecanium canadense* (Cockerell)
9. Female scales larger 1/4 inch long . . . . HICKORY SCALE (*Lecanium caryae*)
10. Scales rather uniform in color 11
10. Scales marbled mottled or distinctly marked with 2 or more colors 15
11. Scales almost globular 12
11. Scales hemispherical or flattened 13
12. Scale with furrow along back; brown, partially covered with powdery or fluffy wax; about 3/8 inch long; occurs only on West Coast; infest many species of hosts . . . . EXCRESCENT SCALE (*Lecanium excrescens* Ferris)
12. Characteristics not as in 12 above; scales very globular gall-like; infest only twigs of oaks . . . . *Kermes* spp.
13. Body heavily covered with powdery, frost-like wax; 1/4 to 3/8 inch long . . . . FROSTED SCALE (*Lecanium pruinosum* Coquillett)
13. Body not covered with powdery wax 14
14. Large scales, 1/4 to 1/3 inch diameter; hosts tulip poplar, magnolia and linden . . . . TULIPTREE SCALE [*Toumeyella liriodendri* (*Gmel.*)]
14. Scales smaller than 1/4 inch long; size, shape and color variable but mostly smooth, shiny brown . . . . EUROPEAN FRUIT SCALE (*Lecanium corni* Bouché)
15. Scales distinctly marked with contrasting colors 16
15. Scales marbled with shades so that they are not very contrasty 17
16. Scales dark brown, regularly marked with white; about 3/16 inch long; occurs in California . . . . CALICO SCALE (*Lecanium cerasorum* Cockerell)
16. Scales oval or pear shaped, color reddish to reddish brown with radiating black bands or vice versa . . . . TERRAPIN SCALE (*Lecanium nigrofasciatum* Perg.)
17. Scales brown with oval mottlings of light yellowish brown; elliptical flattened to elevated; about 1/2 inch long . . . . MAGNOLIA SCALE (*Neolecanium cornuparvum* (Thro.)
17. Scales yellow to brown with a brown marbled effect; body soft and flat . . . . SOFT BROWN SCALE (*Coccus hesperidum* Linn.)

* Those that infest mostly fruit trees are excluded. Characteristics are based on the adult female scales.

FIG. 12.5. A typical soft scale (X4).

PINE TORTOISE SCALES

***Toumeyella numismaticum* (Pettit and McD.) and *T. pini* (King)**
Hemiptera (Homoptera), Coccidae

INJURY. These species severely weaken and sometimes kill young pines.

HOSTS AND RANGE. *T. numismaticum* prefers jack pine, but Scotch pine and red pine also are attacked. *T. pini* infests various hard pines. They occur throughout much of the United States and Canada.

INSECT APPEARANCE. *Adult females:* 3/16 to 1/4 inch long; reddish brown; body convex and hemispherical. *Male scales:* smaller; body flat and elongate; have a transparent glassy appearance.

LIFE CYCLE. There is only 1 generation per year. The overwintering partly grown nymphs become active early in the spring and, at this time, secrete a large amount of honeydew that falls on the needles and twigs where it forms the substrate for the growth of a black, sooty mold. The female nymphs mature by midsummer and lay eggs that hatch within a few hours. The resulting crawlers disperse to new host trees chiefly by means of wind. The males mature, form cocoons, and transform to winged adults in August. As adults they live for only a day or two during which time they mate with the immature female scales. The

FIG. 12.6. Pine tortoise scale (*T. pini*) (X2½).

mated female nymphs continue to develop until fall and then pass the winter as half-grown insects.

CONTROL. Spraying when the crawlers are active during the spring with malathion at a rate of 2 pints of 50 per cent emulsifiable concentrate per 100 gallons of water is practical for killing these scales on small forest trees or on shade trees. One quart of 40 per cent nicotine sulfate per 100 gallons of spray is also very effective.

Orr, L. W. 1931. *Univ. Minn. Agri. Exp. Tech. Bull.* 79.
Rabkin, F. B., and R. R. Lejeune. 1954. *Can. Entomol.* 86:570–575.

ARMORED SCALES

Hemiptera (Homoptera), Diaspididae

Adult armored scales are characterized by having a thick shell-like scale that covers their backs completely. These scales consist mostly of wax secreted from posterior abdominal glands but also contain the molted skins which commonly forms the "nipple." The legs and antennae on the females disappear after the first molt. Male scales are more slender. These scales do not secrete honeydew.

The more common genera of armored scales are presented in Table 12.6. One species, the pine needle scale, is considered below in detail.

TABLE 12.6 COMMON GENERA OF ARMORED SCALES*

1. Infest needles of conifers 2
1. Infest twigs of broad-leaved trees 4
2. Infest junipers . . . . JUNIPER SCALES (*Diaspis carueli* Targ.)
2. Infest hemlock, pines, firs, and spruces 3
3. Scale oblong, dark gray nearly black . . . . HEMLOCK SCALE (*Aspidiotus* spp.)
3. Scale nearly white, elongate to pear shaped (page 363) . . . . PINE NEEDLE SCALE [*Phenacaspis pinifoliae* (Fitch)]
4. Scale circular or oval in shape with the cast nymphal skins forming a nipple-like projection at or near the center 5
4. Scale elongate pear shaped or with anterior end slender and pointed and the posterior end broad and rounded, often resembling an oyster shell with cast nymphal skins located at posterior end 6
5. Nipple in center of each scale . . . . SAN JOSE SCALE and others (*Aspidiotus* spp.)
5. Nipple placed somewhat off center . . . . . GLOOMY SCALES (*Chrysomphalus* spp.)
6. Scale colors light or grayish white; male scales narrow and often ribbed; molten skins usually yellow . . . . SCURFY SCALES (*Chionaspis* spp.)
6. Scale colored dark gray or brown . . OYSTER-SHELL SCALES [*Lepidosaphes ulmi* (L.)]

* Back of adult insects covered with a scale consisting of wax and old skins.

PINE NEEDLE SCALE

Phenacaspis pinifoliae (Fitch) Hemiptera (Homoptera), Diaspididae

INJURY. The pine needle scale causes yellowing and browning of infested needles. When the attacks are heavy and sustained, the twigs, branches, and even small trees are killed.

HOSTS AND RANGE. Pines, spruces, and firs throughout much of North America.

INSECT APPEARANCE. *Adult females:* about 1/8 inch long; scale elongate to pear shape; nearly white with small dark orange nipple (molted skin) at one end; legs and antennae absent. *Male scales:* pure white about 1/32 inch long; adults are minute, 2-winged insects. *Nymphs:* crawlers are pink to pale yellow, legs and antennae well developed; 2nd instar and older female nymphs lack legs and antennae are only small tubercles. *Eggs:* rusty brown; covered with fine powdery wax.

LIFE CYCLE. There is only 1 generation per year. Winter is passed as eggs beneath the parent female scales. These hatch in the spring soon after plant growth starts, and the resulting crawlers disperse before settling. The adult winged males appear later in the summer and mate with females. Oviposition occurs during late summer.

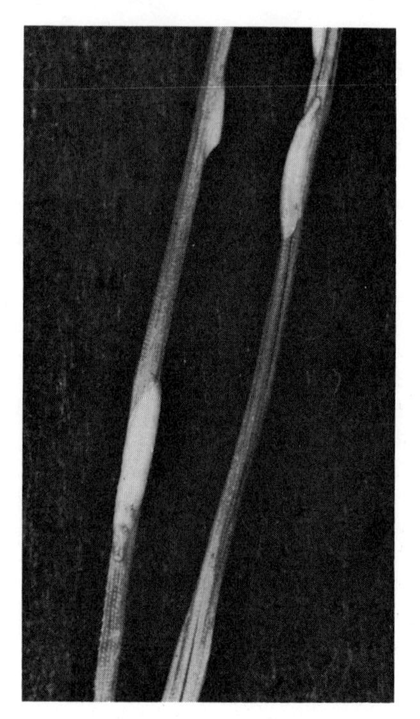

FIG. 12.7. Pine needle scale (*Phenacaspis pinifoliae*) (X8).

CONTROL. Infested ornamental and nursery-grown trees should be sprayed with malathion when the crawlers are active. It should be formulated as an emulsion at the rate of 1 pint of actual insecticide (2 pints 50 per cent emulsion concentrate) per 100 gallons of water. Timing is extremely important, for the effective spray period is only about a week. A dormant spray of 1 part lime sulfur to 9 parts of water applied in the spring before growth starts is also very effective.

Cumming, M. E. P. 1953. *Can. Entomol.* 85:347–352.

MISCELLANEOUS HEMIPTEROUS SAPSUCKERS

SPITTLE BUGS

Aphrophora spp., Hemiptera (Homoptera), Cercopidae

The Saratoga spittle bug, *A. saratogensis* (Fitch), and the pine spittle bug, *A. parallela* Say, are the most common of several species that infest conifers. Of the two, the Saratoga spittle bug has been more injurious.

INJURY. Tree injury is caused chiefly by the adults that feed on the sap extracted from the inner bark of the needle-bearing twigs. Each feeding puncture kills a few square millimeters of inner-bark tissues.

These dead cells together with the underlying wood soon become infiltered with resin which blocks sap movement and causes dehydration and death of the needle-bearing twigs. Light infestations cause only scattered twig dying (flagging), but heavy attacks kill tops, branches, and sometimes whole trees. When severely injured trees are protected from additional attacks they recover quickly.

HOSTS. Chiefly hard pines. Both nymphs and adults of the pine spittle bug infest trees, with the nymphal spittle masses being most conspicuous. Only adults of the Saratoga species feed on the trees. They prefer red pine but also heavily attack jack pine. The nymphs form their spittle masses and feed at the root collars of many species of herbaceous and broad-leaved, woody shrubs. Here they are somewhat concealed by the ground litter. Favored nymphal hosts are willow, sweet fern, and *Rubus* spp. Tree damage by both species occurs mostly to young trees 4 to about 10 feet tall.

RANGE. Both species occur throughout the eastern half of North America. Most damage has been caused by the Saratoga spittle bug in the Lake States region.

INSECT APPEARANCE. *Adults:* small, ⅛ to ⅜ inch long; 2 pairs of wings held roof-like, front pair colored and somewhat thickened; antennae minute, bristle-like, inserted between the eyes; tarsi 3 segmented; hind tibia armed with 1 or 2 stout teeth, tip crowned with short stout spines; Saratoga spittle bugs are tan with lighter markings on forewings forming a subequal diamond-shaped pattern; pine spittle bugs dark tan to reddish brown with 2 narrow oblique lighter lines on each forewing. *Nymphs:* always enveloped in a mucilaginous, white froth (spittle) consisting of anal excretions and glandular secretions; Saratoga spittle bug abdomen is colored bright red during first 4 instars; 5th instar is mahogany brown; abdomen of pine spittle bugs mostly orange with sides red during first 4 instars; 5th instar light to dark brownish with sides reddish. *Eggs:* fusiform shape, laid between the bud scales on trees (*A. saratogensis*) or inserted in the twig tissues of the hosts for *A. parallela.*

LIFE CYCLE. Winter is passed in the egg stage. The nymphs hatch in the spring at about the time plant growth starts. They immediately begin feeding and envelop themselves in masses of spittle. The adults appear about 2 months later and live for a month or two. There is only 1 generation per year.

CONTROL. Spraying with DDT is an effective, practical method for killing the adults, and each year thousands of acres of Lake States pine plantations are sprayed. The spray is applied from the air at the same rate as used against many forest defoliators. This is 1 pound of DDT

FIG. 12.8. Saratoga spittle bug (*Aphrophora saratogensis*). Top: killed stand of young pine; center: bark removed to show undamaged and damaged twigs (X1); lower: adult bugs and a nymph (X2½).

per gallon of spray per acre. The solvents used are fuel oil No. 2 plus accessory naphthalenic solvents.

Large Saratoga spittle bug populations are not likely to develop in young dense pine stands because the alternate hosts will not grow well in the shade; therefore, tree damage can be prevented whenever the young stands can be grown so as to obtain early closure of the tree crowns.

Anderson, R. F. 1945 and 1947. *J. Econ. Entomol.* 38:564–566. 40:26–33. 40:695–701.
Eaton, C. B. 1955. *U.S.D.A. Forest Pest Leaflet* 3.
Speers, C. F. 1941. *N. Y. State College of Forestry Tech. Publ.* 54.

LEAFHOPPERS

Hemiptera (Homoptera), Cicadellidae

INJURY. Leafhoppers feed on the sap they extract from leaves. In so doing they cause small whitish bleached spots to appear on the upper leaf surfaces but seldom is this feeding damage very severe on forest trees. One species (*Scaphoideus luteolus* Van. D.) is of interest because it transmits a serious elm disease known as phloem necrosis.

HOSTS AND RANGE. Foliage of hardwoods wherever the hosts grow.

INSECT APPEARANCE. *Adults:* somewhat similar to spittle bugs but body is more elongate; antennae minute, bristle-like and inserted in front of and between the eyes; tarsi each with 3 segments; hind legs fitted for jumping (femora enlarged); hind tibia with a row of spines along one side; 2 pairs of wings held roof-like when at rest, front wings may be somewhat thickened. *Nymphs:* resemble adults but lack functional wings.

FIG. 12.9. Spittle masses made by spittle bug nymphs (*Aphrophora* spp.). Left: Saratoga spittle bug (*A. saratogensis*) (X¾); right: pine spittle bug (*A. parallela*) (X1).

LIFE CYCLE. Winter is usually passed either as adults or as eggs depending on species. Overwintering eggs are inserted in bark tissues of the host trees, whereas the adults hibernate in the ground litter. The first generation matures about 2 months after plant growth starts, and the resulting adults live for about a month. Commonly there are 2 generations per year.

CONTROL. Insecticidal leafhopper control has been used on trees only for the purpose of preventing the spread of elm phloem necrosis. DDT applied with a mist blower (see page 78) is most practical when many trees are to be treated, but a hydraulic sprayer can also be used. For the latter the usual dosage of 1 pound of actual DDT per 100 gallons of spray is effective. Either emulsions or wettable powder formulations are effective.

Baker, W. L. 1950. *Entomol. Society Washington Proceedings* 52:52.

LACE BUGS

Corythucha spp., Hemiptera (Heteroptera), Tingidae

INJURY. Lace bugs are of only minor importance as tree pests, but inasmuch as they and their damage are frequently encountered, foresters should be able to recognize them. These insects feed on the undersides of the leaves causing small whitish bleached areas to appear on the upper surfaces. This damage is similar to that caused by leafhoppers but can be differentiated by the presence of small blackish excrement spots on the underside of the leaves.

HOSTS AND RANGE. Hardwoods, especially sycamore, cherry, oaks, willows, and elms, wherever the hosts grow.

INSECT APPEARANCE. *Adults:* small, about ⅛ inch long; with net-veined wings; the front pair is held flat when at rest and appears square shaped when viewed from above; prothorax with a lateral, wing-like, net-veined outgrowth on each side. *Nymphs:* resemble adults, but lack the characteristic wings and wing-like lateral prothoracic outgrowths. *Eggs:* flask-shaped; blackish; embedded in the underside of the leaves near the veins.

LIFE CYCLE. Winter is passed either in the adult or the egg stages depending on the species. The females insert the eggs into the underside of the leaves near the veins. The eggs usually hatch within a couple of weeks, and the resulting nymphs mature about a month and a half later; therefore, there may be 2 to 4 generations per year depending on the length of the warm season.

CONTROL. Insecticidal control has never been needed to protect

FIG. 12.10. Cherry lace bug (*Corythucha pruni*). Above: upper leaf surface injury; below: dark excrement spots on under leaf surface; right: adult leaf bug (X6).

forests, and seldom are shade trees sprayed to kill these insects. Should applied control ever be needed, however, any of several insecticides is effective. These include DDT, benzene hexachloride (lindane), and malathion. The concentrations of these poisons that should be used are given on the insecticide containers.

GALL FORMERS

Classes Hexapoda and Arachnida

Most of the gall-producing insects are wasps (Hymenoptera, Cynipidae) or midges (Diptera, Cecidomyiidae), but others are caused by aphids (Hemiptera, Chermidae), mites (Arachnida), and even the larvae of some beetles (Coleoptera, Buprestidae and Cerambycidae). The characteristics of these various groups are presented in Table 12.7. Only the gall wasps and gall midges will be considered here; therefore, the reader is referred to other sections of the book for general discussion of the other groups.

TABLE 12.7 THE COMMON GROUPS OF GALL-FORMING ARTHROPODS

1. Legless 2
1. With 2 or 3 pairs of legs 5
2. With a distinct head; body elongate, cylindrical, and flattened; chewing mouthparts well developed and directed forward; prothorax with plate like areas on upper and lower sides 3
2. Head indistinct; body maggot-like; prothorax lacks plate like areas 4
3. Prothorax flattened and distinctly wider than abdominal segments with distinct equal sized plate-like areas on both the upper and lower sides; upper plate always with a distinct median longitudinal line, V- or Y-shaped lines or grooves; spiracles crescent shaped (page 387) . . . . FLAT-HEADED WOOD BORERS (Buprestidae)
3. Prothorax somewhat flattened with indistinct plate-like areas; the plate beneath less well developed than the back plate; spiracles oval or round (page 387) . . . . ROUND-HEADED WOOD BORERS (Cerambycidae)
4. Usually have an anchor process on the under anterior part of body (p. 370) MIDGES
4. Anchor process lacking on under anterior part of body (page 370) . . . . . . WASPS
5. Animals of microscopic size; 2 pairs legs at front of elongate body; antennae absent (page 372) . . . . MITES
5. With 3 pairs of well-developed legs; antennae well developed and thread-like; mouth a rod-like beak (page 347) . . . . APHIDS

INJURY. Many species of arthropods are able to stimulate plant tissues so that various types of abnormal swellings (galls) are formed on the leaves, twigs, or roots. These vary in size, shape, and structure according to the species of animals involved; therefore, specific identification usually can be made solely from the characteristics of the galls. Insect galls can be differentiated from other types because they contain one or more chambers or depressions in which the insects are or have been. Sometimes the insects become completely enclosed by the plant tissues, whereas for other species there is always one or more openings to the outside.

Galls on trees seldom cause serious damage, but they sometimes make shade trees unsightly. Since they frequently are observed, foresters should be somewhat familiar with them.

HOSTS AND RANGE. All species of plants are attacked, but the oaks, hickories, poplars, and spruces are most commonly heavily infested. They occur wherever the hosts grow.

INSECT APPEARANCE. Gall insects can be most easily identified by host and the characteristics of the galls produced; therefore, the interested reader is referred to the book cited below. The adults are characterized as follows: *Gall wasps:* minute wasps; 2 pairs membranous wings; each with only a few veins; abdomen flattened laterally with the

FIG. 12.11. Insect galls. Upper: twig galls; lower: leaf and needle galls.

basal abdominal segment, usually as long as the remaining segments. *Gall midges:* minute, delicate flies; 1 pair of wings with only a few veins; body and wings clothed with long hairs; antennae long and clothed with whorls of hairs some of which form loops.

LIFE CYCLE. Life cycles of these insects are diverse and sometimes complex; therefore, a generalized statement can not be presented here. Nevertheless, the adults commonly appear early in the spring at the time plant growth starts, so that larval growth and the resulting gall formation occurs during the period of most active leaf and terminal plant growth.

CONTROL. Only rarely is applied control used against gall formers and then only on shade or ornamental trees. The treatment consists of misting or spraying the trees with DDT. For hydraulic spraying the usual rate of 1 pound of actual insecticide (either wettable powder or emulsion) per 100 gallons of spray is effective. Synchronizing the time of application with that of adult appearance is most important. This can be determined by caging many old galls containing larvae and observing when the first adults start to appear.

Felt, E. R. 1940. *Plant galls and gall makers.* Comstock Publishing Co., Ithaca, N. Y.

ARACHNID SAPSUCKERS

MITES (RED SPIDERS)

Paratetranychus spp. and _Tetranychus_ spp., Class Arachnida, Order Acarina

INJURY. Mites are sapsucking arthropods that are serious pests of shade trees, especially in the arid parts of the west. Seldom have they caused serious trouble to forests, but recently some infestations have occurred in the northern Rockies. The forest mite infestations may occur more frequently in the future because of the increasing use of DDT aerial sprays. The DDT kills the mite predators but not the mites. Infested foliage appears bleached, yellowish, blotchy, or bronze. Frequently the leaves also appear to be covered with dust consisting of numerous minute, white, cast skins attached to a loose covering of fine silk webbing.

Eriophyid mites cause the formation of small, brownish blister-like galls, about ⅛ inch in diameter, on the underside of leaves. These are of some importance on fruit trees but never have caused trouble on forest trees.

HOSTS AND RANGE. All species of plants wherever they grow.

MITE APPEARANCE. *Adults:* small, 1/80 to 1/60 inch long; body rather globular; head united with the thorax (cephalothorax); no antennae; 4 pairs of legs; color whitish, green to red; gall mites (Eriophyidae) are of microscopic size; bodies elongate and have only 2 pairs of legs. *Nymphs:* similar to adults but are smaller and have only 3 pairs of legs. *Eggs:* spherical; whitish to red; attached to the silk webbing.

LIFE CYCLE. Depending on species, winter is passed either as eggs or as hibernating adults. The eggs hatch in early spring at the time plant growth starts, and the resulting nymphs mature after 2 to 4 weeks. There are many generations per year, and each female lays 2 to 3 dozen eggs; therefore, large mite populations frequently develop rapidly, especially when natural controls are inoperative.

CONTROL. Although the organic phosphate insecticides such as malathion and demeton are very effective, there are other suitable poisons which are specific acaracides. Two of these are aramite and ovex. Some plants, however, such as dogwood, holly, hawthorn, and flowering crabs are sensitive to ovex. Malathion is commonly applied as an emulsion or wettable powder in a spray at the rate of 1 pint or ½ pound actual insecticide per 100 gallons of water.

CHAPTER 13

Bud-, Twig-, and Seedling-Damaging Insects

Though the more injurious forest insect pests that damage the growing tips are weevils and caterpillars, there are many that belong to other insect groups. See Table 13.1. Most damage caused by these insects is done by borers, but some species eat the bark and girdle the twigs. The females of a few insect species cut small slits in the twigs in which they place their eggs.

TABLE 13.1 THE COMMON GROUPS OF BUD, TWIG AND SEEDLING DAMAGING INSECTS

1. Small trees killed in the vicinity of ant mounds; bark mostly intact, but near ground the bark is sunken and dead around entire stem; no boring nor much bark removal damage (page 395) . . . . ALLEGHENY MOUND ANT
1. Characteristics of damage not as indicated in 1 above 2
2. Insects bore inside the twigs or buds 7
2. Insects eat or otherwise severely damage the bark so that the injury is exposed . . 3
3. Damage consists of rows of short straight or curved slits in the twigs (page 392) . . . . OVIPOSITION INJURY
3. Damage consists of bark eaten from the twigs or a V-shaped girdle cut in the twigs 4
4. Infest conifers 5
4. Infest hardwoods; female beetles cut a V-shaped girdle that encircles twigs; larvae whitish, elongate, rather cylindrical; heads flattened with mouthparts directed straight forward; indistinct plate-like area on underside of prothorax smaller than that on back (page 384) . . . . TWIG GIRDLERS
5. Conspicuous masses of resin formed at the places of attack; small orange-red

maggots present inside pitch masses; see also couplet number 7 (page 394). . . . . PITCH MIDGES

5. Small globular masses of resin sometimes present at edges where bark has been eaten; larvae never present; damage always done by beetles 6
6. Bark eaten from twigs of seedlings and sometimes from twigs of larger trees growing on recently cutover pine lands; bark eaten by adult weevils (page 381) . . . . SEEDLING-DEBARKING WEEVILS
6. Bark eaten from twigs of seedlings and young trees growing near a concentration of fresh coniferous (Pinaceae) logs; bark eaten by long-horned beetles (page 293) . . . . SAWYERS
7. Two types of galleries present; one type is of uniform size (egg galleries) and is made by beetles; the other type (larval galleries) increases in size as the larvae grow; larvae whitish; body cylindrical; curved somewhat in long axis; legless; head capsule globular with mouthparts directed downward; beetles cylindrical; antennae with head-like club (page 203) . . . . BARK BEETLES
7. Only larval galleries present 8
8. Larvae bore mostly in the center of the twigs or in the buds; larvae with 3 pairs small thoracic legs plus 5 pairs fleshy abdominal legs (prolegs) each with groups of hooks (crochets) on ends (page 388) . . . . BUD- AND TWIG-BORING CATERPILLARS
8. Larvae bore mostly in the inner-bark region but may form pupal chambers in pith region; larvae legless but may have locomatory segmental body swellings (ampullae) 9
9. Borings fine and tightly packed in galleries; enlarge swellings (galls)* may form on twigs; larvae whitish; body elongate with head flattened and mouthparts directed forward; prothorax flattened and widened with distinct plate-like areas on both upper and lower sides about of equal size; spiracles cresent shaped (page 387) . . . . FLAT-HEADED BORERS
9. Borings granular or fibrous; larvae various; spiracles shaped oval or round . . 10
10. Borings fine to granular and rather loosely packed in galleries; larvae whitish; body cylindrical, curved somewhat in long axis; head globular with mouthparts directed downward (page 377) . . . . WEEVILS
10. Borings granular or fibrous; larvae whitish; body elongate with head flattened and mouthparts directed straight forward; prothorax with indistinct plates on upper and lower sides; lower plate less well developed (page 383) . . . . ROUND-HEADED BORERS

* See also the section on gall insects on page 369.

TWIG AND TERMINAL BORING WEEVILS

Coleoptera, Curculionidae

The appearance and general characteristics of the more important weevil species that bore in the terminals of trees are presented in Tables 13.2 and 13.3 and in the discussions which follow.

TABLE 13.2 COMMON GENERA OF TWIG AND TERMINAL INNER-BARK FEEDING WEEVILS

1. Pupal cells formed in the wood are lined with shredded wood fibers—"chip cocoons"; but those formed in the twig pith are not so lined; larvae bore and live together in inner bark; feed side by side until mature, at which time they make individual galleries into the wood; older ones score the wood surface; borings, except in pupal chamber, granular . . . . *Pissodes*
1. Pupal cells never lined with coarse shredded fibers; borings fine, sometimes powdery; larvae bore individual galleries throughout developmental period, but they may be very abundant and crowded 2
2. Borings fine and powdery; larvae score sapwood rather deeply; weevils bright green, blue, or black without a groove along breast to received the prominent curved break; hosts include both conifers and hardwoods . . . . *Magdalis*
2. Borings fine and resinous; weevils angular covered with silvery white or bronze scales, have groove along breast in which beak fits when not feeding; hosts pines and firs . . . . *Cylindrocopturus*

TABLE 13.3 COMMON TERMINAL INFESTING *PISSODES* SPP.

1. Occur in eastern North America 2
1. Occur in western North America 3
2. Occur in northeastern North America and south in the mountains to Georgia; infest chiefly eastern white pine, Norway spruce, jack, Scotch, and pitch pines . . . . WHITE PINE WEEVIL
2. Occur in southeastern United States; infest chiefly deodar and Atlas cedars and Cedar of Lebanon, but sometimes may infest pines . . . . DEODAR WEEVIL (*P. nemorensis* Germ)
3. Infest spruces 4
3. Infest pines 5
4. Infests sitka spruce . . . . SITKA SPRUCE WEEVIL (*P. sitchensis* Hopk.)
4. Infests Engelmann spruce . . ENGELMANN SPRUCE WEEVIL (*P. engelmanni* Hopk.)
5. Infest lodgepole pine . . . . LODGEPOLE PINE WEEVIL (*P. terminales* Hopk.)
5. Infest terminals of other species of western pines . . . . *Pissodes* spp.

INSECT APPEARANCE. *Weevils:* beetles with front of head prolonged into a tubular downward projecting snout, which contains small chewing mouthparts at the tip; antennae with a long basal segment connected at a distinct angle (elbowed) with the remaining segments. *Larvae:* body whitish, legless, cylindrical, segmented, elongate to somewhat globular and distinctly curved along the long axis; head small, globular, tan to brown with the mouthparts directed downward. *Pupae:* resemble adults; have a prolonged beak but lack elytra; whitish colored. *Eggs:* usually laid in small circular holes in the bark.

TERMINAL WEEVIL BORERS

Pissodes spp., Coleoptera, Curculionidae

Many species of *Pissodes* infest and destroy the main leaders of various conifers. The most injurious of these is the white pine weevil, *P. strobi* Peck, but several Western species are potentially dangerous and probably will become more troublesome as soon as more planting and regeneration work is undertaken in the West. Some of the more common species are listed in Table 13.3.

INJURY. The white pine weevil has caused much damage to second growth stands of eastern white pine. In saw timber the loss of usable volume often is as much as 30 to 40 per cent.

Each attack kills 2 or 3 years of terminal growth. These destroyed terminals subsequently are replaced by the uppermost living laterals, which bend upwards to become new terminals. This upbending results in worthless, crooked, or multiple boled, bushy trees. Other defects are increased knottiness, cross grain, and openings through which the inoculum of wood-rotting fungi enters. The trees are never killed. The most severe damage occurs to trees less than 25 years old, but all size trees are infested. New shoot growth develops above the parts infested, but this soon wilts and dies.

GALLERY PATTERN. The inner bark of the main leaders on 2 or 3 years growth is completely destroyed leaving only the granular borings. Older larvae also score the wood. Pupal chambers are formed either in the pith or near the wood surface. Those near the surface are tightly packed with fibrous borings, are therefore called chip cocoons.

HOSTS. The chief host of the white pine weevil is eastern white pine but Norway spruce, Scotch pine, and jack pine are also severely attacked. The hosts for some of the other species are presented in Table 13.3.

RANGE. The white pine weevil occurs wherever the hosts grow, but most damage occurs in the Northeast. The general distributions for some of the other species are presented in Table 13.3.

INSECT APPEARANCE. *Larvae:* body cylindrical, elongate, whitish, somewhat curved in long axis, legless; head globular with mouthparts pointed downward, located rather close to neck; about ¼ inch long when full grown. *Weevils:* small about 3/16 inch long; light to dark brown with 3 irregular whitish spots on each elytron; front of head prolonged into a slender beak that is grooved along each side in which the basal antennal segments rest; antennae elbowed; attached on middle of beak and with a compact club; 3rd and 4th abdominal segments much shorter than the 2nd and 5th. *Pupae:* resemble the adult, but are whitish and have wing pads in place of elytra. *Eggs:* small, 1/32 inch long;

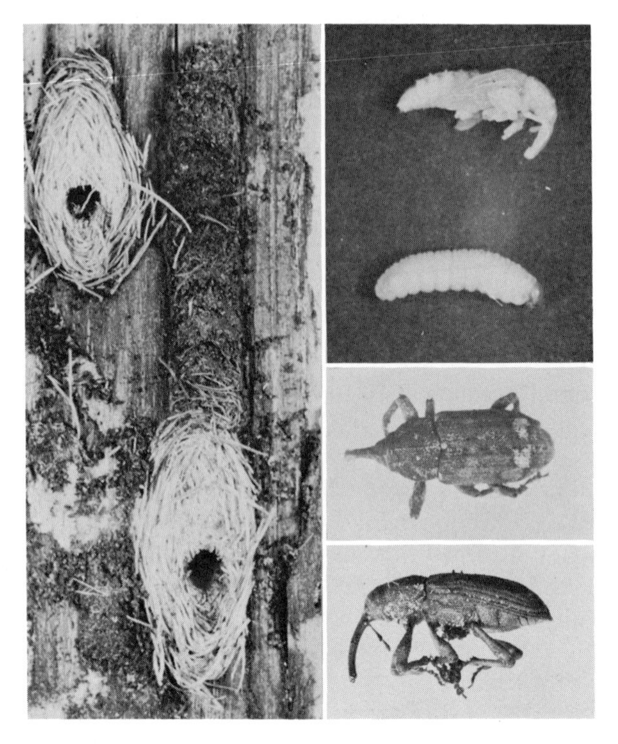

FIG. 13.1. White pine weevil (*Pissodes strobi*). Left: larval galleries and pupal chip-cocoons (X3); right: pupa and larva (X3), and adults (X4).

whitish, oblong, and placed in inner bark through small circular holes cut by female.

LIFE CYCLE. The adult weevils emerge in the fall. They first feed for some time on the inner bark of the terminals by cutting small circular holes through the bark and then go into the ground litter where they hibernate over winter. Weevil activity resumes in the spring when plant growth starts. The females lay their eggs in small niches that they gnaw in the tips of the year-old growth. Each female lays 50 or more eggs and places 2 to 3 in each niche. In heavily infested stands there may be as many as a hundred eggs laid in each leader. The larvae form a ring around the shoot as they feed side by side boring down through the inner bark. As they grow larger, since there is insufficient room and food for all, many get crowded away from the inner-bark food and starve to death. Those that mature first bore into the pith of the twigs to pupate, whereas others make chip cocoons in the surface of the wood. The larval period lasts 5 to 6 weeks, pupation another 2 weeks, and the adults mature and emerge 2 to several weeks later. There is only 1 generation per year.

Life cycles for the western species are similar to that outlined previously but differ for the deodar weevil. For this species the females deposit their eggs in the fall and the resulting larvae kill the terminals during the winter. The larvae mature in the spring and transform to adult weevils. These estivate through the summer in the ground litter.

CONTROL. White pine weevil populations usually do not fluctuate greatly, and natural controls have never been very effective; therefore, the problem is constantly present.

Applied silvicultural control has been advocated for many years but the results have not been wholly satisfactory. The principle appears to be sound, but satisfactory management techniques have not been developed to achieve the desired results. Apparently the female weevils oviposit only within a narrow temperature zone (77° to 85° F) and at relative humidities between 20 to 35 per cent. During the spring, when the weevils are active, these conditions commonly occur in pine stands that are in the open but not in those in the shade. This, of course, suggests that white pine should be grown under a nurse crop. In order to do this successfully the overstory must be carefully regulated so that pines will be protected and yet not greatly suppressed.

Both aerial and hand applications of insecticides are effective for controlling the white pine weevil. Hand spraying consists of applying a poison to the terminals of all trees in young stands during the spring before growth starts. Suitable insecticides are 10 per cent arsenate of lead, (plus 3 per cent linseed oil as an adhesive and 3/10 per cent of any household detergent as a spreader), or emulsions containing 3 per cent of DDT or 1 per cent gamma isomer of benzene hexachloride. The terminal bud cluster and the upper two thirds of the leader must be thoroughly covered from 2 sides. The first spraying should be done when about 5 per cent of the trees in the stand are weeviled during one year. Succeeding applications should be made whenever the infestation reaches 10 per cent per year. Generally only 2 or 3 sprayings are needed to protect the trees until they reach a height of about 16 feet. It is impractical to treat taller trees.

Helicopter applications of DDT applied during the early spring before the weevils emerge also have been successful. The formulation recommended is an emulsion containing 2 pounds DDT in 2 gallons of spray. This also is the dosage applied per acre.

Belyea, R. M., and C. R. Sullivan. 1956. *Forestry Chronicle* 32:58–67.
Connola, D. P., T. McIntyre, and C. J. Yops. 1955. *J. Forestry* 53:889–891.
Crosby, D. 1958. *U. S. Forest Serv., Northeast Forest Exp. Sta. Research Note* 78.

Graham, S. A. 1926. *Cornell Univ. Agri. Exp. Sta. Bull.* 449.
Hopkins, A. D. 1907. *U.S.D.A. Bur. Entomol Circ.* 90.
Jaynes, H. A., and H. J. MacAloney. 1958. *U.S.D.A. Forest Pest Leaflet* 21.
MacAloney, H. J. 1932. *U.S.D.A. Circ.* 221; and 1931 *N. Y. State Coll. Forestry Tech. Publ.* 28.
Ostrander, M. D., and C. H. Stoltenberg. 1957. *U. S. Forest Serv., Northeast Forest Exp. Sta. Research Note* 73.
Waters, W. E., T. McIntyre, and D. Crosby. 1955. *J. Forestry* 53:271–274.
Wright, K. H., and D. H. Baisinger. 1956. *Proc. Soc. Am. Forestry* 1955:64–67. (Sitka spruce weevil.)

REPRODUCTION WEEVILS

Cylindrocopturus spp., Coleoptera, Curculionidae

Although there are many species of reproduction weevils, only two western species have caused serious injury. These are the pine reproduction weevil (*C. eatoni* Buch.) and the Douglas-fir twig weevil (*C. furnissi* Buch.).

INJURY. The larvae of these weevils kill young trees by boring in the inner bark of the stems and twigs. They make individual galleries that cross and run together so that much of the inner bark is destroyed. The borings are of a granular consistency. The mature larvae form pupal chambers in the wood or pith. The adults also cause some injury by their feeding on the needles.

HOSTS AND RANGE. *C. eatoni* attacks chiefly ponderosa and Jeffrey pines in northern California, whereas *C. furnissi* infests Douglas-fir in the Pacific Northwest.

INSECT APPEARANCE. *Larvae:* typical; small, about ⅛ inch long when full grown. *Weevils:* small, about 1/16 inch long; clothed with bronze and silvery scales, which makes them appear gray; beak black, prominent, and curved; beak fits into a groove along the breast; underside of abdomen slopes upward behind thorax. *Eggs:* minute, in small niches in outer bark.

LIFE CYCLE. Winter is generally passed as mature larvae in pupal cells. Pupation occurs during the spring and the adults soon appear. The adults feed on the needles for a couple of weeks after which the females lay their eggs in the outer bark of the stems and twigs. One egg is laid in each niche. These hatch in a couple of weeks and the resulting larvae mature by late fall. There is only 1 generation per year.

CONTROL. Applied control consists of aerial spraying with DDT. The per acre dosage is 1 pound of DDT dissolved in 1 gallon of diesel oil and auxiliary naphthelanic solvent. Application should be made about 3 weeks after the first weevils emerge.

Buchanan, L. L. 1940. *Entomol. Soc. Washington Proceedings* 42:177–181.
Eaton, C. B. 1942. *J. Econ. Entomol.* 35:20–25.
Hall, R. C. 1957. *U.S.D.A. Forest Pest Leaflet* 15.

BARK-EATING INSECTS

The most serious bark feeders are the seedling-debarking weevils. Others include some sawfly larvae (page 178) and sawyer beetles (page 293). The latter frequently injure small pines located close to pulpwood storage yards and sawmills. Only the seeding debarking weevils are considered here.

SEEDLING DEBARKING WEEVILS

Hylobius spp. and *Pachylobius* spp., Coleoptera, Curculionidae

The most injurious debarking weevils are the pales weevil [*Hylobius pales* (Hbst.)] and the pitch-eating weevil (*Pachylobius picivorus* Germ.).

INJURY. Only the adults cause tree injury. They eat irregular areas of bark from the twigs and sometimes almost completely debark the seedlings. Recently planted trees are injured most frequently, but the twigs of older trees also may be heavily attacked, especially when these

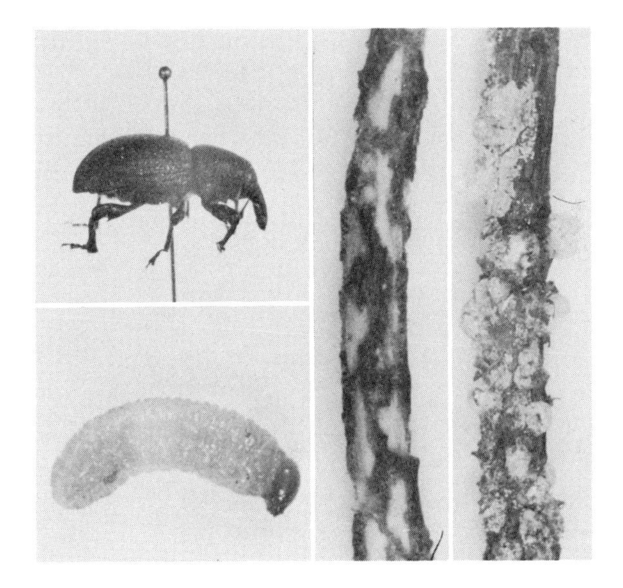

FIG. 13.2. Seedling debarking weevil (*Pachylobius picivorus*) and larva (X2), and damage to the stems of pine seedlings (X1).

trees sparsely stock recently logged pine lands. The weevils are attracted to recently cut-over pine lands where the fresh pine roots provide suitable larval host material. The heaviest damage occurs during June and July.

HOSTS AND RANGE. Various species of pines. The adults eat the bark from the twigs, whereas the larvae bore in the inner bark of the roots. Both species occur throughout eastern North America. The pales weevil is more injurious in the north, whereas the pitch-eating weevil is more common in the South.

INSECT APPEARANCE. *Beetles:* beak stout, at least as long as thorax, grooved for reception of each basal antennal segment (scape); upper lip (labrum) absent; antennae with compact club, elbowed, place of attachment visible when insect viewed from above; eyes not contiguous beneath head; mandibles with teeth at tip; prosternum never triangular shaped and not grooved for reception of beak; tarsi densely hairy underneath, 5-segmented, 3rd segment bilobed, claws simple; tibia with large spine at tip; fore coxae contiguous; thorax wider than long. *Pachylobius*—dark brown, robust, sparsely clothed, short, flattened, white yellowish or reddish brown hairs that form distinct patches on elytra; tibia short and thick with outer end enlarged; 5/16 to 7/16 inch long. *Hylobius*—tibia not thick and outer end not dilated; dark reddish brown with scattered tufts of long gray or yellowish hairs forming 2 oblique cross bars behind middle of elytra; 1/4 to 3/8 inch long. *Larvae:* typical; see page 377. *Pupae:* typical; with characteristic weevil beak.

LIFE CYCLE. Many adults hibernate through the winter in the ground litter. They appear from early spring throughout most of the summer and fall, but there is a peak of beetle abundance in June and July and another again in the fall. In the South, part of the population also winters as partially grown larvae. The females deposit their eggs in niches they cut in the bark of green stumps and exposed roots. Occasionally they attack logs. The resulting larval galleries are mostly in the inner bark of the roots. They are always packed with granular borings. When full grown each grub forms a shallow oval pupal cell in the sapwood, which is covered with fibrous chip-like borings. There is 1 generation in the North but probably 1½ or even 2 in the South.

CONTROL. Injury to planted seedlings can be avoided by delaying the planting until the inner bark of the stumps and roots has died and turned brown. This means that cut-over pine lands must remain unplanted for one full summer in the South and two in the North. Insecticide dips consisting of one of the cyclodienes (2 per cent aldrin emulsion) can also be used for treating the tops of the planting stock. Rubber gloves should be used by all persons handling the treated trees. The advantage of using treated planting stock is that the cut-over pine lands can be replanted immediately so that the competing brush will not get too much of a head start.

Pierson, H. B. 1921. *Harvard Forest Bull.* 3.
Pierson, H. B. 1937. *Mass. Forest and Park Asso. Tree Pest Leaflet* 13.
Beal, J. A., and K. B. McClintick. 1943. *J. Forestry* 36:792–794.
Speers, C. F. 1956. *U.S.D.A. Forest Serv., Southeastern Forest Exp. Sta. Research Note* 96.

ROUND-HEADED TWIG BORERS AND GIRDLERS

Coleoptera, Cerambycidae

Cerambycid larvae that bore in the twigs and smaller branches of trees are not serious forest pests, but sometimes they cause conspicuous damage to shade trees. Even then the injury seldom is serious.

CONTROL. For protecting shade trees it is usually suggested that the fallen, infested twigs should be collected and destroyed. This reduces the insect population the following year. Insecticidal sprays probably would be more effective, but this method has never been tried.

Craighead, F. C. 1950. *U.S.D.A. Misc. Publ.* 657:245–246, 255–257, 264–270.

TABLE 13.4 COMMON GENERA OF ROUND-HEADED TWIG-INFESTING INSECTS*

1.	Twigs each with a V-shaped encircling cut that extends through the bark and part of the wood so as to girdle the twig	2
1.	Twigs not as in 1 above but with tunnels bored on inside	3
2.	Egg niches or punctures adjacent to the buds on the severed portion of the twigs; galleries and larvae also occur in these parts (page 384)	TWIG GIRDLERS
2.	Not as in 2 above; severed portions of twigs never contain tunnels or larvae (page 385)	TWIG BORERS
3.	Infested twigs become swollen and gall-like (page 387)	GALL MAKERS
3.	Do not cause gall formations	4
4.	Larvae cut off the twigs or small branches by boring around inside the twig several times working in concentric circles from the outsides toward the center or vice versa; larvae with heads wider than long†	5
4.	Twigs completely hollowed, with a few to a series of small holes through the bark usually in a straight line	6
5.	Larva with legs; pronotum white and shining (page 385)	TWIG PRUNER
5.	Larvae legless; pronotum brown and velvety (page 386)	BEECH AND BIRCH PRUNER
6.	Larvae with legs; head wider than long;† pronotum shining; hosts oaks and chestnut (page 386)	SEEDLING BORER
6.	Larva legless; head longer than wide;† pronotum rough; hosts poplars, sassafras, and some shrubs (page 385)	TWIG BORERS

* Adapted from F. C. Craighead. U.S.D.A. Misc. Pub. 657 p. 231–2.
† Head must be removed to determine this characteristic.

TWIG GIRDLERS

Oncideres spp., Coleoptera, Cerambycidae

INJURY. Twig girdler attack branches less than about an inch in diameter. The damage consists of V-shaped cuts that encircle the twigs and extend into the wood. Eventually these partially severed twigs break and fall to the ground. Larval galleries occur in the severed portion.

HOSTS AND RANGE. A common eastern species, *O. cingulatus* (Say), infests hickory, persimmon, and many other hardwood species. There are also several Southwestern species that infest shade trees and mesquite. The most common of these is *O. trinodatus* Casey.

INSECT APPEARANCE. *Larvae:* typical cerambycid; see page 272; body cylindrical; head longer than wide with front margin having a transverse row of short ridges; pronotum shining with fine lines; body swellings (ampullae) with 2 or 3 rows of tubercles; legless. *Beetles:* ½ to 1 inch long; gray to brown; body cylindrical; antennae as long or longer than body; prothorax with a spine on each side; pronotum of *O. trinodatus* has 3 spines on middle of back.

LIFE CYCLE. The beetles are present during the late summer and fall. The females first girdle the twigs and then deposit their eggs in small niches adjacent to the buds on the severed portion. The resulting larvae bore in these severed twigs filling the tunnels with loose borings. The pupal cell is tightly packed with fibrous borings. Usually 1 generation develops every year.

FIG. 13.3. Twig-girdler (*Oncideres cingulatus*) beetle and damage (X2).

Essig, E. O. 1958. *Insects and mites of Western North America.* Macmillan Co., N. Y. P. 460.
Sanborn, C. E. 1911. *Okla. Agri. Exp. Sta. Bull.* 91.

TWIG PRUNER

Hypermallus villosus (F.), Coleoptera, Cerambycidae

INJURY. The larvae bore through twigs and small branches that are less than 2 inches in diameter. During late summer each infested branch is further weakened by the galleries then made, which encircle the twig several times. These weakened branches readily break and fall from the trees.

HOSTS AND RANGE. Hardwood species throughout eastern United States.

INSECT APPEARANCE. *Larvae:* body elongate, slender, with shining skin; head wider than long; tip of mandibles rounded; 2 simple eyes (ocelli) on each side; thoracic legs present. *Beetles:* 3/8 to 5/8 inch long; brown; spines present on first few antennal segments and at the tip of abdomen.

LIFE CYCLE. The beetles emerge at about the time the oak leaves appear. After mating, the females deposit their eggs in the leaf axils near the tip of the twigs. The resulting larvae bore down the twigs until late summer. They then bore around the stem several times as described previously. The larvae remain in the severed portions of the branches where they pupate between 2 wads of fibrous borings either in the fall or the following spring.

Chittenden, F. H. 1910. *U.S.D.A. Bur. Entomol. Circ.* 130.

OBEREA TWIG BORERS

Oberea spp., Coleoptera, Cerambycidae

HOSTS AND RANGE. *O. ferruginea* (Casey) infests willow in the Rocky Mountains and Great Plains regions. *O. ruficollis* (F.) infests sassafras and *O. tripunctata* (Swed.) infests poplar. The latter two occur throughout eastern United States. Other species include *O. tripunctata* (Swed.) in dogwood, *O. cingulatus* (Say) in many species of eastern hardwoods, and *O. texanus* Horn in leguminous plants.

INSECT APPEARANCE. *Larvae:* typical cerambycid; see page 272; body elongate, slender, and cylindrical; head longer than wide; posterior of pronotum roughened with point-like elevations, with 2 dark, oblique impressed lines; legless. *Beetles:* body very slender, elongate; yellowish or reddish with black elytra; antennae never longer than body; prothorax without spines on sides but with 3 or 4 black spots on back; last segment of hind tarsus rather long.

LIFE CYCLE. The beetles emerge in the early summer and, after mating, the females first girdle the tips before they deposit their eggs in the living portions below the girdles. The larvae bore down through the center of the twig, and while doing this they make a series of closely placed round holes through which the borings are exuded. Pupation occurs between 2 wads of fibrous borings. There is usually only 1 generation per year.

SEEDLING BORER

Aneflomorpha subpubescens (Lec.), Coleoptera, Cerambycidae

INJURY. The damage done by this insect is very similar to that caused by the *Obera* twig borers, except that generally only the main stems of seedlings are infested. The stem usually is cut off near the ground.

HOSTS AND RANGE. Seedlings (¼ to 1 inch in diameter) of oak and chestnut throughout the South.

INSECT APPEARANCE. *Larvae:* typical cerambycid; see page 272; body elongate, slender; head wider than long; tip of mandible rounded; 6 long curved hairs on each side of head; raised edge on back edge of pronotum; legs present. *Beetles:* ⅝ to ⅞ inch long; color light brown; basal antennal segments and tips of elytra with spines.

LIFE CYCLE. The beetles emerge about the time oak leaves become full grown. The larvae bore in the stems until fall. There is only 1 generation per year.

BEECH AND BIRCH PRUNER

Xylotrechus quadrimaculatus (Hald.), Coleoptera, Cerambycidae

INJURY. The damage these borers do is similar to that caused by the *Hypermallus* twig pruner. One difference is that the larvae bore the concentric circles from the outside in rather than vice versa. The younger larvae feed beneath the bark.

HOSTS AND RANGE. Hardwood species throughout eastern United States.

INSECT APPEARANCE. The generic characteristics for both the larvae and beetles are presented in Tables 11.3 and 11.5. *Beetles:* about ½ inch long; thorax black with 4 yellowish spots; elytra pale brown with whitish marking.

LIFE CYCLE. The beetles appear near midsummer. After mating, the females lay their eggs in wounds or near dead twigs. The resulting larvae bore beneath the bark for a distance of about 10 inches and then make the pupal chamber in the pith. There is only 1 generation per year.

GALL-MAKING ROUND-HEADED BORERS

Coleoptera, Cerambycidae

The maple gall borer, *Xylotrechus aceris* Fisher, and several *Saperda* species are the most common cerambycid gall formers.

INJURY. Boring by the larvae causes gall-like swelling to develop on the infested parts. The borings are mostly packed in the galleries. These are fibrous for the *Saperda* spp. and mostly granular for *Xylotrechus*.

HOSTS AND RANGE. *Saperda* spp. infest poplars, willows, and thorn apples wherever the hosts grow. *Xylotrechus* infests red maple in eastern North America.

INSECT APPEARANCE. The generic characteristics of both the larvae and beetles are presented in Tables 11.3 and 11.5. The *Saperda* beetles are light to dark gray or reddish brown and are about ½ inch long. The beetles of the maple gall borer also are about ½ inch long, but they are colored differently. They have a black pronotum with 4 yellowish spots, and the elytra are pale brown with whitish markings.

LIFE CYCLE. The *Saperda* beetles emerge in the early summer. After mating, the females deposit their eggs in small niches they cut in the bark. The resulting larvae bore beneath the bark for a year before they enter the wood. The *Xylotrechus* beetles appear during midsummer and the females lay their eggs at the bases of small dead twigs or wounds. The resulting larvae bore directly into the wood. Two years or longer are required to complete 1 generation for all these gall formers.

CONTROL. Under forest conditions these insects have never caused serious damage; therefore, no applied control methods have been developed. Removal and destruction of the affected parts is usually the suggested method for treating infested ornamental trees.

FLAT-HEADED TWIG BORERS

Coleoptera, Buprestidae

Some buprestid inner-bark borers that attack the larger branches and the main tree boles also infest the twigs. These are species of *Agrilus, Chrysobothris, Anthaxia,* and *Chrysophana.* For more information on these the reader is referred to the section in Chapter 10 dealing with flat-headed borers.

TWIG-INFESTING BARK BEETLES

Coleoptera, Scolytidae

There are many species of bark beetles that infest the twigs and smaller branches of trees, but the damage they cause seldom is very serious. The interested reader is referred to the general discussion on bark beetles and to the references listed there.

BUD- AND TWIG-BORING CATERPILLARS

Lepidoptera

Bud- and twig-boring caterpillars and moths are difficult to recognize; therefore, these insects usually must be referred to specialists for identification. Nevertheless, a field guide based on hosts and damage for the important genera that infest conifers is presented as Table 13.5. Usually only young sapling-sized stands are seriously injured.

TABLE 13.5 COMMON TWIG- AND TERMINAL-BORING LEPIDOPTERA CATERPILLARS THAT INFEST CONIFERS

1\. Infest cypress, cedars, redwoods, or junipers (Cupressaceae) . . . TWIG MINERS (*Argyresthia* app.)
1\. Infest pines and spruces . . . 2
2\. Buds are bored and destroyed first; then the mines often are extended into the adjacent parts of the twig tissues; infest only pines (page 388) . . . PINE TIP MOTHS
2\. Entrances are made or most boring is done at places below the buds . . . 3
3\. Tunnels made mostly in the wood or the pith of the twigs . . . 4
3\. Mines made in the inner-bark region of the twigs or stems . . . 5
4\. Globular nodules of resin and borings frass up to about ¾ inch in diameter form over points of attack; tunnels commonly 3 to 4 inches long (page 391) . . . PITCH NODULE MOTHS
4\. Tunnel extends 6 to 8 inches through center of current seasons growth; pitch mass as described above absent; (page 391) . . . SHOOT MOTHS
5\. Caterpillars rather large being ¾ to 1 inch long when full grown; thin, irregular layers of resin exude and cover outside bark; inner bark thick and spongy and outer bark very rough (page 260) . . . PITCH MOTHS (*Dioryctria* spp.)
5\. Caterpillars small being about ⅜ inch long when full grown; only limited amount of resin exudations mixed with frass (page 406) CONE MOTHS (*Laspeyresia* spp.)

PINE TIP MOTHS

Rhyacionia spp., Lepidoptera, Olethreutidae

The more important species are the European pine tip or shoot moth [*R. buoliana* (Schiff.)], the Nantucket pine tip moth [*R. frustrana* (Comst.)], and the Western pine tip moth [*R. f. bushnelli* (Busck)]. There are several others which are less troublesome.

INJURY. Tip moth caterpillars first bore into and destroy the buds. From here they often extend their tunnels into the adjacent parts of the twigs for a distance of only a few inches. Sometimes, however, *R. buoliana* larvae may destroy as much as 6 inches of the tips. The external evidences of injury consist of resinous exudations on and around the buds and the twigs near the needle sheaths, but large globular pitch nodules are never formed. Later the needles next to the buds turn yellow and then brown. When the terminal buds or shoots are killed, the nearby uninjured buds become new leaders. This causes the injured trees to become somewhat crooked. Heavy continued infestations result in bushy trees, but seldom, if ever, are the trees killed completely. Usually the crook that results from a single attack is not as severe as that caused by an attack of terminal weevils because a shorter length of stem is destroyed. Pine reproduction in the South frequently appears seriously injured by *R. frustrana* every year over a period of many years, but each spring the trees recover so that eventually they outgrow the attacks without serious stem deformations. Nevertheless, although the loss of tree height growth may amount to ½ to 2½ feet per year, this loss may be partially recovered by the more vigorous spring shoot growth. This generalization, however, is not correct for the more injurious European pine tip moth and in some areas the Nantucket pine tip moth is most injurious.

HOSTS AND RANGE. *R. buoliana*—infests all species of young pines up to about 20 feet tall, but red pine, Scotch pine, and mugo pine are

FIG. 13.4. Pine tip moth (*R. frustrana*) damage (X1), and moth (X4½). Note the short aborted needles at the tip and the resin mass around the bud.

most severely damaged. This European species now occurs throughout much of northeastern United States west to the Plains and south to Virginia. It is still absent from the northern parts of the northern tier of states. *R. frustrana*—infests all native pines up to about 15 feet tall except longleaf, slash, and white pines. It occurs throughout much of the southern two thirds of the East. A western variety, *R. f. bushnelli,* occurs in the Great Plains region, and *R. rigidana* (Fern.) occurs in the far South. Various species occurring in the far West infest lodgepole, Monterey, and ponderosa pines.

INSECT APPEARANCE. *Caterpillars:* typical, with 3 pairs thoracic legs and 5 pairs prolegs; see page 260; small, ⅜ to ⅝ inch long when full grown; body rather naked, yellowish to brownish; pronotum and head black (*R. buoliana*) or dark brown (*R. frustrana*); spiracles elliptical with 8th pair larger and located higher than the others. *Pupae:* small, up to ⅜ inch long; slender; brown; located inside buds or terminals. *Moths:* small, with no good distinguishing characteristics; *R. frustrana* wingspread about ½ inch, grayish; *R. buoliana* wingspread about 1 inch; color reddish brown with whitish legs and irregular silver gray bands crossing wings. *Eggs:* small and flat; yellow at first then turning reddish brown. These insects can be recognized most easily by means of the damage they cause. Only specialists can identify these insects per se.

LIFE CYCLE. *R. buoliana* has only 1 generation per year where it occurs in the United States. After mating the females lay their eggs from mid-June to mid-July. The newly hatched larvae first bore into the needles via the needle sheaths, but as they grow older they bore directly into the buds. Silk spun by the caterpillars together with the resin exudate cover these entrance points. Winter is passed as 4th instar larvae. Boring commences again in the spring and soon the larvae enter the adjoining twig tissues. Pupation occurs within the galleries and probably lasts about 2½ weeks. *R. frustrana*—mostly hibernate over winter as pupae within the galleries or in soil for the variety *bushnelli.* The moths emerge early in spring and the females lay their eggs for the first generation. There are 1 to 4 generations per year depending on the length of the summer season. The other species of tip moths probably have similar habits.

CONTROL. At the present time there is no practical applied method that can be used against these insects to protect forest trees. DDT sprays applied at the usual dosages several times a summer give good control for high value nursery or ornamental trees.

Afanasiev, M., and F. A. Fenton. 1947. *J. Forestry* 45:127–128. (*R. frustrana bushnelli.*)
Busck, A. 1915. *U.S.D.A. Bull.* 170. (*R. buoliana.*)
Friend, R. B., and A. S. West. 1933. *Yale Univ. School Forestry Bull.* 37. (*R. buoliana.*)

Graham, S. A., and L. G. Baumhofer. 1927. *J. Agri. Research* 35:323–333. (*R. frustrana bushnelli.*)
Miller, W. E., and R. B. Neiswander. 1955. *Ohio Agri. Exp. Sta. Research Bull.* 760. (*R. buoliana.*)
Mortimer, M. F. 1941. *Tenn. Acad. Sci. J.* 16:189–206. (*R. frustrana.*)
Stearns, L. A. 1953. *J. Econ. Entomol.* 46:690–692. (Both R.b. & R.f.)

SHOOT MOTHS

Eucosma spp., Lepidoptera, Olethreutidae

INJURY. Wilted leaders and laterals resulting from attacks by shoot moths are often conspicuous, but seldom is the damage serious. The caterpillars bore a distance of 6 to 12 inches through the pith and adjacent tissues of the new growth. This causes the shoots to wilt and die, but trees are never completely killed. See Table 13.5.

HOSTS AND RANGE. Various species of shoot moths occur in different parts of the continent wherever the host pines grow. Only reproduction and sapling-sized trees are infested.

INSECT APPEARANCE. These insects are very similar to other tortricids; therefore, only specialists can identify the insects.

LIFE CYCLE. The seasonal development of these insects is similar to that of *R. buoliana* except that the moths appear earlier and pupation occurs in the soil.

CONTROL. Applied methods have never been needed in forests. Removal and destruction of the wilted shoots is often suggested as the method to use for treating infested ornamental trees.

Butcher, J. W., and A. C. Hodson. 1949. *Can. Entomol.* 81:161–173.

PITCH NODULE MOTHS

Petrova spp., Lepidoptera, Olethreutidae

INJURY. Although infestations of pitch nodule moths have never caused much trouble, the conspicuous, globular masses of resin that form on infested twigs are commonly observed. The larvae bore for a distance of several inches into the pith of twigs, but they never infest the buds. The injury occurs mostly to reproduction and sapling-sized trees. The trees attacked are never killed.

HOSTS AND RANGE. Various species occur in different parts of the country wherever the pines and spruce hosts grow.

INSECT APPEARANCE. All stages are similar to those for the pine tip moths; therefore, for our purpose, they can be identified on the basis of damage as indicated in Table 13.5.

LIFE CYCLE. Similar to that for the European tip moth, except that the moths appear about a month earlier.

CONTROL. In ornamental plantings the infested twigs can be removed and destroyed. Applied control has never been needed in forests.

Turnock, W. J. 1953. *Can. Entomol.* 85:233–243.

OVIPOSITION INJURY

PERIODICAL CICADA

Magicicada septendecim (L.), Hemiptera (Homoptera), Cicadidae

INJURY. The females lay their eggs in shreddy slits they cut in the twigs. When these ovipositor punctures are numerous the twigs are either killed or severely weakened, but usually only the smaller trees and bushes are injured seriously. Apparently the feeding by either the adults or nymphs is not very injurious.

HOSTS AND RANGE. All species of hardwoods throughout most of eastern United States.

INSECT APPEARANCE. *Adults:* large, about 1½ inches long to tip of wings; other species are often larger; eyes and legs red; female with back of abdomen black, sides and underneath banded with orange brown; males with first 4 or 5 back abdominal segments orange brown and with pair of orange covers over sound-producing organs; 3 simple eyes

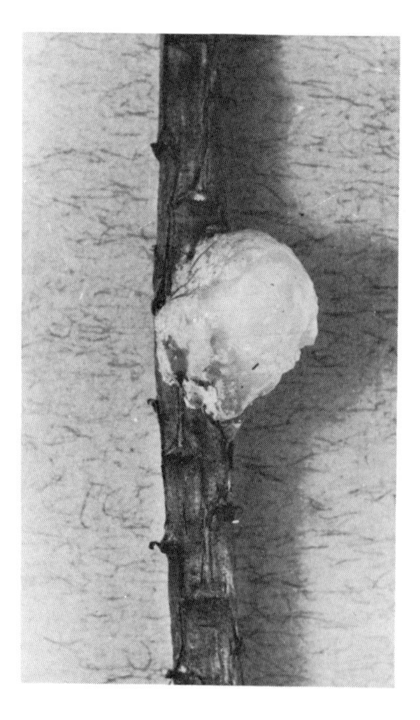

FIG. 13.5. Pitch nodule moth (*Petrova* spp.) resin globule formed on a twig (X1).

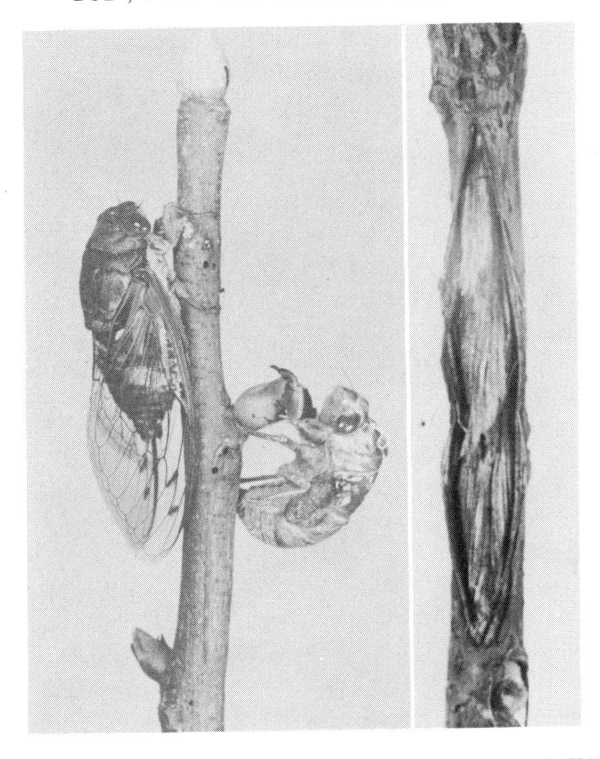

FIG. 13.6. Periodical cicada (*Magicicada septendecim*) adult and nymph (X1¼), and twig damage (X2).

(ocelli); distinct rod-like beak; 4 pairs membranous wings held roof-like when at rest; a distinct W mark on outer part of each forewing. **Nymphs:** seldom seen for they appear above ground for only a short time before the adults emerge, but the molted skins are left on twigs where they are commonly observed; front legs flattened and adapted for burrowing; beak-like mouthparts; somewhat resemble a small crayfish.

LIFE CYCLE. The adults appear in the spring or early summer. They live for about a month during which time they mate and the females deposit their eggs as described above. During this period the males make their characteristic, almost continuous, monotonous song, which is accomplished by vibrating a pair of membranes inside cavities located on the underside of the first abdominal segment. The eggs hatch in 1½ to 2 months, and the resulting young nymphs drop to the ground and immediately burrow into the soil. Here they feed on plant sap extracted from the roots. They remain in the soil and develop slowly for the next 13 or 17 years. The mature nymphs leave the soil and climb up on any plant or object present just before they transform to adults. Other species commonly heard have a shorter life cycle, but these never have been numerous enough to cause noticeable damage.

CONTROLS. Ornamentals can be protected by spraying every few days when the adults are present with an organic phosphate insecticide such as Tetraethyl pyrophosphate (TEPP). One eighth of a pint of actual insecticide should be used per 100 gallons of water. TEPP is very dangerous to humans; therefore, great care must be exercised when using this poison.

1953. *U.S.D.A. Leaflet* 340.

BUFFALO TREEHOPPER

Stictocephala bubalus (F.) Hemiptera (Homoptera) Membracidae

INJURY. The females cut a double row of curved slits in twigs with their ovipositors. This is primarily a pest of fruit trees. Forest trees seldom are injured.

HOSTS AND RANGE. Many species of woody plants throughout United States and southern Canada.

INSECT APPEARANCE. *Adults:* small, about ¼ inch long; light green; pronotum triangular in shape when viewed from above with 1 point extending over the abdomen and the other 2 points forming "horns" on the 2 front corners; antennae attached in front of and between eyes; at most only 1 pair of simple eyes (ocelli). *Nymphs:* resemble adult, but are spiny and wingless.

LIFE CYCLE. Winter is passed as eggs in the twigs. These hatch late in the spring and the resulting nymphs feed on grasses and herbs. The adults appear during late summer. At this time they mate and the females lay their eggs for the next generation. There is only 1 generation per year.

CONTROL. A mineral oil emulsion (4 per cent) applied during the dormant season is effective in killing the eggs but, of course, is practical only for reducing tree hopper populations in valuable plantings such as orchards.

TWIG DAMAGING MAGGOTS

PITCH MIDGES

Retinodiplosis spp. Diptera, Cecidomyiidae

INJURY. Maggots of pitch midges feed on the tender bark of pine twigs and cause globular masses of resin (similar to that shown in Fig. 13.5) to form at the places attached. Inside these masses are small orange-colored maggots. Seldom do they cause serious damage.

HOSTS AND RANGE. Pines and possibly other resinous conifers wherever the hosts grow. *Larvae:* typical; small maggots; see page 402; color orange-red. *Flies:* small 2-winged midges with no good identifying characteristics.

LIFE CYCLE. Probably is as follows: the insects live over winter as pupae in the ground litter. The flies emerge in the spring, mate and the females lay their eggs on the twigs. The resulting maggots mature in late summer. There is only 1 generation per year.

CONTROL. Never has been needed.

TREE KILLING ANTS

ALLEGHENY MOUND ANT

Formica excectoides Forel, Hymenoptera, Formicidae

INJURY. The Allegheny mound building ants kill small trees, shrubs, and other plants in the immediate vicinity (20 to 35 feet) of their nests by injecting formic acid into tissue of the plants. The bark of small killed trees remains rather intact but becomes sunken in a band extending around the entire stem. This sunken bark area occurs just above the ground and extends several inches along the stem. The stem both above and below this girdle is enlarged somewhat. Trees taller than 6 feet seldom are killed. The reason why the ants do this is not known, but one result is to help keep the ant hills from being shaded. The damage they cause seldom is serious.

HOSTS AND RANGE. Though all species of plants are killed, the plant tissues or sap are not used for food. This species occurs throughout northeastern United States south to Virginia.

INSECT APPEARANCE. *Adults:* typical ants; head and thorax reddish brown; legs and abdomen are blackish brown; anal opening round and surrounded by a fringe of hairs; body strongly constricted between thorax and abdomen; pedicle between thorax and abdomen consists of single segment. *Larvae* and *pupae:* similar to the corresponding stages of the carpenter ant. *Mounds:* made of sand, twigs, and straw; often 3 feet tall and 6 feet in diameter.

LIFE CYCLE. Similar to that for the carpenter ant except that this species constructs its nests in the ground and in the adjacent above ground mounds of soil, straw, and twigs. The ants feed chiefly on the sweet exudate (honeydew) produced by aphids and scale insects. They also kill other insects and consume the body fluids.

CONTROL. The applied methods suggested for the control of the town ant would be effective, should conditions ever warrant action.

Andrews, E. A. 1928. *Am. Naturalist* 62:63–75.
Pierson, H. B. 1922. *J. Forestry* 20:325–336.

CHAPTER 14

Root-Feeding Insects

The most serious problems caused by root feeding insects have developed in forest nurseries and in young plantations. Many of these insects consume the fibrous and smaller roots, others bore in the inner bark or woody parts of the larger roots, and some are sapsuckers. The more important of these are presented in Table 14.1.

TABLE 14.1 COMMON GROUPS OF INSECTS THAT INFEST THE ROOTS OF TREES

1. Sapsucking insects; insects with beak-like mouthparts, enclosed needle-like stylets adapted for piercing and sucking; antennae and legs well developed; solid parts of host not eaten	2
1. Insects that consume the solid parts by eating the finer roots or by boring in the roots	3
2. Forelegs enlarged and adapted for digging; large insects, up to about 1 inch long (page 392)	CICADAS
2. Front legs not adapted for digging; insect bodies small, globular to pear shaped and soft, may be covered with woolly wax; frequently with a pair of gland-like or tube-like structures (cornicles) on back of 6th abdominal segment (page 347)	PLANT LICE
3. The fine roots are eaten; when seedlings and small trees are infested frequently, only parts of the tap roots remain	4
3. Root borers; tunnels formed either in the inner bark or in the wood	7
4. Larvae with 3 pairs well-developed thoracic legs	6
4. Larvae legless	5
5. Larvae without a distinct head (page 401)	SEED MAGGOTS
5. Larvae with a distinct globular head capsule (page 400)	WEEVILS*
6. Body cylindrical elongate; tan to brown; with surface rather hard and leathery legs small; up to about 1½ inches long (page 398)	WIRE WORMS

6. Body whitish, soft, cylindrical in shape with long axis strongly curved so that the insect appears C-shaped; posterior end enlarged and blunt (page 397) . . . . WHITE GRUBS
7. Borers in the woody parts; elongate, whitish, cylindrical with 3 pairs of short peg-like thoracic legs; head somewhat flattened with mouthparts directed forward (page 400) . . . . PRIONID ROOT BORERS
7. Inner bark borers; larvae cylindrical, legless, and with globular head 8
8. Both adults and larvae bore in the inner bark, hence both uniform width beetles galleries and variable width larval galleries present (page 401) . . . BARK BEETLES
8. Only larvae and larval galleries present in the inner bark (page 399) . . . . ROOT COLLAR WEEVIL

* If the head is flattened, check the characteristics for both round-headed and flat-headed borers presented on pages 247 and 272.

WHITE GRUBS

Phyllophaga spp. *Serica* spp., Coleoptera, Scarabaeidae

INJURY. White grubs have caused serious losses in both forest nurseries and in young plantations. At times the tree losses in nurseries have been as high as 30 to 40 per cent, and in some places field plantings have been completely destroyed. The damage is caused by the soil-inhabiting grubs which eat the roots. The adults, known as May or June beetles, sometimes cause defoliation to hardwood trees surrounding grassy fields or in wood lots. This aspect is discussed on page 196.

HOSTS AND RANGE. White grubs are general feeders attacking the roots of grasses, woody plants, and many herbs. They occur throughout North America.

INSECT APPEARANCE. *Larvae:* body large, often up to 1 inch long, segmented, soft, whitish, cylindrical; the long axis sharply curved so grubs appear C-shaped; posterior of abdomen enlarged, smooth, shining, and translucent so that the dark intestinal contents are visible; head tan or brownish, globular with mouthparts directed downward; 3 pairs of thoracic legs. *Beetles:* well-known May or June beetles; large, up to 1 inch long; brown to blackish brown; oval; antennae elbowed each with a club consisting of several flattened, plate-like segments that can be closely appressed together and that extend out from only one side of main antennal axix (lamellate); abdomen with 6 visible segments underneath.

LIFE CYCLE. Both beetles and larvae live in the soil during winter. The beetles emerge in the spring from March to July depending on when the soil becomes warm. They are nocturnal. Each morning they return to the soil where the females lay their eggs. These eggs hatch in 3 to 4

FIG. 14.1. White grub and the adult June beetle (*Phyllophaga* spp.) (X1½).

weeks, and the resulting larvae feed on decaying vegetation and on roots. The period of time required for each generation to develop varies from 2 to 4 years, depending on the length of the summers. Pupation occurs in the soil during the late summer preceding the spring the beetles emerge.

CONTROL. Any of the cyclodiene insecticides are effective against white grubs. The actual amounts of the poisons that should be applied per acre are 10 pounds of chlordane or half as much of the aldrin, dieldrin, or heptachlor. They are usually applied with a fertilizer drill as 5 to 10 per cent dusts and then disked into the soil. Of course, this type of treatment is practical only in nurseries.

Luginbill, P., Sr., and T. R. Chamberlin. 1953. *U.S.D.A. Farmers' Bull.* 1798.
Shenfelt, R. D., and H. G. Simkover. 1950. *J. Forestry* 48:429–435.
Shenfelt, R. D., H. R. Liebig, and R. C. Dosen. 1955. *Tree Planters Notes* 20:14–17.

WIRE WORMS (CLICK BEETLES)

Coleoptera, Elateridae

INJURY. Some wire worms are borers in rotting wood, others are predators, and few occur in the soil where they feed on and damage the roots of various plants. They have not been serious tree pests.

HOSTS AND RANGE. Wire worm larvae are general feeders that are widely distributed.

INSECT APPEARANCE. *Larvae:* see characteristics presented in Table 14.1. *Beetles:* tan to brown, elongate, resemble buprestid beetles but have prothorax and mesothorax joined with a dorso-ventral hinge-like connection. This characteristic makes it possible for the beetles to throw themselves in the air whenever they fall on their backs. In doing this they make a clicking sound; therefore, they are called click beetles.

LIFE CYCLE. Similar to that for the white grubs, but the length of development varies from 2 to 6 years. The beetles are strongly attracted to sweets.

CONTROL. Similar to that recommended for white grubs.

ROOT COLLAR WEEVILS

Hylobius radicis Buchanan, Coleoptera, Curculionidae

INJURY. The larvae bore in the inner bark at or below the ground line and frequently girdle the trees. Trees usually attacked are between 5 and 20 years old, but trees of all ages may be infested. The soil adjacent to infested trees becomes blackened and soaked with resin.

HOSTS AND RANGE. Many species of pines are attacked, but the introduced European species such as Scotch, Austrian, and Corsican pines are preferred. *H. radicis* occurs in northeastern United States west to Minnesota. Another species, *H. warreni* Wood, attacks the roots of lodgepole pine in western Canada.

INSECT APPEARANCE. The insects are very similar to the pales weevil; therefore, only specialists can distinguish them.

LIFE CYCLE. Similar to that of *H. pales.* See page 381.

CONTROL. Ornamental trees can be treated with benzene hexachloride (⅓ pound gamma isomer per 100 gallons) or with chlordane (1 pound per 100 gallons). Apply enough of either poison around the base of each tree so as to thoroughly wet the bole and the soil.

Schaffner, J. V., Jr., and H. L. McIntyre. 1944. *J. Forestry* 42:269–275.
Shenefelt, R. D. 1950. *J. Econ. Entomol.* 43:684.

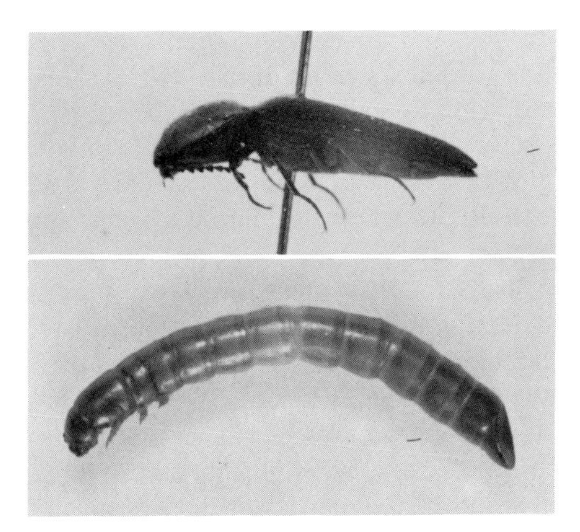

FIG. 14.2. Wire worm and the adult click beetle (Elateridae) (X2½). Note that the beetle is pinched together between the pronotum and elytra and that the larva has a hard pigmented body wall.

ROOT WEEVILS

Brachyrhinus ovatus (L.) Say, and *B. sulcatus* (F.), Coleoptera, Curculionidae

INJURY. The fine roots of seedlings are eaten by the larvae. The damage most frequently occurs in nurseries but is not as common as that caused by white grubs.

HOSTS AND RANGE. Various species of plants throughout northern United States and Canada.

INSECT APPEARANCE. *Larvae:* typical; see page 376; about ⅜ inch long when full grown. *Beetles:* wing covers (elytra) grown together so insects cannot fly; about ¼ inch long, color shining black with reddish brown antennae and legs; surface coarsely punctured and somewhat hairy; antennae clubbed; upper lip (labrum) absent; beak short and stout, about as long as head, dilated and notched at tips; tarsi dilated, spongy pubescent beneath, 3rd segment deeply notched (bilobed).

LIFE CYCLE. Winter is passed in both the adult and larval stages. The weevils hibernate in the ground litter or within the crowns of herbaceous plants such as strawberry. They appear late in the spring and each female lays about 50 eggs during a period of 1 to 2 weeks. The adults feed on the leaves of many species of both woody and nonwoody plants. The larvae feed on roots. There are 1 or 2 generations per year.

CONTROL. The same control suggested for white grubs is effective against the root weevils.

Schread, J. C. 1951. *Conn. Agri. Exp. Sta. Circ.* 174 (revised).

PRIONID ROOT BORERS

Prionus spp. Coleoptera, Cerambycidae

INJURY. The seriousness of these root borers is difficult to determine even though evidence of their activity is commonly seen. The larvae first feed in the inner bark of living roots but soon enter and bore in the wood. The galleries are filled with coarse fibrous borings, pellet-like excrement, and soil. Mature trees in open growing stands are preferred.

HOSTS AND RANGE. Living oaks are preferred, but other hardwoods are also attacked. Different *Prionus* spp. occur in various parts of the country.

INSECT APPEARANCE. *Larvae:* typical cerambycids, see page 272; body elongate, rapidly tapering toward posterior; head wider than long, retracted into prothorax; front of head with a transverse ridge divided at the middle; small radially lined disk below spiracles on each of first 6 abdominal spiracles; 3 pairs thoracic legs present. *Beetles:* see Table 11.5.

LIFE CYCLE. The adults appear in late spring or early summer. After mating the females lay their eggs on the roots and the resulting larvae

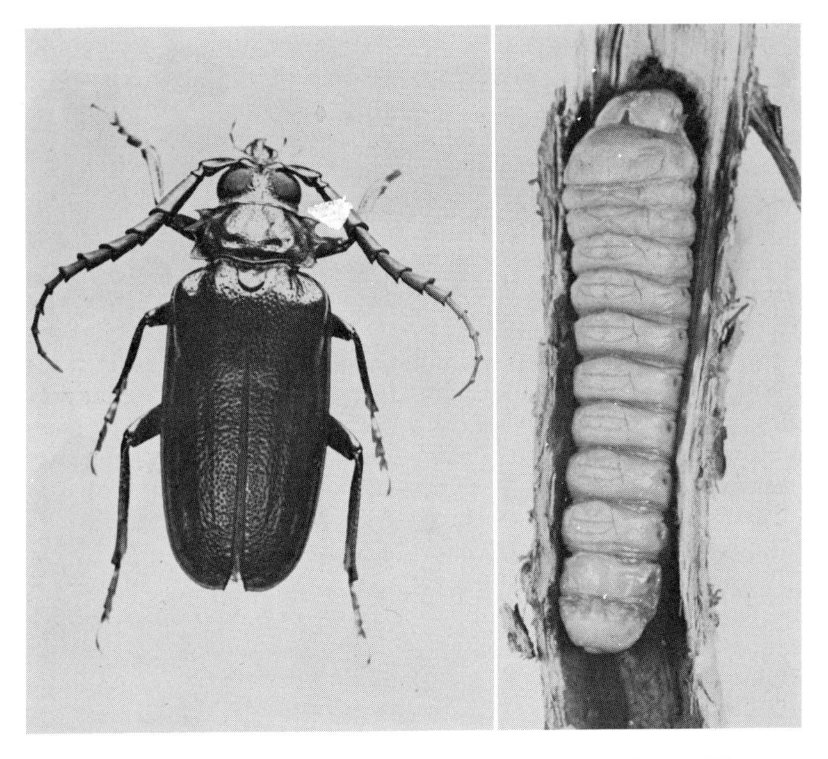

FIG. 14.3. *Prionus imbricornis* root borer, beetle (X1⅔) and larva (X1).

bore for 4 or 5 years before they mature. Pupation requires 1 or 2 months.

CONTROL. There is no known method for controlling these root borers.

ROOT BARK BEETLES

Coleoptera, Scolytidae

Many species of bark beetles infest the roots of trees. The more common of these are *Dendroctonus terebrans, D. valens, Hylastes* spp., and *Hylurgops* spp. The latter two genera seldom cause serious trouble. For additional information on these insects the reader is referred to the section on bark beetles in Chapter 10.

ROOT MAGGOTS

Hylemya spp., Diptera, Anthomyiidae

INJURY. Occasionally maggots damage young seedlings during cool, wet springs.

HOSTS AND RANGE. These general feeders mostly attack the roots of crop plants, but tree seedlings sometimes are attacked. They occur everywhere.

INSECT APPEARANCE. *Maggots:* body soft, segmented, whitish; about ¼ inch long when full grown; head indistinct with head end pointed; a pair of dark-colored mouth hooks occur inside; these can be seen through the translucent body wall; posterior end large and blunt with 2 spiracles. *Pupae:* inside hard, brown, shell-like, segmented, oval cases (puparia); about ⅕ inch long. *Flies:* small, grayish brown, 2-winged insects; ⅕ inch long.

LIFE CYCLE. Pupae hibernate in the soil over winter. The flies emerge early in the spring, and after mating the females lay their eggs in the soil. The resulting larvae eat the roots, seeds, or other organic matter present in the soil. Development is rapid; consequently, there may be 2 to 5 generations per year.

CONTROL. Three pounds of aldrin or 2 pounds of heptachlor per acre worked into the surface soil will kill the maggots.

Breakey, E. P., C. J. Gould, and C. E. Reynolds. 1945. *J. Econ. Entomol.* 38:121.

CHAPTER 15

Cone- and Seed-Destroying Insects

Much interest has developed recently in cone- and seed-destroying insects primarily because of the large amount of forest regeneration work being done and because of the increased interest in forest tree improvement. The most injurious of these insects are caterpillars, cone beetles, weevils, wasps, and maggots. Others of lesser importance include some of the small powder post beetles (Anobiidae), round-headed borers and flat-headed borers. The characteristics of the seed and cone infesting stages are presented in Table 15.1 and in the discussions that follow. The best compilation of information on the western species is contained in the publication listed below.

Keen, F. P. 1958. *U.S.D.A. Tech. Bull.* 1169.

TABLE 15.1 CONE- AND SEED-FEEDING INSECTS*

1. Thoracic legs present 2
1. Thoracic legs absent 3
2. Caterpillars with 3 pairs short thoracic legs and 2 to 5 pairs of fleshy abdominal legs (prolegs); webbing as well as borings often present in galleries (page 406). . . . . CONE AND NUT MOTHS
2. Larvae with only 3 pairs of short thoracic legs; body elongate, cylindrical; head flattened with mouthparts directed forward . . . . ROUND-HEADED BORERS (Cerambycidae)
3. Larvae with distinct heads that are hardened and darker colored (tan or brown) than whitish body 4
3. Larvae apparently headless or head indistinct and not colored darker than rest of body 6

4. Infest cones of coniferous trees 5
4. Infest larger fruits of broad-leaved trees such as nuts and acorns; larval bodies cylindrical and C shaped; heads globular with mouthparts directed downward; legless (page 404) . . . . WEEVILS
5. Prothorax flattened and much wider than other body segments with plate-like areas on back and underneath about equal size; head flattened with mouthparts directed forward; only larval galleries present . . . . FLAT-HEADED BORERS (Buprestidae)
5. Both uniform sized adult galleries and larval galleries present; larvae similar to but smaller than weevil larvae (see No. 4 above and general description for bark beetles) (page 408) . . . . CONE BEETLES
6. Grubs; mouthparts, if present, are opposable jaws or points; body soft, whitish, pointed at both ends, often C shaped; 7 to 11 inconspicuous spiracles (page 407) . . . . SEED WASPS
6. Small maggots; body spindle shaped or peg like with head end pointed and posterior end large and blunt; mouthparts 2 parallel, dark hooks inside pointed front end; 1 or 2 pairs spiracles present with one large pair at blunt posterior end; some species with spatula shaped hardened "breast bone" on thorax (page 409) . . . . SEED MAGGOTS

* Insects that feed in old dead cones are not included.

NUT WEEVILS

Curculio spp. and *Conotrachelus* spp., Curculionidae, Coleoptera

INJURY. The larvae bore and feed in nuts or acorns and, in so doing, they often destroy large proportions of the seed crops.

HOSTS AND RANGE. All large seeded nut-like fruits such as those produced on oaks, hickories, walnuts, butternuts, chestnuts, and hazelnuts are attacked. The insects occur wherever the hosts grow.

INSECT APPEARANCE. *Larvae:* typical weevils, see page 376; body very stout and plump, cylindrical, long axis strongly curved; head brownish, globular, with mouthparts directed downward, legless. *Weevils:* see page 376 for a general description of weevils. *Curculio* spp.—body robust, densely clothed with grayish, yellowish, or brownish hairs; beak very long and very slender, grooved on sides for reception of antennae, beak longer than body on females but shorter on males; mandibles move in a vertical direction; antennae elbowed, attached well back from middle of beak on females; hind angles of prothorax not pointed; legs long; ventral abdominal segments, except 1st, about of equal length. *Conotrachelus* spp.—typical weevils with breast grooved for reception of beak; eyes somewhat covered by ocular lobes; prothorax much narrower than elytra; outer front angles of elytra rather square shaped; body scaled

FIG. 15.1. Acorn weevil (*Curculio* spp.) (X2), larva (X2½), and in weeviled acorn showing exit hole.

on upper surface; front coxae close to each other; tibia each with a strong spine.

LIFE CYCLE. *Curculio* spp.—winter is passed as larvae in the soil. Pupation occurs during early summer, and weevils emerge during July and August. After mating the females cut small circular holes in the developing nuts and deposit their eggs in these niches. Oviposition continues until the nuts are fully formed. The eggs hatch within a week or two and the larvae mature in the fall. At this time mature larvae cut small, circular, exit holes through which they emerge and drop to the ground. There is only 1 generation per year. *Conotrachelus* spp.—similar to *Curculio* spp. except that the adults appear in the fall; these hibernate over winter in the ground litter.

CONTROL. Under forest conditions there is no practical method for controlling these insects. In nut orchards a DDT spray is effective and practical. It consists of using 3 pounds actual insecticide, either as an emulsion or a wettable powder, per 100 gallons of water. The first application should be made as soon as the weevils appear in late July or early August, and the second application after the first heavy rain in late August or early September. Presence of the weevils can be determined by jarring the trees and catching the insects that fall on a sheet placed underneath. Weevil larvae inside the nuts can be killed if the nuts are immersed in water at 120° F for 30 to 45 minutes or are fumigated with methyl bromide at the rate of 1 ounce per 25 cubic feet.

Asburn, M. R., A. M. Phillips, W. C. Pierce, and J. R. Cole. 1954. *U.S.D.A. Farmers' Bull.* 1829 (revised).
Lotti, T. 1959. *J. Forestry* 57:923.

CONE AND NUT MOTHS

Lepidoptera, Pyraloidea and Tortricoidea

The various species of caterpillars that infest cones and nuts are not easily identified; therefore, their specific characteristics will not be presented here.

INJURY. The boring caterpillars eat the bracts, scales, and seeds of cones, acorns, or nuts. Frequently masses of resin mixed with excrement pellets occur both inside and on the outside of infested cones. Some silk webbing also may be present in the galleries. *Laspeyresia* spp. commonly bore through the axis of cones and then into the seed, whereas *Eucosma* spp. bore mostly in the cone scales. The others bore irregularly through the cones or nuts.

HOSTS AND RANGE. The seed of all species of plants wherever they occur.

The general hosts of a few of the more common genera are as follows: *Dioryctria* spp.—pine, fir, and spruce cones. *Heinrichia* spp.—cypress and Sitka spruce cones. *Barbara* spp.—true fir and Douglas-fir cones. *Laspeyresia* spp.—pine and spruce cones, hickory nuts. *Eucosma* spp.—pine cones. *Eupithecia* spp.—fir, Douglas-fir and hemlock cones.

INSECT APPEARANCE. *Caterpillars:* see table 15.1 for the general characteristics. Small; full grown length ⅜ to ½ inch for smaller *Laspeyresia* and *Eucosma* spp., ⅝ inch or longer for the larger *Dioryctria* and *Bar-*

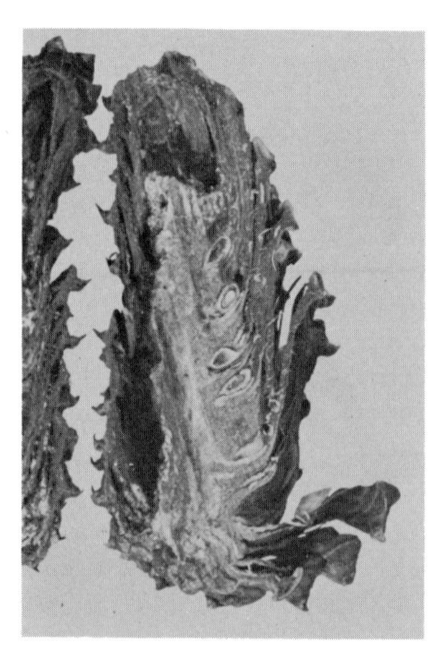

FIG. 15.2. Cone moth (*Eucosma* spp.) damage (X¾).

bara spp.; naked, soft bodied, cylindrical; 3 pairs thoracic legs plus 5 pairs fleshy abdominal legs (prolegs) (3 pairs for *Eupithecia* spp.); body color gray, pinkish, reddish, or greenish; yellow white for *Barbara* spp. *Moths:* small and inconspicuous; they are seldom noticed and can be identified only by specialists.

LIFE CYCLE. These insects usually pass the winter as pupae inside cocoons formed within or on the surface of the cones or in the ground litter. The moths emerge, mate and the females lay their eggs during the spring or summer at the time the host fruits are suitable for the larvae. The eggs hatch within a week or two and the resulting caterpillars bore into the host. Pupation occurs in the fall so there is usually only 1 generation per year.

CONTROL. There is no practical way to control these insects in forests. For protecting the cones in high value seed orchards or on experimental seed trees a DDT spray can be used. A dosage of 1 pound of the actual insecticide per 100 gallons of spray should be effective if the application is made so as to have the poison on the cones or nuts when the eggs hatch.

SEED WASPS

Megastigmus spp., Hymenoptera, Chalcidoidea

INJURY. Each larva hollows out a single seed, but the injury cannot be detected externally.

HOSTS AND RANGE. The seeds of many species of plants including both conifers and angiosperms are attacked by seed wasps. These insects are widely distributed wherever the hosts grow.

INSECT APPEARANCE. *Larvae:* body very minute; 1/16 to 3/16 inch long; body soft, whitish, cylindrical, plump, with long axis of body sharply curved, often C shaped; head indistinct, neither hardened nor colored darker than body; with 7 to 11 pairs inconspicuous spiracles. *Wasps:* very small, 1/16 to 3/16 inch long; color yellow and/or black, usually with an iridescent metallic sheen; antennae with many segments; pronotum small, each wing with only a single vein; ovipositer long, attached far anterior to tip of abdomen.

LIFE CYCLE. The adult wasps emerge through small circular holes in the seed during the spring or early summer. After mating the females deposit their eggs directly inside the developing seeds by means of their long ovipositers. The larvae develop to maturity before fall and remain inside the seeds until the following spring or summer. Often 10 to 50 per cent do not emerge but remain in a diapause through the next summer. These pupate and the wasps emerge the following year. Thus, the life cycle is completed in either 1 or 2 years.

CONTROL. Insecticidal control is impractical under most conditions. The seed should be fumigated if it is to be exported to areas where the insects are absent. An effective fumigant consists of a mixture of 4½ pounds ethylene dichloride and 1½ pounds carbon tetrachloride. This amount is adequate to treat 1,000 cubic feet of space. After the poison is applied, the fumigation chamber should be kept closed for a period of 24 hours. Temperatures should be above 75° F, and the seed treated should not have a water content of more than 12 per cent. It is possible that seed produced in seed orchards can be protected by spraying with DDT, provided the treatment is applied at the right time.

CONE BEETLES

Conophthorus spp., Coleoptera, Scolytidae

INJURY. Both adults and larvae of cone beetles bore in small immature pine cones and cause them to wither and die. Masses of resin (pitch tubes) usually form on the cones or stems at the places where the beetles enter. The egg galleries made by the beetles extend through the axis of the cones, whereas the larvae bore and feed on the scales and seeds. These insects often cause heavy losses sometimes killing 25 to 75 per cent of the cones.

HOSTS AND RANGE. Almost all species of pines are attacked. Different species occur and infest different hosts wherever they grow.

INSECT APPEARANCE. *Beetles:* head concealed from above by prothorax; front tibia widened toward the tip and serrated on the outer margin; antennal funicle (part between club and basal segment) with more than 3 segments, club without cross lines (satures); body stout; pronotum with a fine raised line along back border; mouthparts densely clothed with stiff hairs. *Larvae* and *pupae:* typical, see general characteristics for bark beetles.

LIFE CYCLE. Young adults pass the winter in the cones in which they were reared. The beetles emerge early in the summer and immediately attack young year old cones. The eggs are laid in individual niches along the sides of the egg galleries, and the larvae bore into the scales and seeds. They mature within a couple of months and then pupate. The resulting adults remain in the cones until the following spring. There is only 1 generation per year.

CONTROL. No practical control methods have been developed for protecting forest cone crops. It is possible, however, that sprays containing insecticides such as benzene hexachloride might be effective if applied at the right time. The cost of such treatments probably would limit their use to high value cone crops such as experimental trees or seed orchards.

CONE AND SEED MAGGOTS

Diptera, mostly Cecidomyiidae

INJURY. These insects mine the seed and cone scales, but usually they cause little damage.

HOSTS AND RANGE. Various coniferous species wherever the hosts grow.

INSECT APPEARANCE AND LIFE CYCLE. See the general description given for root maggots.

CONTROL. Applied methods are not practical under forest conditions.

Host Index

In this book it is impossible to record all the known hosts for each insect species considered; therefore, only a few of the more commonly infested trees are listed for each insect. In many cases, especially for the foliage eating Lepidoptera and for many wood borers, the insects exhibit little, if any, host preference. Insects with these diverse food habits are grouped under the host headings of either "Hardwoods" and/or "Conifers." The reader must always keep in mind, however, that many of the insects for which specific hosts are listed also attack other tree species. Thus, host lists such as this have only limited use.

In using this index the following page key should be useful:

	PAGES
Foliage-eating insects	95–199
Inner-bark borers	200–266
Wood borers	267–344
Sapsucking insects	345–373
Twig-, bud-, and seedling-damaging insects	374–395
Root-infesting insects	396–402
Cone and seed borers	403–409

General Index

(Bold face type indicates illustrations)